Microbial Waterborne Pathogens

Microbial Waterborne Pathogens

Edited by
T.E. Cloete, J. Rose, L.H. Nel and T. Ford

IWA
Publishing

Published by IWA Publishing, Alliance House, 12 Caxton Street, London SW1H 0QS, UK

Telephone: +44 (0) 20 7654 5500; Fax:+44 (0) 20 7654 5555; Email: publications@iwap.co.uk
Web: **www.iwapublishing.com**

First published 2004

Edited and typeset by Techbooks, India
Printed by TJ International (Ltd), Padstow, Cornwall, UK.

British Library Cataloguing in Publication Data
A CIP catalogue record for this book is available from the British Library

Library of Congress Cataloging-in-Publication Data
A catalog record for this book is available from the Library of Congress

ISBN: 1 84339 055 8

Contents

Preface

The year 2003 was a year to focus on raising awareness, protecting and respecting our water resources. Because 70% of the Earth is covered by water, it is called the 'Blue Planet'. However, only 2.5% of the world's water is fresh, while 97.5% is in oceans. And only 0.3% of this freshwater is available from rivers, lakes and reservoirs, 30% from groundwater, while the rest is stored in glaciers, ice sheets and mountain peaks, making it difficult to extract.

In the developed world, the connection between water, hygiene and health is taken for granted. However, for the less fortunate majority, access to potable water is non-existent and remains a daily struggle. Bacteria, viruses and parasites in contaminated water cause waterborne disease. Of concern are the so-called new emerging pathogens, contributing to waterborne disease, one of the biggest human tragedies, killing more than 5 million people each year. About 2.3 billion people suffer from diseases linked to contaminated water, and some 6,000 people die daily as a result of this. Some 60% of all infant mortality worldwide is linked to water-related infectious and parasitic diseases.

Treating water before use can eliminate most of these waterborne pathogens. Water supply and sanitation is therefore essential for meeting basic human needs. The importance of promoting and protecting health has risen to the top of the political agenda and much progress has been made. However, millions of people are still deprived of water supply and sanitation, resulting in waterborne diseases. Inadequate water supply and sanitation reinforces the cycle of poverty, slowing down the ability of societies to develop. If we fail to react, this situation will only get worse considering that the world population will increase by 5 billion people by the year 2050. In order to meet the demand, new sewerage facilities must be provided to 383,000 people a day. In the shorter term (2004–2015) the United Nations Millennium Development Goals of reducing the proportion of people without access to potable water by 50% implies giving access to clean water to an additional 27,400 people each day. To achieve this, water supply and sanitation services will have to overcome major constraints, including limited financial resources, insufficient cost recovery of services and inadequate operation and maintenance capacities. Unfortunately, many politicians and decision makers do not realise that providing access to sanitation facilities will halve the death toll of those who do not currently enjoy access to potable water and sanitation. However, the challenge we face goes beyond political decisions. In many cultures, norms and beliefs also have a negative impact on water supply and sanitation provision. The promotion of fundamental behaviour changes is key to integrating the appropriate use of services into daily routine and needs to start in childhood. School health and hygiene education programmes should therefore be an essential part of every water supply and sanitation programme. Relatively low-cost water management solutions that can make a significant impact in preventing waterborne disease exist, we need only apply them. However, more difficult seems to be finding the will to do so, which will be one of the most important challenges that the international community will have to face over in the future. An integral part of this is knowledge of the disease-causing organisms, detection techniques and epidemiology, which is the focus of this book.

T.E. Cloete
Pretoria, June 2004

Contributors

E. Dei-Cas
Institut Pasteur de Lille
BP 245
59019 Lille cedex
France

Tim Ford
Professor and Head
Department of Microbiology
Montana State University
Bozeman, MT 59715
USA

Elizabeth J. Hrudey
Department of Public Health Sciences
10-102 Clinical Sciences Building
University of Alberta
Edmonton, Alberta T6G 2G3
Canada

Steve E. Hrudey
Professor of Environmental Health Sciences
Department of Public Health Sciences
10-102 Clinical Sciences Building
University of Alberta
Edmonton, Alberta T6G 2G3
Canada

Henri Leclerc
Professor of Microbiology
Faculté de Médecine de Lille, pôle recherche
1, place de Verdun
59045 Lille cedex
France

Erin K. Lipp
Assistant Professor
Dept. of Environmental Health Science
The University of Georgia
206 Environmental Health Science Bldg.
Athens, GA 30602
USA

W. Markotter
Department of Microbiology and Plant Pathology
University of Pretoria
Pretoria 0001
South Africa

LH Nel
Professor and Head of the Microbiology Division
Department of Microbiology and Plant Pathology
University of Pretoria
Pretoria 0001
South Africa

Joan B. Rose
Homer Nowlin Chair in Water Research
Department of Fisheries and Wildlife
13 Natural Resources Building
Michigan State University
East Lansing, MI 48824
USA

L. Schwartzbrod
Faculté de Pharmacie
5,rue Albert Lebrun
54000 Nancy cedex
France

Troy M. Scott
Director
BCS of North Florida, Inc.
4641 NW 6th Street, Suite A Gainesville,
FL 32609
USA

J. Weyer
Department of Microbiology and Plant
Pathology
University of Pretoria
Pretoria 0001
South Africa

1

Microbial agents associated with waterborne diseases

H. Leclerc, L. Schwartzbrod and E. Dei-Cas

1.1 INTRODUCTION

Four epidemiological categories were initially devised for classifying water-related infectious 'diseases' concerning the engineering technologies required to prevent or control them. The categories defined for water transmission of infectious agents were waterborne (classical and other); water-washed (intestinal and body surface); water-based (dependent on intermediate aquatic host(s) and water-related insect vectors (breed in/bite near water) (Bradley, 1977; Craun, 1986; Moe, 1997). Waterborne diseases are those that are transmitted through the ingestion of contaminated water, with water acting as the passive carrier of the infectious agent.

The use of operative words such as 'waterborne' and 'disease' is justified for reports of outbreaks or cases of disease associated with drinking water because traditional epidemiological investigation relies on the occurrence of disease, which for drinking water is primarily waterborne. The use of 'waterborne' as a generic term encompassing all infections arising from water use, however, is too simplistic and poses difficulties when attempting a theoretical analysis of water-associated infections, because not all develop from ingestion of water. Likewise, 'disease' should not be used synonymously with 'infection,' which may be either asymptomatic (without clinical expression) or symptomatic (clinically observable syndrome and therefore a 'disease,' when the causal agent was identified). It could be argued that if 'disease' is not the end result of a water-related infection with a pathogenic agent, then the role of water in its epidemiology is not important. However, many of the bacteria and viruses, which may be spread in low levels by the water route, produce asymptomatic infections. This seeding of susceptible individuals could result in an endemic situation with the potential for an epidemic to occur from direct person-to-person, food or water spread (through contamination with greater levels of pathogenic agents released from infected carriers). Therefore, for purposes of clarification and general discussion, the use of the word 'infection' instead of 'disease' is preferred.

This chapter attempts to describe present knowledge about waterborne diseases, their epidemiology and the microbial agents commonly associated with them.

1.2 WATERBORNE DISEASES AND AETIOLOGIC AGENTS

A wide variety of bacterial, viral and protozoan pathogens excreted in faeces are capable of initiating waterborne infections, although the potential for this, predicted from latency, survival and infective dose data, which are usually higher for viruses and protozoa, remains unknown in many cases. Waterborne spread of infection by pathogenic agents depends on factors such as pathogen survival in water and the dose required for establishing infection in susceptible individuals. In addition to pathogen survival, latency (the period between pathogen excretion and acquisition of actual infectious power) and ability of the pathogen for multiplication in the environment are the factors influencing the infective dose. The minimum infective dose has been determined only for some bacteria, viruses or protozoa that are excreted with faeces and, therefore, potentially water-transmitted.

Therefore, it is very important to distinguish different classes of pathogens:

(1) Viruses may only remain static in number or die off. They cannot grow in the receiving waters; they cannot, indeed, multiply outside of susceptible living cells. The infective dose of these agents is low, typically in the range of 1 to 10 infectious units.

(2) Enteric protozoa such as *Giardia* and *Cryptosporidium* are highly resistant to most disinfectants commonly used in water treatment and survive for long time periods in the aquatic environment. Like viruses, they cannot multiply in the receiving waters. The infective dose of *Cryptosporidium* in humans is highly variable in terms of the strain virulence (Okhuysen *et al.*, 1999) and, probably, of the host susceptibility.

(3) The recognized waterborne bacterial pathogens include enteric and aquatic bacteria. The persistence of enteric bacteria, including *Salmonella* spp.,

Shigella spp. and *Escherichia coli*, in the aquatic environment depends on various parameters, including nutrients present and temperature. Although enteric bacteria are usually assumed to exist under starving conditions, there is evidence that some can grow in fresh water. Other bacterial infectious agents such as *Legionella* spp., *Aeromonas* spp., *Pseudomonas aeruginosa* and *Mycobacterium avium* are indigenous aquatic organisms, which can both survive and proliferate in drinking water. An additional mode of water-related transmission is through direct inhalation of aerosols, specifically with *Legionella pneumophila*, the aetiologic agent of Legionnaires' disease, and probably with non-tuberculous mycobacteria such as *Mycobacterium avium*. The infectious dose of many enteric bacteria is in the range of 10^7 to 10^8 cells, but is much lower with some species, including *Shigella* spp., *Campylobacter* spp. and enterohaemorrhagic *E. coli* O157:H7.

This discussion concerns the epidemiology of waterborne diseases, that is to say the rates of infection and characteristics of the diseases principally associated with waterborne transmission in developed countries. The situation is markedly different in developing countries. In many countries worldwide, more illnesses and deaths result from the sheer lack of availability of water in quantities sufficient for personal and household hygienic uses than from impurities in drinking water. These deficiencies in quantity or availability, along with malnutrition and lack of medical care, are responsible for the millions of deaths ascribed annually to diarrhoea and enteritis, which are water-associated, but much less commonly waterborne. In 1997, diarrhoea ranked first in the WHO's report on assessment of causes of morbidity (4×10^9) and sixth in causes of mortality (WHO, 1998; Ford, 1999). It can be suggested that up to 70% of diarrhoeal illness could happen from contaminated food, and therefore 30% could result from polluted water. It should also be stated that food responsible for outbreaks could be contaminated by water.

The most complete data on waterborne disease are those reported for the United States by the Center for Disease Control and Prevention (CDC) and the Environmental Protection Agency (EPA). Two or more persons must show the same clinical symptoms before the incident is included in the statistics. Water systems are classified as community water systems, non-community systems and private or individual water systems (CDC, 2000). National statistics dating back to 1920 on outbreaks associated with drinking water were analyzed by Craun (1986, 1992) in an attempt to estimate the effects and adequacy of public health programmes, regulations, treatment technologies and microbial monitoring. The epidemiological course of waterborne disease in the United States (Figure 1.1) is marked by a clear reduction in bacterial diseases during the period 1920 through 1998. From the 1970s, shigellosis outbreaks have continued to occur and campylobacteriosis and *E. coli* O157:H7 haemorrhagic colitis have emerged. At the same time, there has been an increase in incidence of protozoan outbreaks, including the largest documented incident in Milwaukee in 1993, since record keeping began in

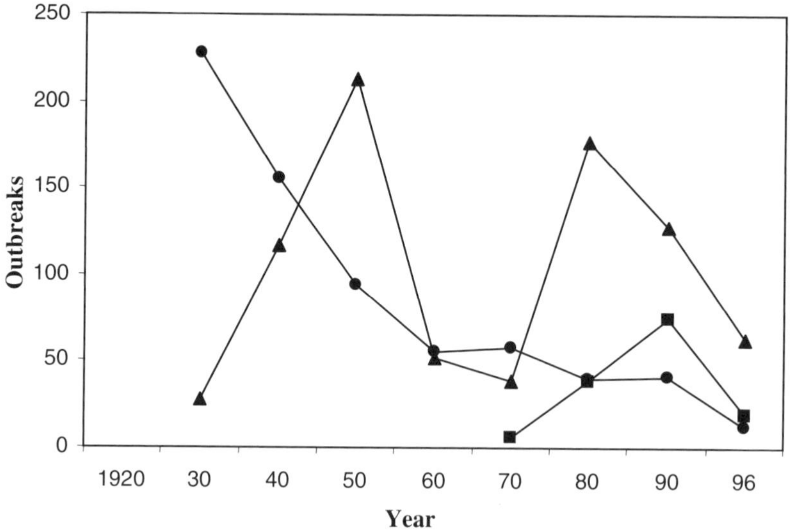

Figure 1.1 Etiology of waterborne outbreaks, United States of America, 1920–1996. ●, bacterial outbreaks; ▲, acute gastrointestinal illness of unknown etiology (AGI); ■, protozoan outbreaks.

1920. During the past decade, the aetiology of acute gastroenteritis (AGI) was not determined in almost half of the outbreaks. However, in many of these outbreaks, a viral aetiology (Norwalk agents, human rotaviruses and adenoviruses) has been suspected and was supported by the detection of Norwalk agents in the United States from 1980 to the 1990s (Kaplan *et al.*, 1982; Moe, 1997). Several reports and reviews have been published recently in the United States on this subject (Anon., 1999; Ford, 1999; LeChevallier *et al.*, 1999a, b, c; Szewzyk *et al.*, 2000b).

Unfortunately, there is a paucity of data available on the occurrence of outbreaks of water-related infections in Europe. This may be because water-related outbreaks are rare in Europe, but it is more likely attributed to the problems associated with epidemiological investigations of such outbreaks. Some reports on European waterborne disease incidence have been published, including reports for the United Kingdom (Galbraith, 1994) and Nordic countries (Stenström, 1994). Hunter (1997) has conducted a comprehensive review of waterborne disease worldwide.

It is generally agreed that waterborne outbreak reports are incomplete. A growing number of reports (Craun, 1992; Morris and Levin, 1995; Ford, 1999) indicate that waterborne disease is far more prevalent than reported outbreaks. Craun suggests that only one-half to one-third, or even one-tenth, of waterborne outbreaks occurring in the United States are detected, investigated and reported. According to Morris and Levin (1995), the annual incidence in the United States could be 7–8 million cases of illness and 1200 deaths attributable to waterborne infectious disease.

1.3 BACTERIAL ENTEROPATHOGENS

1.3.1 *Salmonella–Shigella*

Waterborne outbreaks of bacterial origin in the United States have declined dramatically from 1920 through 1960–1970. This applies particularly to typhoid fever, which has been seen rarely since the 1970s, as a result of the widespread application of disinfection and filtration treatments, introduced in the early 1900s. Deaths from waterborne infections were almost entirely attributed to typhoid fever from 1920 to around 1950 (Craun, 1992). On the other hand, the number of reported outbreaks of salmonellosis has been relatively stable over the years, whereas outbreaks of shigellosis have increased (Figure 1.2). Since 1961, *Shigella sonnei* has become the most prevalent causative agent, replacing *S. flexneri* as the dominant species. The change in *Shigella* species causing the majority of cases of illness can probably be explained by a similar change in incidence of *Shigella* serotypes in patients.

1.3.2 *Campylobacter*

Waterborne outbreaks caused by *Campylobacter jejuni* in the United States (3 outbreaks 1971–1980, 10 outbreaks 1981–1990, 3 outbreaks 1991–1996) (Craun, 1986, 1992) demonstrate that *Campylobacter* is now a leading cause of bacterial gastroenteritis in the community. It is isolated as commonly as *Salmonella* and *Shigella* from patients with diarrhoea; it is the most frequently identified bacterial cause of diarrhoea in the United States and the United Kingdom. Human campylobacteriosis is a severe disease, often leading to serious sequellae and sometimes resulting in death. Following the introduction of an improved surveillance system in England and Wales, the Communicable Disease Surveillance Centre observed

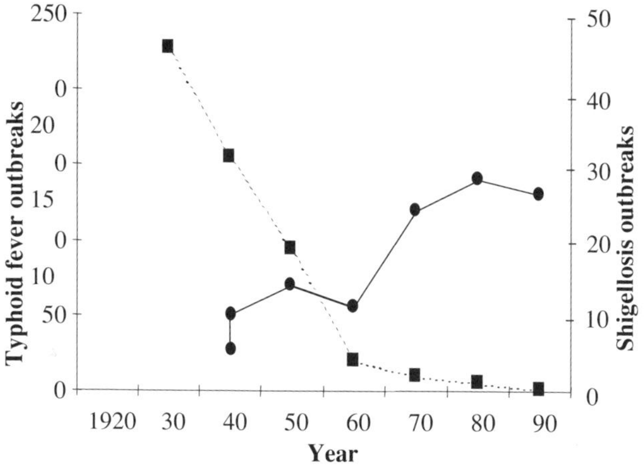

Figure 1.2 Comparison of typhoid fever and shigellosis occuring in the waterborne outbreaks (USA, 1930–1990). ■, typhoid fever outbreaks, ●, shigellosis outbreaks.

26 outbreaks between 1992 and 1995 in which there was evidence of waterborne transmission of infectious agents (Furtado *et al.*, 1998). In Nordic-European countries, campylobacters have been known as important human pathogens since the late 1970s (Rautelin and Hanninen, 2000). More recently, they were associated with presumed waterborne outbreaks (Melby *et al.*, 1991, 2000; Merritt *et al.*, 1999; Godoy *et al.*, 2002; Frost *et al.*, 2002); in Finland, a waterborne epidemic in a hospital caused nearly a hundred persons to become ill (Rautelin *et al.*, 1990). Other outbreaks have been reported in Canada (Alary and Nadeau, 1990; Millson *et al.*, 1991) and in New Zealand (Stehr-Green, 1991). Some investigations demonstrate the potential for outbreaks involving more than one pathogen (Maurer and Sturchler, 2000; Bopp *et al.*, 2003).

Campylobacter species appear to be particularly well adapted to the avian intestinal tract, thus causing poultry to be the primary vehicle of transmission to humans (Tauxe *et al.*, 1988). *Campylobacter* enteritis is a zoonosis, and animals used for food production, such as poultry, account for a large reservoir of environmental pollution. Virtually all surface waters contain campylobacters, even in remote areas where contamination stems from wild birds. Waterborne outbreaks typically follow consumption of untreated surface water, which may be contaminated with bird faeces (Taylor *et al.*, 1983).

The occurrence of thermophilic campylobacters was studied in river and lake waters, some of which were used as drinking water sources. *C. jejuni* was recovered frequently from samples of fresh water (Brennhovd *et al.*, 1992; Popowski *et al.*, 1997; Obiri-Danso and Jones, 1999). However, no campylobacter-type bacteria were found in treated water because of their susceptibility to chlorination (Megraud and Serceau, 1990). The first isolation of *C. jejuni* from groundwater (Stanley *et al.*, 1998) supports the theory that groundwater may be a vehicle for campylobacter transmission. The use of large sample volumes (several litres) increases the chance of detecting the pathogen *C. jejuni* in water (Hanninen *et al.*, 2003). In water microcosm experiments, factors such as temperature, oxygenation, anaerobic conditions, nutrient and biofilms influence their survival times (Buswell *et al.*, 1998; Thomas *et al.*, 1999). Survival was enhanced by the presence of autochthonous microflora and nutrients, and decreasing temperature and anaerobic conditions. The existence of viable but non-cultivable campylobacter, and their power for retaining pathogenicity and virulence (Rollins and Colwell, 1986; Jones *et al.*, 1991; Medema *et al.*, 1992; Tholozan *et al.*, 1999) was a debatable point until demonstrated by Cappelier *et al.* (1999) and Talibart *et al.* (2000).

The spread of *Arcobacter* organisms (previously designated as aerotolerant *Campylobacter*) via the drinking water path must be suspected (Jacob *et al.*, 1998; Rice *et al.*, 1999). Some cases of acute diarrhoea, however, were associated with *Arcobacter butzleri* (Lerner *et al.*, 1994), although very little is known about the clinical significance of these infections in humans.

1.3.3 *E. coli* O157:H7

E. coli O157:H7, first recognized as a pathogen in 1982 (Riley *et al.*, 1983), is now known as an important cause of bloody diarrhoea (haemorrhagic colitis) and renal failure (haemolytic uraemic syndrome) in humans. Its pathogenicity has been attributed in part to the production of a toxin cytotoxic for Vero cells (a fibroblastic green monkey kidney line cell). The toxin was appropriately named Verotoxin (VT), and the group of *E. coli* that produce VT became known as the Verotoxin-producing *E. coli* (VTEC) (Chart 2000). Subsequently, other verotoxin-producing *E. coli* strains have been associated with haemorrhagic colitis and haemolytic uraemic syndrome. *E. coli* O157:H7, however, is recognized as the most common cause of VTEC-related human illness. VTEC, including *E. coli* O157:H7, are strongly associated with cattle, and they can clearly pass through the stomachs of ruminants. The transmission of VTEC O157 is often foodborne, particularly from contaminated ground beef or raw milk, or person to person, and contact with farmed animals has also resulted in human infection. Farmed and wild animals grazing in water catchment areas, however, are a potential source of faecal contamination, and therefore of waterborne VTEC O157 infection. VTEC O157 is particularly well adapted for surviving in the aquatic environment (Kogure and Ikemoto, 1997), as suggested by outbreaks associated with swimming in lakes (Rice *et al.*, 1992; Wang and Doyle, 1998). It is capable of surviving for many days, especially at cold temperatures.

The waterborne route of VTEC O157 infection was first clearly demonstrated by an unusually large outbreak through unchlorinated municipal water in the Missouri Community (Swerdlow *et al.*, 1992). Since then, waterborne VTEC O157 has been described in sporadic cases and in outbreaks of illness. Chalmers *et al.* (2000) have analyzed published outbreaks of VTEC O157 associated with recreational waters, private and municipal supplies. Despite the potential for large contamination of the environment with VTEC O157, waterborne infection is relatively rare because VTEC O157 is as susceptible to chlorination as bacterial indicators.

Epidemiological investigations have elucidated the mechanisms by which *E. coli* O157:H7, *Shigella* spp. and campylobacters have become a source of concern recently. The most striking feature is the low inoculum of organisms that may trigger disease. As few as 10 to 100 organisms of the most virulent *S. dysenteriae* type 1 are sufficient for causing clinical dysentery, while the other species may require a 10–100 times higher dose (DuPont *et al.*, 1989). The dose required to trigger campylobacteriosis is also low, probably not more than a few hundred bacteria (Robinson, 1981; Black *et al.*, 1988). *E. coli* O157:H7, like *Shigella* and *C. jejuni*, appears to have a low infectious dose, approximately 100 organisms, or less (Griffin and Tauxe, 1991).

1.3.4 *Yersinia enterocolitica*

Bacteria of the genus *Yersinia* have been frequently isolated from the environment, including water supplies of various types. Environmental strains should be differentiated from serotypes O:3, O:9, O:5, O:27 and O:8 of *Y. enterocolitica*, which are the ones most frequently associated with human infections in Europe, Japan, Canada and the United States (Schieman, 1990). These pathotypes are psychrotrophic, hence can multiply in fresh waters and could constitute a major hazard to drinking water. Epidemiological data concerning waterborne yersiniosis, however, are scarce (Schieman, 1990) and involve only a small number of individuals. A large outbreak of intestinal illness at a Montana ski resort in 1977 was suggested to be waterborne yersiniosis (Eden *et al.*, 1977), but the many strains isolated from water sources were later identified as non-pathogenic, leaving the cause of this outbreak uncertain (Hunter, 1997). Three prospective studies from Norway (Ostroff *et al.*, 1994; Saebo *et al.*, 1994) and from New Zealand (Satterthwaite *et al.*, 1999), however, seem to demonstrate that *Y. enterocolitica* could be waterborne in these areas.

Y. enterocolitica became the repository for a large number of 'atypical' or *Y. enterocolitica*-like strains. Usually isolated from terrestrial and fresh water ecosystems, they are often referred to as environmental strains, including non-pathogenic serotypes of *Y. enterocolitica* and the related species such as *Y. aldovae*, *Y. bercovieri*, *Y. frederiksenii*, *Y. intermedia*, *Y. kristensenii*, *Y. mollaretii* and *Y. rohdei*. These strains invariably lack the characteristic virulence determinants of the enteropathogenic O:3, O:9, O:5, O:27 and O:8 serotypes of *Y. enterocolitica*. These aquatic environmental strains were, however, relatively uncommon in the water supplies examined (Kapperud, 1977; Langeland, 1983; Aleksi and Bockem, 1988).

1.3.5 *Vibrio cholerae*

Cholera caused by toxigenic *Vibrio cholerae* is a major public health problem plaguing developing countries, where outbreaks occur in a regular seasonal pattern and are associated with poverty and poor sanitation. Among the 193 currently recognized O serogroups of *V. cholerae*, only serogroups O1 and O139 have been associated with epidemics of cholera. The other serogroups, usually referred to as non-O1 and non-O139, can cause sporadic diarrhoea (Chakraborty, 2000). This sharp distinction between serogroups is related to virulence-associated genes. The strains belonging to serogroups O1 and O139 (more than 95%) produce cholera toxin (CT) and a colonization factor known as toxin-coregulated pilus (TCP), which regulated in coordination with CT production. In contrast, the strains of non-O1 and non-O139 (more than 95%) serogroups do not possess these two virulence genes.

From 1817 to 1994, seven distinct pandemics of cholera occurred and all of these were caused by *V. cholerae* O1. During the first six pandemics, cholera remained principally confined to South and Southeast Asia, whereas the seventh pandemic reached West and East Africa and South America. In late 1992, *V. cholerae* belonging to a non-O1 serogroup (now referred to as O139 Bengal) caused explosive epidemics of cholera through India, Bangladesh and neighbouring countries. Systematic surveillance of cholera in these countries from 1992 to 2000 revealed either temporal alternation or coexistence of *V. cholerae* O1 and O139 in different regions (Faruque *et al.*, 1998; Reeves and Lan, 1998; Basu *et al.*, 2000). A clone of serogroup O37 demonstrated epidemic potential in the 1960s (Beltran *et al.*, 1999). The close evolutionary relationships among O1, O139 and O37 epidemic clones indicates that new cholera clones likely arose from a lineage that was already epidemic or closely related to such a clone (Faruque *et al.*, 1998, 2000; Beltran *et al.*, 1999; Jiang *et al.*, 2000). An endemic focus of cholera was reported in the United States from 1970 to 1990. Most cases were associated with consumption of undercooked crabs or shrimp, or raw oysters, from the Gulf of Mexico (Craun *et al.* 1991). In Europe, several countries were plagued by cholera in the 1990s (Onishchenko *et al.*, 1995; Maggi *et al.*, 1997; Clark *et al.*, 1998; Cavalieri d'Oro *et al.*, 1999).

V. cholerae is now recognized as an autochthonous member of the microflora in many aquatic environments such as in riverine and estuarine areas. The importance of water ecology is suggested by the close association of *V. cholerae* with surface water and the human population interacting with the water. Food also plays an essential role, although in many instances water is the source of contamination of foods (Albert *et al.*, 1997). Many ecological aspects of the seasonal appearance of epidemic *V. cholerae* strains and outbreaks of cholera remain unknown. Recent data suggest that the following *Vibrio* characteristics may be involved: (1) the aquatic reservoir of *V. cholerae* O1 or O139 is capable not only of surviving in water but also of forming a complete component of the ecosystem (Franco *et al.*, 1997). It has been postulated that under stress conditions vibrios can be converted to a viable but non-culturable form (VBNC) (Colwell and Huq, 1994; Borroto, 1997; Jesudason *et al.*, 2000), which can revert back to live infectious bacteria. (2) The major pathogenic gene in toxinogenic *V. cholerae* are clustered in several chromosomic regions (CTX genetic element and TCP pathogenicity island) which are capable of being propagated horizontally to non-O1 and non-O139 strains by lysogenic conversion (Faruque *et al.*, 1998). More recently, Chakraborty *et al.* (2000) have demonstrated the occurrence and expression of critical virulence genes in environmental strains of *V. cholerae* that appear to constitute an environmental reservoir of virulence genes. These new data on the ecology of *V. cholerae* appear to be of great significance. Likewise, the fact that non-O1 and non-O139 strains are more commonly detected than O1 and O139 strains, in fresh water and estuarine areas, is also relevant (Faruque *et al.*, 1998; Uchiyama, 1998; Thomson *et al.*, 1998).

In outbreaks with a bacterial aetiology, the CDC and the EPA make efforts to correlate waterborne disease with process deficiencies. Thus, it has been established that untreated surface or groundwater, inadequate disinfection procedures or distribution system deterioration were the major causes of waterborne outbreaks. However, control of such deficiencies is usually within reach. These conclusions substantiate the basic concepts of management of enteric diseases by multiple barriers and monitoring by use of indicators (Ford, 1999).

1.4 OPPORTUNISTIC PATHOGENS THAT GROW IN WATER SUPPLIES

Similar to faecal pathogens, which may have access to the aquatic environment and survive or remain viable but non-culturable, other bacterial pathogens are able to multiply in water even with low levels of organic nutrients. They are opportunistic pathogens in humans, but concern about their occurrence in drinking water is disputed. The most important organisms to consider are *Legionella, Pseudomonas aeruginosa, Aeromonas* and *Mycobacterium avium* complex.

1.4.1 *Legionella*

The genus *Legionella* has at least 48 named species (Fields *et al.*, 2002) amongst which *L. pneumophila* serogroup 1 is most frequently related to human disease. In addition to *L. pneumophila*, 19 *Legionella* species have been documented as human pathogens on the basis of their isolation from clinical material (Muder and Yu, 2002). *L. micdadei* is the second most frequent species causing legionellosis in the United States and *L. longbeachae* infections are frequent in Australia (WHO, 2002a). *Legionella* infections can lead to two forms of disease, namely legionellosis or Legionnaires' disease, an acute purulent pneumonia and Pontiac fever, a self-limiting non-pneumonic disease consisting of fever and mild constitutional symptoms. Inhalation of *Legionella*-containing aerosols is regarded as the usual mode of infection. Aspiration following ingestion of contaminated water is another route of infection considered as major by some authors (Yu, 1993). There has been no proved person-to-person transmission. The epidemiology of Legionnaires' disease is dominated by the frequency of sporadic cases. However, during the past 3 years several large outbreaks were reported from Australia, the Netherlands and Spain, which involved cooling towers and whirlpool spas (Fields *et al.*, 2002). Travel-associated Legionnaires' disease has caused concern, particularly among European countries because of the threat for the tourist industry.

Legionella is a common inhabitant, usually in low numbers, of natural aquatic environments and water supplies that meet drinking water standards (States *et al.*, 1990). A number of abiotic factors, of which temperature is the most important one, significantly influence *Legionella* survival and growth (Bates *et al.*, 2000). *L. pneumophila* multiplies at temperatures between 20°C and 50°C with a maximal growth between 30°C and 40°C (Lee and West, 1991). *Legionella* are difficult to grow in the laboratory requiring a specific combination of nutrients in the medium. These nutritional requirements seem to contradict the widespread distribution of *Legionellae* in freshwater environments where nutrients levels are low. However, there are many other aquatic bacteria that cannot be grown in simple media, which can grow on nutrients in the water. The growth of *Legionella* has been shown to be substantially affected by the occurrence of algal–bacterial mat communities in thermal effluents (Tison *et al.*, 1983). Therefore, hot-water tanks, cooling systems and towers (Martinelli *et al.*, 2000), because of their heat-exchanging function, serve as bacterial 'amplifiers'. *Legionella* spp. are able to colonize all parts of internal distribution systems of hospitals, hotels or buildings (Kool *et al.*, 1999; Rangel-Frausto *et al.*, 1999; Yu and Stout, 2000). These bacteria can potentially infect susceptible persons through aerosols created in showers, sinks, whirlpool spas, etc. In addition to temperature and nutrients, several other factors in aquatic environments can increase the risk of Legionnaires' disease:

(1) First, there is now ample evidence that amoebae such as *Acanthamoeba*, *Hartmanella*, *Valkampfia*, *Naegleria* and other protozoa may be natural hosts and 'amplifiers' for *Legionellae* in the environment. Invasion and subsequent intracellular replication of *L. pneumophila* within protozoa could play a major role in the transmission of Legionnaires' disease (Rowbotham, 1980; States *et al.*, 1990; Fliermans, 1996; Winiecka-Krusnell and Linder, 1999). The intracellular multiplication cycle of *Legionella* in amoeba or in macrophages, indeed, has many notable similarities, as well described by Swanson and Hammer (2000). During the course of amoebic infection, the *Legionella* acquire the ability to survive and multiply in a phagosome surrounded by rough endoplasmic reticulum, which constitutes a store of nutrients and at the same time forms an additional cell compartment. Within this protected vacuole, *L. pneumophila* converts into a replicative form, tolerant to acids but no longer expressing virulence factors, including those which block lysosomal fusion. In the lysosome compartment, a nutrient-rich replication niche, the *Legionella* multiply abundantly. Once the pool of amino acids become scarce, accumulation of ppGpp signals entry into the stationary phase and the bacteria then express a number of virulence traits that promote transmission to a new host. As its natural reservoir, protozoan phagocytes amplify the *L. pneumophila* population in fresh and potable water supplies.

Recently, it has been shown that viability and infectivity of environmental non-cultivable *L. pneumophila* can be restored by intracellular replication within protozoa (Steinert *et al.*, 1997). Intracellular growth even enhances the infectivity of *L. pneumophila* to human-derived cells (Fritsche *et al.*, 1998). At some stage after the ingestion of bacteria, amoebae produce vesicles that contain high numbers of *Legionellae*, which may reach the lung alveolus and constitute an infectious dose (Berk *et al.*, 1998).

(2) Second, aquatic biofilms which are widespread not only in water environments but also in medical devices provide comfortable ecological niches in which *Legionella* become established and proliferate (Lee and West, 1991; Rogers *et al.*, 1994; Flierman, 1996; Atlas, 1999, Fields *et al.*, 2002). Bacteria, as colonial organisms in biofilms, elaborate systems of intercellular communication to facilitate their adaptation to changing environmental conditions (Wimpenny *et al.*, 2000; Davey and O'Toole, 2000). Therefore, biofilms provide an ideal environment for the etablishment of syntrophic relationships in which two metabolically distinct types of bacteria depend on each other to utilize specific substrates, typically for energy production.

Only a few studies have attempted to characterize the interactions between water bacteria and *Legionella* in such diverse habitats as free water and biofilms (Fields *et al.*, 2002). According to some authors (Rogers and Keevil, 1992; Rogers *et al.*, 1995) it might be that the biofilms support the survival and growth of *Legionellae* outside a host cell. The studies of Murga *et al.* (2001) using a biofilm reactor suggest that *L. pneumophila* may persist in the absence of amoebae, but in the model system, the amoebae were required for multiplication of the bacteria. Similar findings have been reported by Szewzyk *et al.* (2000a). Of particular importance is the potential protection, granted not only by biofilms but also by amoebic hosts to *Legionella*, from the effects of desinfectants, low pH and heat (Lee and West, 1991; Fliermans, 1996).

(2) When nutrient conditions of aquatic habitats become unfavourable, both sessile and planktonic bacterial cells are sharply reduced in size to form very small (+0.3 μm) spherical ultramicrobacteria by a process that is now well documented as starvation survival (Kjelleberg, 1993; Morita, 1997). Thus, the overall strategy of bacteria in oligotrophic environments is to grow in biofilms on surfaces where nutrients are locally available and to persist in nutrient-deprived zones in the starved state as floating, dormant minicells, with the capability of returning to the vegetative state when nutrients again become available (Costerton *et al.*, 1995). On exposure to nutrient limitation, bacteria synthesize a large number of new proteins (stress proteins) that increase their resistance to multiple stresses (Morita, 1997). In low-nutrient environments, *L. pneumophila* is able to enter a non-replicative viable but non-culturable state (VBNC) (Oliver, 1993). It can result not only from nutrient deprivation but also due to temperature, antibiotics, salt, and chlorine stress (Roszack and Colwell, 1987). The addition of amoebae to the dormant bacteria appears to result in the resuscitation of *Legionellae* to a culturable state (Steinert *et al.*, 1997).

Legionella outbreaks and single cases have been traced to cooling towers, evaporative condensers, hot water from showers, spas and whirlpools, nebulizers and other aerosol-producing medical devices. It is not feasible to prevent the contamination of water supply systems and consequently hot water systems with *Legionellae* over extended periods of time, given the possible entry of a single bacterium or bacteria phagocytosed by amoebic hosts from the water distribution system or during repair and construction. Colonization of domestic hot water systems that deliver water below 50°C is particularly frequent. To avoid this, it is recommended that the water should be stored and distributed at a temperature of at least 50°C, or greater (WHO, 2002a) and measured at the tap after less than a minute of flow. The drinking water system should be free of significant *Legionella* concentration (i.e., below the detection limit of 50 per litre) when the water temperature does not exceed 20°C, and assuming a concentration of 0.2 mg free chlorine per litre within the water mains and additional measures have been taken to prevent stagnation of water. Unfortunately, all of these conditions are difficult to meet in technical terms.

The risk of infection following exposure to *Legionella* cannot be assessed and remains open to speculation (O'Brien and Bhopal, 1993). Therefore, risk management strategies should be introduced to control *Legionella* at locations where a health risk is recognized, for example, in water supply systems of hospitals. Persons who are severely immunosuppressed or who have chronic underlying illnesses, such as haematologic malignancy or chronic lung disease, are at a markedly increased risk for legionellosis. It would be advisable to propose threshold levels of *Legionella* in water systems, related to immunocompromised states of patients

(Simon *et al.*, 2001). Health care facilities, including hospital and other health centres, should have general safety plans within the framework of a general infection control strategy (WHO, 2002b). Such plans should concern the control of water contamination by *Legionella*, including medical devices by which these organisms enter a patient's respiratory tract.

1.4.2 *Pseudomonas aeruginosa*

P. aeruginosa is a ubiquitous micro-organism, which is an inhabitant of fresh waters, soil and plants. This bacterium has been isolated from numerous vegetables, such as tomatoes, radishes, cucumbers, onions and lettuces (Stiles, 1989) at rates capable of reaching 10^3/g. Its presence is constant and abundant in wastewaters (Leclerc et Oger, 1974, 1975) and consequently in surface waters, which receive polluted effluents. Its growth in water is not directly linked to the organic matter content, because it can develop abundantly in the purest of waters. *P. aeruginosa* is a species of considerable versatility and a significant pathogen which can cause infection in a variety of plants, insects and warm-blooded animals. In humans, it is an opportunist pathogen, well known in the hospital environment; it seems likely to be the cause of 10 to 20% of nosocomial infections. Its extreme resistance to antibiotics explains why these ubiquitous bacteria are able to readily colonize the skin and mucous membranes of patients.

As some *P. aeruginosa* strains are capable of producing enterotoxins, the enteropathogenicity of this species is understandable. Since 1894, many later publications have recognized this bacterium as an enteric pathogen and causative agent of diarrhoea in infants and children (Williams and Cameron, 1897; Hunter and Ensign, 1947; Florman and Schifrin, 1950; Henderson *et al.*, 1969). Enteric disease associated with septicaemia was described earlier by Dold in 1918 as 'Shangai fever', a prolonged febrile illness that affected both children and adults. As expounded by Hardalo and Edberg (1997) the colonization of children by *P. aeruginosa* quickly ceases once the environmental sanitary conditions are corrected. Moreover, each of these 'infections' was diagnosed before there were adequate means of precluding a viral or protozoan aetiology. Community acquired *P. aeruginosa* gastrointestinal disease with sepsis rarely occurs in healthy infants, that is, those who lack identified underlying immunological or haematological problems (Lepow, 1994). On the other hand, *P. aeruginosa* can be a colonizer of the gastrointestinal tract in immunocompromised and hospitalized adults and children. There have been no significant outbreaks reported in recent decades, possible as a result of better hygienic control mesures and diagnostic techniques. Moreover, a study by Buck and Cook in 1969 demonstrated that the ingestion of up to 10^6 viable *P. aeruginosa* did not lead to infection or colonization, but only to a brief period of recovery of the organism from the stool.

It should be noted that *P. aeruginosa* is primarily an opportunistic pathogen (see Section 1.10) and that only patients who are at significant risk are susceptible. The host immune defects for which *P. aeruginosa* is a major challenge are largely confined to specific groups such as (1) burn patients with destroyed non-specific defenses, resulting in local and systemic immunosuppression, on both the humoral and cellular levels; (2) patients undergoing treatment for cancer and leukaemia who may remain profoundly neutropenic for extended periods of time; (3) hospitalized patients in intensive or special care units subject to instilled medical devices (e.g., catheters, endotracheal tubes and medical interventions such as antibiotics, chemotherapy); (4) patients with cystic fibrosis who suffer from chronic infections of *P. aeruginosa* biofilms in the lungs. Other specific conditions must be met for *P. aeruginosa* to result in infection. The infecting clone must be able to produce certain virulence factors: attachment facilitated by the production of pili and alginate (mucoid strains), tissue invasion with cellular destruction factors such as exotoxin A, elastase, alkaline protease, haemolysin, leucocidin, and anti-inflammatory activity, destroying immunoglobulins and complement components. Normal drinking water is not always suitable for all such individuals for all uses, e.g., medical devices containing water, respiratory care equipment, wound irrigation, etc. (Boukadida *et al.*, 1991; Römling *et al.*, 1994; Foca *et al.*, 2000) and it is essential that it is carefully monitored and controlled. Consequently, advice should be provided to at-risk groups (WHO, 2002b).

Other than these well-defined groups the general population is relatively resistant to infection with *P. aeruginosa*. This bacterium is predominantly an environmental organism and fresh surface water is an ideal reservoir. *P. aeruginosa* reaches relatively high numbers in food and on moist surfaces. Daily, substantial numbers of the species are ingested with our food, particularly with raw vegetables, while our body surfaces also are in continuous contact with the organism. The clinical history of a *P. aeruginosa* infection in relation to drinking water has never been reported. There is no evidence of association between the presence of these bacteria in drinking water and gastrointestinal infection among the general population (Hunter, 2002; Fricker, 2002; Allen *et al.,* 2002). Therefore it is reasonable to consider that *P. aeruginosa* in drinking water does not currently pose a problem for public health. Based on these observations, attemps to regulate *P. aeruginosa* in drinking water would not yield public health protection benefits.

1.4.3 Aeromonas

Traditionally, the genus *Aeromonas* has been divided into a psychrotrophic group and a mesophilic group. The first group contains the fish pathogen, *A. salmonicida*, while the second one belongs to a mesophilic group, initially represented by the *A. hydrophila* species. Adoption of DNA–DNA hybridization allowed a substantially improved grouping of *Aeromonas* spp., to define 16 hybridization groups (HGs), all of which have assigned species names (Janda *et al.*, 1998; Edberg and Allen, 2003). Only a few species are recognized as having an association with human disease: *A. hydrophila*, *A. caviae*, *A. jandei*, *A. shubertii*, *A. veronii* subsp. *veronii*, *A. veronii* subsp. *sobria*.

Aeromonas spp. are widely associated with environmental waters; most available strains are not from human faecal origin, but represent the natural habitat of the organism. The concentration of mesophilic aeromonads range from 10^6 to 10^8/ml in crude sewage to 10 to 10^3/ml in clean river water and surficial groundwater (Hazen *et al.*, 1978; Schubert, 1991; Kersters *et al.*, 1995; Pettibone, 1998). Aeromonads readily multiply in domestic or industrial wastewater and are also found in siphons, sinks and drainage systems (Schubert, 1991). Temperature is an important factor for enhancing colonization by aeromonads in wastewater. That is probably the reason why *Aeromonas* spp., which are psychrotrophic organisms, outgrow coliforms in sewage (Seidler *et al.*, 1980; Araujo *et al.*, 1989; Schubert, 1991). Hence aeromonads are natural residents in wastewater and fresh surface water, and their concentration is closely related to organic load and temperature (WHO, 2002a). On the other hand, it appears that aeromonads survive poorly in nutrient-poor waters in comparison with other autochthonous oligotrophic bacteria (Schubert, 1991; Kersters *et al.*, 1996). Since 1962, we have demonstrated that 30% of drinking water samples found positive for 'faecal coliforms' contained strains of *Aeromonas* which would have falsely indicated that the sample was positive in the faecal coliform test (Leclerc and Buttiaux, 1962). These observations have since been confirmed by many teams (Burke *et al.*, 1984; Havelaar *et al.*, 1990; Schubert, 1991). *Aeromonas* indeed may colonize the distribution system biofilm, as do many species of coliforms including *Enterobacter*, *Klebsiella* or *Citrobacter* but they appear to be unable to compete successfully with other heterotrophic bacteria in a pipe sediment suggesting they constitute only a small part of the total biofilm population.

Aeromonads are occasionally responsible for wound infections following contact with water (bathing, fishing, canoeing, etc.) (Delhaes *et al.*, 2001). However, they are primarily involved in human gastroenteritis which can be due to food and drinking water. Many experimental, clinical and epidemiological data tend to lend credence to the assertion that *Aeromonas* may be aetiologically involved in diarrhoeal illness (Janda, 1991; Janda and Abbott, 1998). Some authors are more cautious and consider that only some strains are likely to be pathogenic, a situation similar to that with *E. coli* and *Y. enterocolitica* (Farmer *et al.*, 1992). *Aeromonas* spp. may produce several toxins, cytolytic as well as cytotonic (Stelma *et al.*, 1986; Janda, 1991; Deodhar *et al.*, 1991; Kirov, 1997). The beta hemolysin called aerolysin is the major virulence factor of *Aeromonas hydrophila* (Stelma *et al.*, 1986; Xu *et al.*, 1998; Buckley and Howard, 1999). The isolation of *Aeromonas* in stools of individuals with diarrhoeal diseases (in particular in children less than 2 years) has been reported in several studies (Ashiru *et al.*, 1993; Albert *et al.*, 2000; Huys *et al.*, 2002). Beyond any doubt, *Aeromonas* may be isolated as many times from the faeces of patients without diarrhoea as persons with diarrhoea, suggesting that *Aeromonas* would as a rule, be a non-pathogenic 'fellow traveller'. The most striking argument against the role of *Aeromonas* in human diarrhoea emerged from studies of Morgan *et al.* (1985) with human volunteers; despite the fact that high challenge doses were used, this investigation failed to establish *Aeromonas* spp. as an enteropathogen. However, the pathogenicity of aeromonads may be strain- and even pathovar-related and the value of these data has been challenged by Gosling (1996), considering the poor characterization of the strains. Ultimately although there have been conflicting studies concerning *Aeromonas* as a potential pathogen, it is generally assumed that this species can cause self-limiting diarrhoea, with children being the most susceptible population (Isonhood and Drake, 2002; Edberg and Allen, 2003).

The frequent presence of *Aeromonas* in drinking water raised the question of its role as a waterborne pathogen, because production of enterotoxins and/or adhesins had been demonstrated (Krovacek *et al.*, 1992; Kühu *et al.*, 1997). However, exposure to *Aeromonas* is not only through drinking water. In recent years, there has been increasing interest in *Aeromonas* spp. as a possible cause of food-borne gastroenteritis (Wadström and Ljungh, 1991; Kirov, 1993, 1997). Foods that are consumed without cooking, or improperly cooked foods, would be the ones most likely to pose a threat. Concerning drinking water, Burke *et al.* (1984) first observed that *Aeromonas* spp.-associated gastroenteritis was closely correlated with the mean number of *Aeromonas* spp. in water samples within distribution systems. Typing methods have shown, however, that there is little similarity between the majority of water-related *Aeromonas* strains and diarrhoea-associated isolates (Holmberg *et al.*, 1986; Kuijper *et al.*, 1989; Havelaar *et al.*, 1992; Hänninen, 1994). With regard to the epidemiological relationship with drinking water, in contrast to other waterborne pathogens, no clearly defined outbreaks of diarrhoeal illness due to *Aeromonas* have ever been reported, although this bacterium is frequently isolated from water (Schubert, 2000). In fact, all of the mentioned studies are consistent with people who have become colonized by *Aeromonas* from drinking water and been infected with an intestinal disease from another source (Hunter, 2002).

Edberg and Allen (2003) considered the specific criteria that find that a micro-organism poses a health risk. Based on these observations there is insufficient evidence that *Aeromonas* spp. can be considered as enteric pathogens when found in drinking water. Therefore it would be inappropriate to suggest analytical regulations and monitoring for this category of micro-organisms.

1.4.4 Non-tuberculosis mycobacteria (NTMS)

The *Mycobacterium tuberculosis* complex (*M. tuberculosis*, *M. bovis*, and *M. africanum*) includes the species pathogenic for humans and animals, and 'non-tuberculosis mycobacteria' (NTMs) or 'mycobacteria other than *M. tuberculosis*', include the formerly called 'atypical' mycobacteria. In a benchmark review (Wolinsky, 1979), the evidence was summarized that some non-tuberculosis mycobacteria were capable of causing disease. The most common among these include the *Mycobacterium avium* complex (MAC) composed of *M. avium* and *M. intracellulare*, clearly different species based on DNA–DNA homology and 16S rRNA sequences (Shinnick and Good, 1994). On the other hand, on the basis of DNA homology and phenotypic relatedness, it has been proposed that *M. avium*, *M. paratuberculosis*, and the wood pigeon bacillus be placed in one species with the designations *M. avium* subsp. *avium*, *M. avium* subsp. *paratuberculosis*, *M. avium* subsp. *silvaticum*, respectively (Thorel *et al.*, 1990). It should be noted that a "MAIS" complex, including *M. avium*, *M. intracellulare*, and *M. scrofulaceum* has been occasionally described. However, *M. scrofulaceum* is genetically and phenetically distinct from *M. avium* and *M. intracellulare*, and is not a member of the *M. avium* complex (Wayne and Kubica, 1986).

In contrast to tuberculous bacteria that live and grow in human tissue, non-tuberculous bacteria are free-living saprophytes that are widely distributed in the environment: water, soil, dust and aerosols (Grange *et al.*, 1990; Jenkins, 1991; Falkinham, 1996; Yajko *et al.*, 1995). They have been recovered from many piped and treated drinking waters throughout the world (Collins *et al.*, 1984; du Moulin and Stottmeier, 1986; du Moulin *et al.*, 1988; Von Reyn *et al.*, 1993, 1994; Glover *et al.*, 1994; Montecalvo *et al.*, 1994). NTMs are not contaminants picked up from another source but residents capable of surviving and growing in water.

The physiological characteristics of non-tuberculous mycobacteria have provided an understanding of their ecological distribution (Falkinham, 1996); (i) they grow best at low pH values, that is, between 5 and 5.5, (ii) they grow microaerobically; (iii) *M. avium* strains grow at 45°C, unlike *M. intracellulare* which can grow to only 42°C; (iv) they grow equally well in water with high level of salt and are found in large numbers in brackish swamps and estuaries; (v) they are hydrophobic and collect largely at air–water interfaces; (vi) they are relatively resistant to heavy metals and may be isolated from water with high levels of metal pollution; (vii) it has been shown that they may be responsible for degradation of a wide variety of xenobiotic organic coumpounds; they are capable of transforming compounds not readily biodegradable such as humic and fulvic acids (Kirschner *et al.*, 1999); (viii) their high resistance to disinfection by chlorine undoubtedly contributes to their persistence in drinking water systems. Experimental data suggest a role for protozoa as hosts for pathogenic mycobacteria, such as *M. avium* (Cirillo *et al.*, 1997; Steinert *et al.*, 1998; Miltner and Bermudez, 2000). There have been few systematic studies of drinking water systems over long periods of time and of the effects of treatment plants. Within the Paris water distribution system (Le Dantec *et al.*, 2002), saprophytic mycobacteria accounted for 41% of all positive samples and potential pathogens for 16%. *M. intracellulare* was detected in only one sample. In the studies of Falkingham *et al.* (2001) and LeChevallier *et al* (2001) mycobacterial recovery from drinking water distribution systems throughout the United States was low (15% of samples, 3% yielding *M. avium* and 1% *M. intracellulare*). Water treatment substantially reduced the number of mycobacteria in raw water by 2 to 4 log units and biofilms in the distribution pipework appear to be rich sources of these organisms. Although the data confirm that members of the genus *Mycobacterium* are isolated in drinking water, the numbers and frequencies of recovery of *M. avium* and *M. intracellulare* are low.

The concern about non-tuberculosis mycobacterial disease has been radically changed by emergence of the AIDS epidemic throughout the world. Before the AIDS epidemic, and still today in immunocompetent people, non-tuberculous mycobacterial disease was primarily pulmonary and the major pathogens were *M. kansasii*, *M. avium* and *M. intracellulare*. In the absence of evidence of person-to-person transmission, it was suggested that man is infected from environmental sources via aerosols (Wolinsky 1979). *M. scrofulaceum* was considered as the causative agent of cervical lymphadenitis in children and *M. marinum* was associated with skin infections originating from aquaria or swimming pools. However, with the advent of the AIDS epidemic in the United States and Europe, in immunodeficient individuals, non-tuberculous mycobacterial disease is usually systemic with acid-fast organisms, principally *M. avium*, isolated commonly from either blood or stool. In Africa and other areas of the developing world, where the incidence of tuberculosis is high, the rate of non-tuberculous mycobacterial disease in AIDS people is low (Morrissey *et al.*, 1992). In all probability, AIDS patients die of other infections before they reach a stage at which slow onset *M. avium* disease develops.

Limited data exist to evaluate the risk of *M. avium* complex infection from water. Most interest relates to those in people with HIV disease. Von Reyn *et al.* (1994) were among the first to document a relationship between infections in AIDS patients and water as the source of the *M. avium* complex. Examination of isolates from patients and from waters by PFGE showed identical patterns. A study from Aronson *et al.* (1999) also supports the possibility that potable water is a source of nosocomial spread of *M. avium* infections in hospitals. Recirculating hot water systems, which are used in many institutions such as hospitals, hotels, apartments, and office buildings, may allow thermotrophic and chlorine-resistant mycobacteria to persist and colonize, once they have been introduced from municipal systems (du Moulin *et al.*, 1986, 1988). Infection with *M. avium* complex is thought to occur from colonization of the gastrointestinal tract (Bodmer *et al.*, 2000), although respiratory access has also been documented (Mc Fadden *et al.*, 1992; Horsburgh, 1992). Hot water showers may be a source of infection, but drinking water cannot be excluded as a possible source, because a common tap may deliver both hot and cold water (Squier *et al.*, 2000). The major case–control study (Horsburgh *et al.*, 1994; Ristola *et al.*, 1999) does not support a waterborne hypothesis for MAC infections of AIDS patients. From a public health point of view, the occurrence of bacteria of the *M. avium* complex in drinking water systems does not appear to pose any particular risk for the general population. Epidemiological characteristics of these mycobacterial infections in HIV patients remain poorly understood, especially in terms of the mode of transmission and measures for preventing clinical disease, as discussed by Singh and Yu (1994) and Montecalvo *et al.* (1994) commenting on the work by Van Reyn *et al.* (1994).

Before the emergence of the AIDS epidemic, *M. kansasii* infections were generally more common than *M. avium* complex infections. The AIDS epidemic has had, in most countries, a striking effect on the incidence of disease caused by the *M. avium* complex but not *M. kansasii*. Indeed, between 1984 and 1992 there has been a 10-fold increase in the number of *M. avium* complex infections compared with *M. kansasii* infections in England (Yates *et al.*, 1993). *M. kansasii* disease in immunocompetent patients (e.g., those with AIDS) can be disseminated or exclusively pulmonary. There have been a number of reports relating the occurrence of *M. kansasii* in drinking water and shower heads (Engel *et al.*, 1980; Kaustova *et al.*, 1981; Collins *et al.*, 1984; Falkinham, 1996). Therefore, as with *M. avium* complex infections, the water exposure route may be possible.

M. avium subsp. *paratuberculosis* (MAP) is the agent of Johne's disease in cattle. A number of microbiological and epidemiological studies have suggested that it may be also implicated in Crohn's disease in humans, which is a chronic, incapacitating inflammatory disease of the gastrointestinal tract (Falkinham, 1996). Water might be one of a number of sources through which people could be exposed to MAP (Herman Taylor *et al.*, 2000; Nichols and Lee, 2002). If it is shown that MAP or wood pigeon strains of *M. avium* (McFadden *et al.*, 1992) are responsible for disease, human infection could also occur through exposure to contaminated water (e.g. drinking or aerosols).

1.5 *HELICOBACTER PYLORI*

The assumption that *Helicobacter pylori* is waterborne needs to be substantiated. Half of the world's population is infected with *H. pylori*, making it a pathogen of potentially great significance. Although in the majority of cases, infection is harmless, many infected people develop chronic gastritis, peptic ulcer disease or gastric cancer (Ernst and Gold, 2000). Although all evidence suggests that *H. pylori* is well suited to attach to the gastric mucus and the gastric epithelium, it is difficult, however, to establish whether this ecological niche is the only one. Studies suggest four ways of transmission (Tursi *et al.*, 1997; Leung *et al.*, 1999; Parsonnet *et al.*, 1999; Deltenre and de Koster, 2000): (1) by the faecal–oral route, the bacterium, excreted by faeces, might colonize water sources, becoming available to be transmitted to humans; (2) by oral–oral route, *H. pylori*, which colonizes dental plaque and saliva, may be transmitted by saliva to other individuals (person to person transmission); (3) by gastric–oral route, the typical modality of transmission in childhood, by contaminated vomitus; (4) finally, by gastric–gastric route, the bacterium might be transmitted by endoscopic procedures.

The natural route of transmission is by gastric juice, specifically as a result of epidemic vomiting in childhood (Leung *et al.*, 1999; Parsonnet *et al.*, 1999). The different modalities of transmission may be contemporaneously involved, however, because no one *per se* is able to explain the widespread occurrence of *H. pylori* infection. Many studies have examined the possibility that *H. pylori* is waterborne. *H. pylori*-specific DNA was detected in sewage (Vincent, 1995), surface water (Sasaki *et al.*, 1999), and water supplies (Hulten *et al.*, 1996, 1998; McKeown *et al.*, 1999; Baker and Hegarty, 2001; Imanischi *et al.*, 2003) including biofilms (Park *et al.*, 2001; Bunn *et al.*, 2002) although the organisms should be readily inactivated by free chlorine (Johnson *et al.*, 1997). Actively respiring bacteria have been found by testing with monoclonal antibodies, in the majority of surface and shallow groundwater samples tested in the United States (Hegarty *et al.*, 1999). The survival capacity of *H. pylori* is related to the non-cultivable coccoid form, which may persist up to 20 to 30 days in water (Mizoguchi *et al.*, 1999; Nilsson *et al.*, 2002) and also in food (Jiang and Doyle, 1998). It has also been reported that under experimental conditions free-living amoebae promote growth and survival of this bacteria (Winiecka-Krusnell *et al.*, 2002).

Studies of prevalence or seroprevalence suggest that drinking water might play some role in infection with *H. pylori* (McKeown *et al.*, 1999), which has not been confirmed by other work (Parsonnet *et al.*, 1992; Begue *et al.*, 1998). An increasing body of data show that *H. pylori* DNA can be detected by PCR from faecal samples of infected individuals or patients with peptic ulcer (Watanabe *et al.*, 1998; Makristathis *et al.*, 1998; Russo *et al.*, 1999; Gramley *et al.*, 1999) strongly suggesting faecal-to-oral transmission. The detection of *H. pylori* antigen in stools (Hp SA) brings new promise for diagnosing *H. pylori* infections (Vaira *et al.*, 1999). The results obtained from *H. pylori* infection in an experimental murine model also seem to support an oral–faecal route as the primary mode of transmission (Cellini *et al.*, 1999).

However, many characteristics make *H. pylori* a special bacterium in the world of human pathogens. Considerable work remains to fully understand the epidemiology of transmission.

1.6 PATHOGENIC PROTOZOA

The most prevalent enteric protozoa associated with waterborne disease include *Giardia lamblia* and *Cryptosporidium parvum* (Casemore, 1990; Current and Garcia, 1991; Meinhard *et al.*, 1996; Perz *et al.*, 1998). *Cyclospora cayetanensis*, *Isospora belli* and many microsporidian species can be considered emerging pathogen protists (Marshall *et al.*, 1997; Curry and Smith, 1998; Herwald, 2000; Mota *et al.*, 2000). Microsporidian organisms were recently classed as fungi on the basis of molecular data (Lecointre and Le Guyader, 2001). However, they will be treated in the present chapter, as detection methods for these pathogens are similar to those used for protists.

1.6.1 *Giardia, Cryptosporidium*

As a result of proliferation and evolution in the intestinal tract, cysts (*Giardia*), oocysts (*Cryptosporidium*, *Cyclospora*, *Isospora*) or infective spores (Microsporidia) are produced and excreted in faeces in a fully infective form (*Giardia, Cryptosporidium*, Microsporidia), or as immature stages (*Cyclospora, Isospora*), which will shortly complete their development in the environment and acquire their infectious form. Giardiasis has become the most common cause of human waterborne disease in the United States over the past 30 years.

The first outbreak was documented in 1965 at Aspen, CO (Craun, 1986). From then until 1996, 122 outbreaks involving 27,000 patients have been reported (Table 1.1).

The apicomplexan *Cryptosporidium* has only more recently been recognised as a cause of waterborne outbreaks. The first occurred in 1985 in Texas, with sewage-contaminated groundwater as the source of the infection (D'Antonio, 1985). The outbreak in Milwaukee was the largest documented waterborne incident ever in

Table 1.1 Involvement of enteric eukaryotic micro-organisms in waterborne disease outbreaks

Enteric protozoa	Waterborne disease outbreaks	High risk for AIDS
Prevalent pathogens		
Giardia lamblia	From 1961: 122*	–
Cryptosporidium parvum	From 1985: 10*	+++
Emerging pathogens		
Cyclospora cayetanensis	1994: 1*	?
Microsporidia	Not reported	+++
Isospora belli	Not reported	+++

*In U.S.A.

United States since 1920. An estimated 403,246 cases of diarrhoea were recorded over a span of 2 months, 4400 persons were hospitalized and 69 deaths were recensed among immunocompromised patients (Kaminski, 1994; Haas and Rose, 1994; Mac Kenzie *et al.*, 1995). From 1985 to 1996, 10 outbreaks occurred in the United States that resulted in 419,914 cases of illness. Many outbreaks have also been reported in the UK (Hunter, 1997). The size and severity of waterborne cryptosporidiosis outbreaks have prompted in-depth investigations into their aetiology and transmission.

Many outbreaks of giardiasis and cryptosporidiosis have been associated with potable water, even when processing systems had been operated in accordance with conventional standards of water treatment, and while current microbiological standards were met (Meinhardt *et al.*, 1996; Goldstein *et al.*, 1996; Yamamoto *et al.*, 2000). Relying on *Cryptosporidium* as a model protozoan pathogen, investigations by Casemore (1990), Rose (1990), Current and Garcia (1991), Meinhart *et al.* (1996) and Guyot *et al.* (2001, 2002) attempted to clarify the transmission patterns for these diseases, and the problems to be overcome in their management and monitoring for compliance. These are briefly reviewed below.

(1) A broad variety of animal reservoirs, including farm livestock, pets and wildlife will spread oocysts to humans and the water environment, particularly through agricultural and human effluents, compounded by abundant rainfall. In addition, zoonotic and person-to-person transmission occurs frequently. Fully infective oocysts excreted by humans and animals in large numbers constitute probably the main risk factor for the spread of *Cryptosporidium* in aquatic environments (Rose, 1990, 1993; Meinhard *et al.*, 1996). The contamination of water sources may lead to penetration of water treatment plants and drinking water supplies (Meinhardt *et al.*, 1996; Payment *et al.*, 2000).

(2) *Cryptosporidium* oocysts possess a substantially elevated resistance to disinfectants used in water treatment, with the exception of ozone and chlorine dioxide (Casemore, 1995; Chauret *et al.*, 1998). The physical removal of

the parasite by coagulation and filtration should, in theory, address the risks. All outbreaks of waterborne cryptosporidiosis that occurred in the United States between 1985 and 1993, however, hit communities where surface water supplies had been filtered and where processing equipment met federal and state standards. According to Gale (1996), available evidence suggests that many pathogenic micro-organisms, including cryptosporidia, are clustered to some degree, even within small volumes, exposing some drinking water consumers to much higher doses than others. By assuming that enteric pathogens are randomly dispersed, current models underestimate the risk of contracting infectious diseases. In order to model pathogen densities in drinking water obtained from source water, and develop performance criteria for treatment (for example, removal of efficiency), requires information on the degree to which treatment processes (e.g., filtration and coagulation) affect pathogen clustering. The practice of recycling filter backwash water is thought to lead to breakthrough of oocysts into finished water (Lisle and Rose, 1995). Oocysts of *Cryptosporidium* are thought to attach to and persist in biofilms. Their subsequent release into the water system by sloughing of biofilm material could result in exposure of some individuals to infectious clusters (Lisle and Rose, 1995).

(3) The infective dose of *Cryptosporidium* in humans is not clearly known. Experiments in human volunteers (Dupont *et al.*, 1995; Chappel *et al.*, 1996) obtained dose–response curves indicating a median infective dose (ID_{50}) of 132 oocysts and a minimum infective dose of less than 30. A more recent study (Okhuysen *et al.*, 1999) reported that the minimal infective dose for immunocompetent volunteers can vary between 10 and 1000 oocysts. Studies using animal models showed that as few as 1 to 10 oocysts may initiate an infection (Miller *et al.*, 1986; Kwa *et al.*, 1993; Meinhardt *et al.*, 1996). Using *Cryptosporidium* isolates from human origin, Chappell *et al.* (2000) observed that 100% of immunocompetent volunteers developed an actual infection with inocula of 100 oocysts. Although IDs vary with the viability of oocysts that in turn depends on the degree of environmental stress and on oocyst virulence, the risk assessment model developed by Perz *et al.* (1998) suggests that transmission of oocysts may occur at low levels of the pathogen in drinking water (Gibson *et al.*, 1998). With respect to the Milwaukee outbreak, mathematical modelling combined with epidemiological data suggests that persons might indeed become infected after exposure to only one oocyst (Haas *et al.*, 1994). However, waterborne cases have been inferred, in the absence of detectable oocyst levels, as in the Las Vegas outbreak in 1994 (Haas, 2000).

(4) Molecular studies on *Cryptosporidium* genetic diversity are basically changing our current perspective on the epidemiology of cryptosporidiosis. Until recently, *C. parvum* was the species recognized as responsible for cryptosporidiosis in both animals and humans. Consequently, human cryptosporidiosis was usually considered as a typical zoonosis. In fact, on the basis of isoenzymatic and molecular studies, two genotypes (1 and 2) have been described

in the species *C. parvum*. Genotype 1 was found only in humans, whereas genotype 2 was found in both infected humans and livestock. Moreover, cross infection studies have shown that the bovine genotype readily infects mice and cattle, while the human genotype does not (Peng *et al.*, 1997; Awad-El-Kariem *et al.*, 1998; Widmer *et al.*, 1998). Multilocus analysis indicated that the two genotypes are genetically isolated (Spano *et al.*, 1998; Caccio *et al.*, 1999). These results suggest the existence of two separate *C. parvum* transmission cycles: one that may occur in humans or animals, and another that exclusively occurs in humans. In addition, more recent experimental studies performed in gnotobiotic pigs have revealed marked differences between human and bovine *C. parvum* genotypes (Pereira *et al.*, 2002). The infection with *C. parvum* bovine genotype was clearly more severe than the infection induced by the human *C. parvum* genotype. Bovine *C. parvum* isolates had shorter prepatent and patent periods, and caused more severe symptoms or intestinal lesions than parasites from human origin. Subsequent studies have revealed nine other genotypes in *C. parvum* isolates from domestic or wild mammals (dog, mouse, pig, ferret, marsupial, monkey, deer and bear). Furthermore, the species *C. meleagridis* (from birds), *C. felis* and *C. muris* have been identified in human patients (Guyot *et al.*, 2001). Indeed, as only molecular methods allow species or genotype identification of human, animal or environmental *Cryptosporidium* isolates (Guyot *et al.*, 2001, 2002), new *Cryptosporidium* taxonomy and transmission patterns are likely to continue to emerge.

(5) In immunocompromised persons, particularly AIDS patients, cryptosporidiosis symptoms are of greater severity than in immunocompetent people, and indeed may be fatal (Meinhardt *et al.*, 1997; Perz *et al.*, 1998; Byleveld *et al.*, 1999). Most waterborne cryptosporidiosis outbreaks have caused the more fulminant reaction in AIDS patients. Among the 69 deaths in the Milwaukee outbreak, the majority of victims were AIDS patients (MacKenzie *et al.*, 1995).

1.6.2 *Cyclospora, Isospora, Microsporidia*

In addition to *Giardia* and *Cryptosporidium*, Apicomplexa protozoa such as *Cyclospora*, *Isospora* and many microsporidian species are emerging as opportunistic pathogens and may have waterborne routes of transmission. In general, these species have received attention recently because of their high infection rates in AIDS patients (Sobottka *et al.*, 1998; Leder *et al.*, 1998; Curry and Smith, 1998; Mota *et al.*, 2000; Brasil *et al.*, 2000).

Before 1995, the parasite *C. cayetanensis* was primarily described in gastroenteritis among children living in poor sanitary conditions and in travelers who had visited developing countries. Several outbreaks of *Cyclospora* infection (or isolated cases) have been linked to waterborne transmission (Huang *et al.*, 1995; Koumans *et al.*, 1998) but in only one case (Rabold *et al.*, 1994) have *Cyclospora* oocysts been demonstrated in drinking water. In 1996, the largest ever reported outbreak of cyclosporiasis, affecting more than 1,400 persons in North America (Herwald *et al.*, 1997; Bern *et al.*, 1999) was associated with eating fresh raspberries from Guatemala. Most likely, faecally-polluted water, used for spraying biocides on fruit was the indirect source. Notwithstanding technologic limitations, *Cyclospora* have been recovered in limited numbers from water sources and vegetables (Sterling and Ortega, 1999; Mota *et al.*, 2000). Unlike *Cryptosporidium*, which has many known animal hosts (Dumoulin *et al.*, 2000), *Cyclospora*-like organisms have only been recovered from ducks, chickens, dogs, and primates (Sterling and Ortega, 1999).

The apicomplexan *Isospora belli* is among the protozoa that most commonly cause gastroenteritis in immunocompromised hosts, such as patients with AIDS (Lindsay *et al.*, 1997). This species is restricted to the human host. Waterborne transmission of this parasite is considered possible (Karanis *et al.*, 1993), and limited epidemiological data support this route of transmission (Ford, 1999).

Microsporidia protists are obligate, intracellular, spore-forming parasitic organisms recently classified with the fungi. Their host range is extensive and includes most invertebrates and all classes of vertebrates. Five microsporidian genera have been associated with human disease, primarily in immunocompromised persons. These are the *Enterocytozoon*, *Encephalitozoon*, *Trachipleistophora*, *Vittaformae* and *Nosema*. *Enterocytozoon bieneusi* and *Encephalitozoon intestinalis* are the most prevalent microsporidian parasites that cause gastrointestinal infection in AIDS patients (Weber *et al.*, 1994; Brasil *et al.*, 1998; Gumbo *et al.*, 1999). Microsporidia could also be a possible cause of traveller's diarrhoea (Lopez-Velez *et al.*, 1999). Microsporidia may disseminate to cause systemic infection as well as cornea, conjunctiva, liver, and biliary tract infections. They have been recognized as a group of pathogens that have potential for waterborne transmission (Sparfel *et al.*, 1997; Wolk *et al.*, 2000), and recently, methodologies have been developed for the detection of human pathogenic Microsporidia in water (Dowd *et al.*, 1998). However, the study of Fournier *et al.* (2000) showed a low rate of water contamination by *E. bieneusi*, suggesting that the risk of waterborne transmission to humans is limited.

1.7 VIRAL PATHOGENS

The modern era in environmental virology is characterized by recognition of hepatitis E virus (HEV) and more recently identified enteric viruses capable of producing waterborne outbreaks (Metcalf *et al.*, 1995; Koopmans *et al.*, 1997, 1998; Grabow *et al.*, 1999; Van der Poel *et al.*, 2000; Lopman *et al.*, 2002). More than 15 different groups of viruses, encompassing more than 140 distinct types can be found in the human gut. They are excreted by patients and discharged to the aquatic environment in sewage.

These viruses can be divided into three categories with respect to their epidemiological significance (Table 1.2). (1) Some are either non-enteropathogenic, or cause illness unrelated to the gut epithelium; these include poliovirus, coxsackie viruses A and B, echovirus, Hepatitis A and E and some other human enteroviruses.

Table 1.2 Human viral pathogens or putative pathogens that can be transmitted by faecally polluted water. Adapted from Metcalf *et al.* (1995)

Virus	Drinking water	Recreational water	Shellfish
Pathogen viruses			
Enteroviruses:			
Polioviruses	+/−	−	−
Coxsackie A viruses	−	−	−
Coxsackie B viruses	−	−	−
Echoviruses	−	−	−
Enteroviruses 68-71	−	−	−
Hepatitis A viruses	+	+	+
Hepatitis E-like viruses	+	−	+
Gastroenteritis virus			
(enteropathogenicity proved)			
Noroviruses:	+	+	+
Norwalk virus			
Sapoviruses:	+	+	+
Sapporo virus	+	−	+
Astroviruses	+	−	−
Rotaviruses	−	−	−
Enteric adenoviruses 40, 41			
Putative enteropathogens			
Coronaviruses	Toroviruses		
Enteroviruses	Parvoviruses		
Picobirnaviruses	Pestiviruses		

(2) A relatively small group of viruses has been linked to acute gastroenteritis in humans, and some have been shown to be true aetiologic agents; they include noroviruses (NV) and saporoviruses (SV), rotaviruses, astroviruses and some enteric adenoviruses. (3) A third category encompasses putative enteropathogens, such as coronavirus, enterovirus, torovirus, parvovirus and reovirus, for which causal relationships are still unproven.

The occurrence of a virus in a faecal specimen from a patient with gastroenteritis does not prove that the virus is indeed the causative agent of diarrhoea. The first step in an attempted confirmation procedure relies on establishing the identity of the virus by molecular biology (PCR), immunological assessment, or after it has been grown in culture. The second step in demonstrating causality is based largely on guidelines developed from epidemiology (Glass, 1995).

1.7.1 Hepatitis

Hepatitis A virus (HAV) and Hepatitis E virus (HEV) are associated with epidemics and sporadic cases of hepatitis; they are transmitted by the faecal–oral route and therefore sewage-polluted water may constitute a source.

The most established virus group associated with waterborne outbreaks is HAV. It results in widespread epidemics and constitutes an important cause of asymptomatic infections in young children. Mosley (1967) summarized reports of 50 outbreaks from 1895 to 1964. Twenty-nine of these outbreaks show features supporting transmission by water.

The highest incidence of waterborne infectious hepatitis in the United States occurred between the 1950s and the 1970s with 47 outbreaks reported (1653 cases of illness). Since 1971 incidence has declined with only 27 outbreaks between 1971 and 1990 (Craun, 1986, 1992). Since the early 1990s, very few outbreaks of HAV were reported to the CDC (CDC, 1994, 1996; Winn, 1999; De Serres *et al.*, 1999). The largest documented epidemic of shellfish associated HAV occurred in 1988 in Shanghai, China: 250,000 people had Hepatitis A after consumption of contaminated clams (Halliday *et al.*, 1991).

Hepatitis E is much less widespread and mostly confined to tropical and subtropical areas. Infection can be more severe than with hepatitis A, with a high incidence of cholestasis and increased mortality in pregnant women (20%). In 1991 the largest waterborne outbreak of viral Hepatitis E yet reported occurred in Kasspur, India, with an estimated total of over 79,000 cases of disease (Naik *et al.*, 1992). Another waterborne outbreak occurred in Delhi from 1955 to 1956, affecting approximately 29,300 persons (Vishwanathan, 1957). Such large-scale hepatitis epidemics are invariably waterborne and caused by HEV (Bradley, 1992; Aggarwal and Krawczynski, 2000). Recent evidence indicates that HEV might also be prevalent at a very low level in Europe (Zaaijer *et al.*, 1993; Winn, 1999; Widdowson *et al.*, 2003).

1.7.2 Viral gastroenteritis

The recognized viral agents of gastroenteritis include rotavirus, calicivirus (norobreak virus and sapovirus), astrovirus and some enteric adenovirus. Rotaviruses are classified by antigenic groups A, B and C, subgroup and serotype. Group A rotaviruses are the most common cause of diarrhoea disease in infants and young children. Group C rotaviruses, like group A strains, principally cause diarrhoea in young children, age 4 months to 4 years. In contrast to other human rotaviruses, group B rotaviruses are reponsible predominantly for adult diarrhoeal disease, and are therefore designated as adult diarrhoea rotavirus or ADVR.

The most notorious of caliciviruses identified as a cause of gastroenteritis was the agent termed Norwalk virus. This originated from an outbreak of epidemic gastroenteritis in an elementary school in Norwalk, OH, in 1968. Many other related viruses were later classified by immunoelectron microscopy (IEM) as small round structured viruses (SRSV). These included the Hawaï, Snow Mountain, Montgomery County, Taunton, Otofuke, Sapporo and the Osaka agents. Most of these have been identified as calicivirus, together with 50 other SRSVs. The family *Caliciviridae* includes the following four genera: *Vesivirus*, *Lagovirus*, *Norovirus* (NV) and *Sapovirus* (SV) (Pringle 1998). Viruses in the NV and SV genera have been found only in humans, but recent data suggest that calves and pigs may be reservoir hosts of NVs (Van der Poel *et al.*, 2000). Genetically, the human caliciviruses (Hu CVs) contain at least three genetic clusters or genogroups (GC), i.e., GGI, GGII NV and GGIII SV (Jiang *et al.*, 1996, 1997; Green *et al.*, 2000). Recently, accumulated sequences of HuCV strains indicate these genogroups can be further divided (Jiang *et al.*, 1997; Noel *et al.*, 1997; Wright *et al.*, 1998; Parks *et al.*, 1999; Vinje *et al.*, 2000).

Some SRSVs isolated from stools of infants with diarrhoea were named astrovirus, for the characteristic five or six pointed star configuration that was evident on the surface of the viral particles. To date, seven serotypes of human astroviruses (HAstV) have been identified by enzyme immuno-assays, immuno-electron microscopy or reverse transcriptase–polymerase chain reaction (RT–PCR) (Koopmans *et al.*, 1998; Matsui *et al.*, 1998). Phylogenetic analysis identified two genogroups A and B correlated with serotypes (Belliot *et al.*, 1997).

There are currently 51 serotypes of human adenoviruses (WHO 2002), subdivided into six subgroups, A through F. Many types of adenoviruses have been isolated from stool samples, but only types 40 and 41 have been consistently associated with gastroenteritis (Grimwood *et al.*, 1995; Herrmann and Blacklow, 1995; Barnes *et al.*, 1998). The two serotypes from the F subgroup are referred to as enteric adenoviruses (EAd).

Rotaviruses are one of the most important causes of infantile gastroenteritis in the world. If untreated, rotavirosis can result in severe dehydration and death, especially in developing countries. Calicivirus and astrovirus gastroenteritis are generally mild, typically explosive but self-limited, with symptoms lasting 1 to 4 days after a mean incubation time of 2 days. The clinical picture of adenovirus gastroenteritis is similar to that of rotavirosis, but the incubation period lasts from 8 to 10 days.

In North America, viral enteritis outbreaks from drinking water occur infrequently: 10 between 1971 and 1980, 15 between 1981 and 1990, and 1 from 1991 to 1996. Because of their low incidence they have not been included in Figure 2.1. However, it is more than likely that a marked fraction of gastroenteritis of unidentified origin (Acute Gastro-intestinal Infection, AGI) is indeed associated with a viral aetiology (section 2). In recent years, the predominant role of SRSVs, especially NVs, was revealed in several waterborne outbreak reports (Kapikian *et al.*, 1996; Brugha *et al.*, 1999; Kukkula *et al.*, 1999; Hardy, 1999; Glass *et al.*, 2000; Hafliger *et al.*, 2000; Boccia *et al.*, 2002; Anderson *et al.*, 2003; Pedalino *et al.*, 2003). In the Netherlands, approximatively 80% of outbreaks of gastroenteritis that were reported in the past 7 years were attributed to NV (Koopmans *et al.*, 2002). In the United States, 96% outbreaks of non-bacterial acute gastroenteritis reported to the CDC between January 1997 and June 1998 were caused by NV infection. Epidemiological studies of seroprevalence confirm that a large number of infections from Norwalk viruses occur throughout the year (Payment, 1991). The sensitive new molecular techniques based on the reverse transcriptase–polymerase chain reaction (RT–PCR) available today, have made it possible to detect low levels of a wide variety of enteric viruses. The results obtained to date (Grabow *et al.*, 1999; Abbaszadegan *et al.*, 1999; Yates *et al.*, 2000) also support epidemiological data that suggest low-level transmission of viruses by drinking water supplies.

1.7.3 Other candidate viruses

A relatively small group of viruses has been incriminated as causes of acute gastroenteritis in humans, but while most of them could be identified in faecal specimens from patients with diarrhoea, not all were necessarily the aetiologic agent. Coronaviruses have been associated with human gastroenteritis (Mortensen *et al.*, 1985), and with neonatal necrotising entero-colitis (Chany *et al.*, 1982) and have been found regularly in the faeces of patients with non-bacterial gastroenteritis (Gonzalez *et al.*, 1997). Despite two decades of research to assess their causality, the results so far have been inconclusive. Other viruses such as picobirnaviruses, toroviruses, enteroviruses, reoviruses and parvoviruses were also identified in faecal specimens from children and adults with gastroenteritis, and HIV-infected patients (Glass, 1995; Koopmans *et al.*, 1997, 2002; Giordano *et al.*, 1999). However, their role as an aetiologic agent of diarrhoea in the human population remains to be determined because many of these are also found in healthy controls. More recently immune response has been introduced as a parameter in such investigations.

1.7.4 Factors affecting viral infections

Certain characteristics of enteric viruses point to their potentially important role in waterborne outbreaks. (1) More than 140 types of human or animal enteric viruses can be found in sewage (Table 1.2). Some of them have proved enteropathogenicity, others are just putative enteropathogens in humans. Most of the novel viral agents (caliciviruses, astroviruses) found in human diarrhoea specimens are known to cause gastroenteritis in different animal species. (2) Viruses, found in the faeces of symptomatic or asymptomatic hosts attain numbers up to 10^{11} infectious units per gram, excreted for 1 to 4 weeks. The number of viruses currently detectable in raw sewage can reach 10^2 to 10^3 infectious units per litre. (3) The wide distribution of enteric viruses is demonstrated by sero-epidemiological surveys of North American populations, with most of the adults testing positive (Payment, 1991). (4) Many methods have been applied for eliminating viruses from water; all complicated by their small size (20 to 100 nm), resistance to treatment (much higher than that of bacteria), variation in microbial load of raw water, seasonal effects and no random dispersion of pathogens within large volumes and various analytical shortcomings. After drinking water treatment, viruses are still occasionally detectable (Payment and Armon, 1989). The major limitation of water treatment is that it may promote further clustering of pathogens, exposing some consumers to much higher doses than others. (5) The number of viruses required to initiate infection can be estimated in animal models or human volunteers. Under appropriate conditions as little as one infectious particle of rotavirus can trigger disease in animal (Graham *et al.*, 1987) or human models (Ward *et al.*, 1986). From human volunteer studies with Norwalk and related viruses (Cubitt, 1989; Kapikian and Chanock, 1990), astroviruses (Kurtz and Lee, 1987), and rotaviruses (Ward *et al.*, 1986), it appears that the number of virions needed to infect humans or animals is low in terms of infectious units: Payment (1993) rightly stated "The cell culture infectious unit is only a fraction of the real number of infectious viruses". For rotavirus the proportion of culturable particles among electron microscopically detectable viruses can be as high as 1:1000 or more (Payment and Morin, 1989). In the most enteropathogenic bacteria, the ID_{50}s are generally high, of the order of 10^6 to 10^8, with some exceptions (cf. section 1.2). (6) In some waterborne outbreaks of gastroenteritis, viruses were isolated from water that met current bacteriological standards and contained an adequate chlorine level (Goyal, 1983; Craun, 1986; Gerba and Rose, 1990; Schwartzbrod, 1991; Payment, 1993). The isolation of viruses from samples of 'bacteriologically safe' potable water indicates that bacteriological indicators are not adequate to monitor the occurrence of viruses in water.

1.8 GASTROENTERITIS OF UNDETERMINED AETIOLOGY

In the majority of outbreaks of acute gastroenteritis reported in the United States, the aetiology could not be determined. Between 1940 and 1950, 213 outbreaks were recorded and 177 in 1970 to 80. Because waterborne outbreak reporting is voluntary, the differences observed in the frequency of outbreak occurrence may be largely attributed to poor recording by State and Local Health Departments (Craun, 1986). Although in principle, many incidents could have a bacterial or protozoan aetiology, many of these outbreaks might also be of viral origin.

A bacterial cause is relatively easy to identify, while filtration and disinfecting procedures effectively eliminate most of these agents. Some of the outbreaks were clinically and epidemiologically consistent with Norwalk-like virus infections (Kaplan *et al.*, 1982); however, the detection of enteric viral pathogens was impossible at that time. Even in 1970 to 1980 *Giardia* and *Cryptosporidium* were only rarely recognized as aetiologic agents.

1.9 ENDEMIC GASTROENTERITIS

A limited number of studies carried out in Canada and later in France seem to point at endemic infections associated with drinking water complying with bacteriological standards. Payment *et al.* (1991a, 1991b) in Montreal, followed a randomised trial design, in order to quantify the health hazard remaining after a standard, well-conducted treatment of bacteriologically contaminated surface water has been completed. Payment's study showed that municipal tap water drinkers had a significantly higher rate of acute gastrointestinal illnesses (35%) than those provided with a domestic water filter. The treatment used in the community water system included predisinfection, flocculation, sand filtration, ozonation and final chlorine or chlorine dioxide disinfection procedures meeting the North American microbiological and physicochemical water quality standards. Payment *et al.* (1994) suggest that heterotrophic, potentially pathogenic agents, for example, *Bacillus* spp. could be responsible for these incidents. This assumption has never been confirmed, however, while a disproportionately high percentage of *Bacillus* in the studied water supplies has not been found by other authors (Edberg *et al.*, 1997). An epidemiological study was conducted in France by Ferley *et al.* (1986) and Zmirou *et al.* (1987) in order to directly assess the residual health risk after disinfection of piped water processed from groundwater. The crude incidence rate of diarrhoea was 1.4 times more frequent among children drinking treated water than among controls (Zmirou *et al.*, 1995). The morbidity observed in this study comprises small sporadic outbreaks, along with an endemic occurrence of acute gastrointestinal infection. Even when days with 'epidemics' were excluded from the analysis, the routine occurrence of acute gastroenteritis events was still greater in children drinking treated water than controls, suggesting that microbiological contamination is continuous. Information on the association of specific HPC (Heterotrophic Plate Count) bacteria with health effects may be derived from other more recently conducted epidemiological studies (Hellard *et al.*, 2001; Colford *et al.*, 2002). Opportunistic pathogenic bacteria may be recovered among HPC, including strains of *P. aeruginosa* and *Aeromonas* spp. There is no evidence, however, of a relationship of any of these opportunistic pathogens with acute gastrointestinal illness through the waterborne route among the general population (WHO, 2002b).

In Vermont, where giardiasis has been the most commonly diagnosed gastrointestinal disease, with annual incidence rates higher than in other states, waterborne transmission has been suggested to be an important cause of non-outbreak related cases. Rates of infection were reported to be highest in persons that drank non-filtered city water (Birkhead and Vogt, 1989). These observations have been confirmed by Fraser and Cook (1991) in New Zealand and by Dennis *et al.* (1993) in New Hampshire, where drinking untreated surface water and recreational water exposure were strongly associated with endemic giardiasis.

Cryptosporidium spp. probably present a greater endemic risk than *Giardia* spp., because of their more frequent occurrence in surface water and higher resistance to water purification technology, including filtration. Perz *et al.* (1998) used a risk assessment model to examine the potential role of tap water in endemic infections with *Cryptosporidium* spp. In comparing the model output with surveillance data for cryptosporidiosis, the analysis showed that low-level transmission via tap water can represent an important exposure route for endemic *Cryptosporidium* infection.

Endemic transmission of protozoan infections results from the low infectious dose of these organisms for humans, and their heterogeneous distribution in clusters in treated waters. Similarly, enteric viruses might be involved in outbreaks of endemic waterborne gastroenteritis. On the other hand, a similar role for bacterial enteropathogens seems less likely because their infectious dose is usually higher and their resistance to water disinfection lower.

Management of endemic transmission of waterborne infections requires considerable improvements in existing epidemiological and analytical databases. Prospective studies on populations that have been exposed to different levels of pathogens, including control subjects, are needed. The outcome of such investigations may well result in future improvements in water treatment technologies aimed at an unconditionally safe drinking water supply.

1.10 PATHOGENICITY AND AT INCREASED RISK POPULATIONS

Drinking waters should be safe for lifetime use, taking account of differing sensitivities across life stages. However, all drinking waters are not necessarily suitable for people suffering from immunodeficiency (WHO, 2002b). Pathogenicity of micro-organisms reflects the host–parasite equilibrium, governed by very dynamic ecological conditions (Isenberg, 1988). Historical definitions of pathogens were based on their ability to cause disease as an invariant trait. Nosocomial diseases and the complications of acquired immune deficiency syndrome (AIDS) illustrate why the pathogenicity and virulence concepts are no longer sufficient to explain fully the harmful interactions between the micro-organisms and the human host. Even in 1934, Smith recognized the importance of the host. Defining microbial pathogenesis in terms of host damage permits the inclusion of many variables which affect the host–pathogen relationship. In this view, virulence is a property of the pathogen, but it is modulated by host susceptibility and resistance (Casadevall and Pirofski, 1999, 2000). Therefore, it is neither easy nor appropriate to make an absolute distinction between pathogenic micro-organisms (often qualified as 'true' or 'real'), those considered as only 'potential' pathogens and members of the normal human microbiota. Many of these potential pathogens are found in both raw sources and in drinking water, including representatives from the following genera: *Pseudomonas*, *Acinetobacter*, *Stenotrophomonas*, *Aeromonas*, *Legionella* and *Mycobacterium*. In this context, infections that arise with increasing frequency in hosts with generalized or local defects in antimicrobial defense mechanisms have been called 'opportunistic infections'. An opportunistic micro-organism, as defined by Von Graevenitz (1977), is one that utilizes the opportunity offered by weakened defence mechanisms to inflict damage to the host.

There are certain sensitive groups in the general population (Gerba, 1996) such as pregnant women and newborn children, young children (up to 5 years old) and elderly people. The mortality rate is significantly greater in young children and the elderly compared with adults. Common clinical observations show us that microbial infections are more serious and more frequent during pregnancy, and in particular in the third quarter of gestation. This global sensitivity is still subject to major variations according to groups. So, within the general population, children and the elderly are at highest risk for waterborne gastrointestinal disease

(Schwartz *et al.*, 2000). In contrast, children and young people are rarely affected with legionellosis and then only when severely immunologically impaired, while immunocompromised patients, especially transplant recipients, run a very high risk of disease.

The human immuno-deficiency virus epidemic has resulted in a marked increase in infections by organisms that rarely cause disease in healthy hosts. In all waterborne cryptosporidiosis outbreaks it has been observed that immunocompromised persons, such as AIDS patients, are at greater risk than immunocompetent persons for developing severe illness, and for a potential lethal outcome. In the Milwaukee epidemic (Hass and Rose, 1994; Kaminski, 1994; Mac Kenzie *et al.*, 1994), at least 69 fatalities were recorded among immunocompromised patients. In the outbreak that occurred in 1994 in Clark County (Nevada) (Roefer *et al.*, 1996; Goldstein *et al.*, 1996) 63 of the 78 laboratory confirmed cases were in HIV-infected adults, among whom 32 died. Not surprisingly, more recently reported cases of cyclosporiasis, and diarrhoea with *Isospora* or Microsporidia in stools, include over-representation of immunocompromised patients with AIDS (Marshall *et al.*, 1997; Cotte *et al.*, 1999). The increased susceptibility to protozoan infection, and particularly to *Cryptosporidium*, prompted public health authorities to launch an information campaign targeted at recommendations for specific measures aimed at reducing the risk for waterborne cryptosporidiosis, including boiling water before use, using water filters or bottled water and avoiding swimming in lakes, rivers or public swimming pools. This does not exempt these authorities, however, from the obligation to manage health risks associated with low-level oocyst contamination of fully treated drinking water. Such efforts should be guided by surveillance systems and by epidemiologic studies designed to assess the public health significance of very low levels of *Cryptosporidium* oocysts.

Health care facilities including hospitals and health centres represent a specific area of concern for emerging and nosocomial infections, given increased numbers of hosts with immune deficiencies. Organ transplantation, invasive surgery, implantation of prosthetic devices, and the use of immunosuppressive therapies prolong survival for some diseases, but also result in compromised immunity and render previously immunocompetent individuals susceptible to microbes formerly considered to be common saprophytes. It is important to emphasise that the majority of infections due to opportunistic bacteria are from non-water sources (endogenous flora, cross-infection, invasive exploration, multi-resistant bacteria). The most frequent opportunistic and/or nosocomial infections are, for hospital patients, the urinary tract infections, pneumonia, bacteriurias on catheters, and upper respiratory infections. It is only when the microbe has particular virulence factors for sites in the gastrointestinal tract, and the specific host defence mechanisms in the gastrointestinal tract are breached, that infection of this organ system occurs (Edberg, 1996; Duncan and Edberg, 1995). Bacteria typically described as 'opportunistic pathogens' that may regrow in water supplies or hot water systems include strains of *Legionella*, *P. aeruginosa*, *Aeromonas* spp., and non-tuberculous Mycobacteria. The hot water plumbing system of many hospitals is colonized by *Legionella* species, and many hospital outbreaks of Legionnaires' disease have been attributed to *Legionella* contaminated potable water systems. The other opportunistic bacterial species have been identified in a number of hospital outbreaks, where investigations have implicated the water supply (WHO, 2002b).

1.11 CONCLUSION

Many classes of pathogens excreted in faeces are capable of initiating waterborne infections. There are bacterial pathogens, including enteric and aquatic bacteria, enteric viruses and enteric protozoa, which survive in the water environment and are strongly resistant to most disinfectants. The infection dose of viral and protozoan agents is lower than bacteria, in the range of one to 10 infection units.

Waterborne outbreaks of bacterial origin (particularly typhoid fever) in developing countries have declined dramatically through the 1900s. However, some bacterial agents such as *S. sonnei* remain prevalent. New pathogens of faecal origin such as zoonotic *C. jejuni* and *E. coli* O157:H7 may contaminate pristine waters through wildlife or domestic animal faeces. The common feature of these bacteria is the low inoculum (a few hundred cells) that may trigger disease. The emergence in early 1992 of serotype O139 of *V. cholerae* with epidemic potential in Southeast Asia, suggests that other serotypes (other than *V. cholerae* O1) could also have epidemic potential.

New pathogens include environmental bacteria that are capable of survival and proliferation in water distribution systems. Other than specific susceptible hosts, the general population is relatively resistant to infection with ingested *P. aeruginosa*. The significance of *Aeromonas* spp. in drinking water and its relation to acute gastroenteritis remains a debatable point and has to be evaluated in further epidemiological studies. *Legionella* and *Mycobacterium avium* complex (MAC) are environmental pathogens that have found an ecologic niche in drinking and hot water supplies. Numerous studies have reported Legionnaires' disease caused by *L. pneumophila* occurring in residential and hospital water supplies. *M. avium* complex causes disseminated infections in AIDS patients and drinking water is strongly suggested as a source of infection.

An increasing number of reports show that *Helicobacter pylori* DNA can be amplified from faeces samples of infected patients, strongly suggesting faecal-to-oral transmission. It is therefore possible that *H. pylori* infection is waterborne, but this assumption needs to be substantiated.

Giardiasis has become the most common cause of human waterborne disease in the United States over the past 30 years. However, as a result of the massive outbreak of waterborne cryptosporidiosis in Milwaukee, Wisconsin, affecting an estimated 403,000 persons, there is increasing interest in the epidemiology and prevention of new infections caused by *Cryptosporidium* spp., as well as monitoring of water quality. The transmission of *Cryptosporidium* and *Giardia* through treated water supplies that meet water quality standards, demonstrates that water treatment technologies have become inadequate, and that a negative coliform no longer guarantees that water is free from all pathogens, especially from protozoan agents. Substantial concerns persist that low levels of pathogen occurrence may be responsible for the endemic transmission of enteric disease. In addition to *Giardia* and *Cryptosporidium*, some species of the genera *Cyclospora* and *Isospora* and of the Microsporidia group are emerging as opportunistic pathogens and may use waterborne routes of transmission.

More than 15 different groups of viruses, encompassing more than 140 distinct types can be found in the human gut. Some cause illnesses unrelated to the gut epithelium, such as HAV and HEV. Numerous large outbreaks have been documented in the United States between 1950 and 1970 and the incidence has strongly declined in developed countries since the 1970s. Hepatitis E is mostly confined to tropical and subtropical areas but recent reports indicate that it can occur at a low level in Europe. A relatively small group of viruses have been incriminated as causes of acute gastroenteritis in humans and have proved to be true aetiologic agents, including rotavirus, calicivirus (norovirus and saporovirus), astrovirus and some enteric adenoviruses. These enteric viruses have not been frequently identified as the aetiologic agents of waterborne disease outbreaks, because of inadequate diagnostic technology. However, many outbreaks of unknown aetiology, currently reported, are likely to be due to viral agents. Today, noroviruses are recognized as the major causes of waterborne illnesses worldwide.

The global burden of infectious waterborne disease is considerable. Reported numbers dramatically underestimate the real incidence of waterborne diseases. The most striking concern is that enteric viruses, such as caliciviruses, and some protozoan agents, such as *Cryptosporidium*, are the best candidates to reach the highest levels of endemic transmission. They are ubiquitous in water intended for drinking, and are highly resistant to relevant environmental factors, including chemical disinfection procedures. Other concerns are the enhanced risks for the classical group of debilitated subjects (very young, old, pregnant and immunocompromised individuals) and the basic requirement to take specific measures aimed at reducing the risk of waterborne infectious diseases in this growing, immunologically weaker population.

REFERENCES

Abbaszadegan, M., Stewart, P. and LeChevallier, M. (1999) A strategy for detection of viruses in groundwater by PCR. *Appl. Environ. Microbiol.* **65**, 444–449.

Aggarwal, R. and Krawczynski, K. (2000) Hepatitis E: an overview and recent advances in clinical and laboratory research. *J. Gastroenterol. Hepatol.* **15**, 9–20.

Alary, M. and Nadeau, D. (1990) An outbreak of *Campylobacter enteritis* associated with a community water supply. *Can. J. Public. Health* **81**, 268–271.

Albert, M.J., Neira, M. and Motarjemi, Y. (1997) The role of food in the epidemiology of cholera. *World Health Stat.* **50**, 111–118.

Albert, M.J., Ansaruzzaman, M., Talukder, K.A., Chopra, A.K., Kuhn, I., Rahman, M., Faruque, A.S., Islam, M.S., Sack, R.B. and Mollby, R. (2000) Prevalence of enterotoxin genes in *Aeromonas* spp. isolated from children with diarrhea, healthy controls, and the environment. *J. Clin. Microbiol.* **38**, 3785–3790.

Aleksic, S and Bockem, H.L.J. (1988) Serological and biochemical characteristics of 416 *Yersinia* strains from well water and drinking-water plants in the Federal Republic of Germany: lack of envidence that these strains are of public health importance. *Zentralbl. Bakteriol. Mikrobiol. Hyg. B* **185**, 527–533.

Allen, M.J., Edberg, S.C. and Reasoner, D.J. (2002) Heterotrophic plate count (HPC) bacteria. What is their significance in drinking water? Presented at the *NSF International/World Health Organization Symposium on HPC Bacteria in Drinking Water.* Geneva, 22–24 April.

Anderson, A.D., Heryford, A.G., Sarisky, J.P., Higgins, C., Monroe, S.S., Beard, R.S., Newport, C.M., Cashdollar, J.L., Fout, G.S., Robbins, D.E., Seys, S.A., Musgrave, K.J., Medus, C., Vinje, J., Bresee, J.S., Mainzer, H.M. and Glass, R.I. (2003) A waterborne outbreak of Norwalk-like virus among snowmobilers-Wyoming, 2001. J. *Infect. Dis.* **187**, 303–306.

Anon. (1999) Waterborne pathogens. *AWWA Manual of Water Practices*, M48. A.W.W.A, Denver, CO.

Araujo, R.M., Arribas, R.M., Lucena, F. and Pares, R. (1989) Relation between *Aeromonas* and faecal coliforms in fresh waters. *J. Appl. Bacteriol.* **67**, 213–217.

Aronson, T., Holtzman, A., Glover, N., Boian, M., Froman, S., Berlin, O.G., Hill, H. and Stelma, G. Jr. (1999) Comparison of large restriction fragments of *Mycobacterium avium* isolates recovered from AIDS and non-AIDS patients with those of isolates from potable water. *J. Clin. Microbiol.* **37**, 1008–1012.

Ashiru, J.O., Salau, T. and Rotilu, I.O. (1993) Incidence of *Aeromonas* species in diarrhoeic stool University College Hospital Ibadan, Nigeria. *Comp. Immunol. Microbiol. Infect. Dis.* **16**, 51–54.

Atlas, R.M. (1999) *Legionella*: from environmental habitats to disease pathology, detection and control. *Environ. Microbiol.* **1**(4), 283–293.

Awad-El-Kariem, F.M., Robinson, H.A., Petry, F., McDonald, V., Evans, D. and Casemore, D. (1998) Differentiation between human and animal isolates of *Cryptosporidium parvum* using molecular and biological markers. *Parasitol. Res.* 84, 297–301.

Baker, K.H. and Hegarty, J.P. (2001) Presence of *Helicobacter pylori* in drinking water is associated with clinical infection. *Scand. J. Infect. Dis.* **33**(10), 744–746.

Barnes, G.L., Uren, E., Strevens, K.B. and Bishop, R.F. (1998) Etiology of acute gastroenteritis in hospitalized children in Me from April 1980 to March 1993. *J. Clin. Microbiol.* **36**(1), 133–138.

Basu, A., Garg, P., Datta, S., Chakraborty, S., Bhattacharya, T., Khan, A., Ramamurthy, S., Bhattacharya, S.K., Ymasaki, S., Takeda, Y. and Nair, G.B. (2000) *Vibrio cholerae* O139 in Calcutta, 1992–1998: incidence, antibiograms and genotypes. *Emerg. Infect. Dis.* **6**, 139–147.

Bates, M.N., Maas, E., Martin, T., Harte, D., Grubner, M. and Margolin, T. (2000) Investigation of the prevalence of *Legionella* species in domestic hot water systems. *N. Zl. Med. J.* **113**, 218–220.

Begue, R.E., Gonzales, J.L., Correa-Gracian, H. and Tang, S.C. (1998) Dietary risk factors associated with the transmission of *Helicobacter pylori* in Lima, Peru. *Am. J. Trop. Med. Hyg.* **59**, 637–640.

Belliot, G., Laveran, H. and Monroe, S.S. (1997) Detection and genetic differentiation of human astroviruses: phylogenetic grouping varies by coding region. *Arch. Virol.* **142**, 1323–1334.

Beltran, P., Delgado, G., Navarro, A., Trujillo, F., Selander, R.K. and Cravioto, A. (1999) Genetic diversity and population structure of *Vibrio cholerae*. *J. Clin. Microbiol.* **37**, 581–590.

Berk, S.G., Ting, R.S., Turner, G.W. and Ashburn, R.J. (1998) Production of respirable vesicles containing live *Legionella pneumophila* cells by two *Acanthamoeba* spp. *Appl. Environ. Microbiol.* **64**, 279–286.

Bern, C., Hernandez, B., Lopez, M.B., Arrowood, M.J., de Mejia, M.A., de Merida, A.M., Hightower, A.W., Venczel, L., Herwaldt, B.L. and Klein, R.E. (1999) Epidemiologic studies of *Cyclospora cayetanensis* in Guatemala. *Emerg. Infect. Dis.* **5**, 766–774.

Birkhead, G. and Vogt, R.L. (1989) Epidemiologic surveillance for endemic *Giardia lamblia* infection in Vermont. The roles of waterborne and person-to-person transmission. *Am. J. Epidemiol.* **129**, 762–768.

Black, R.E., Levine, M.M., Clements, M.L., Hughes, T.P. and Blaser, M.J. (1988) Experimental *Campylobacter jejuni* infection in humans. *J. Infect. Dis.* **157**, 472–479.

Boccia, D., Tozzi, A.E., Cotter, B., Rizzo, C., Russo, T., Buttinelli, G., Caprioli, A. Marziano, M.L. and Ruggeri, F.M. (2002) Waterborne outbreak of Norwalk-like virus gastroenteritis at a tourist resort, Italy. *Emerg. Infect. Dis.* **8**, 563–568.

Bodmer, T., Miltner, E. and Bermudez, L.E. (2000) *Mycobacterium avium* resists exposure to the acidic conditions stomach. *FEMS Microbiol. Lett.* **182**, 45–49.

Bopp, D.J., Sauders, B.D., Waring, A.L., Ackelsberg, J., Dumas, N., Braun-Howland, E., Dziewulski D., Wallace, B.J., Kelly, M., Halse, T., Musser, K.A., Smith, P.F., Morse, D.L. and Limberger, R.J. (2003) Detection, isolation and molecular subtyping of *Escherichia coli* O157:H7 and *Campylobacter jejuni* associated with a large waterborne outbreak. *J. Clin. Microbiol.* **41**(1), 174–180.

Borroto, R.J. (1997) Ecology of *Vibrio cholerae* serogroup O1 in aquatic environments. *Rev. Panamer. Salud Publica* **1**, 3–8.

Boukadida, J., de Montalembert, M., Gaillard, J.L., Gobin, J., Grimont, F., Girault, D., Véron, M. and Berche, P. (1991) Outbreak of gut colonization by *Pseudomonas aeruginosa* in immunocompromised children undergoing total digestive decontamination: analysis by pulsed-field electrophoresis. *J. Clin. Microbiol.* **29**, 2068–2071.

Bradley, D.J. (1977) Improvements of rural domestic water supplies. *Proc. R. Soc. Lond.* B **199**, 37–47.

Bradley, D.W. (1992) Hepatitis E: epidemiology, aetiology and molecular biology. *R. Med. Virol.* **2**, 19–28.

Brasil, P., de Paiva, D.D., de Lima, D.B., da Silva, E.J., Peralta, J.M., da Silva, A.J., Sodre, F.C., Villela, E.V. and Moura, H. (1998) A 3-year follow-up of a Brazilian AIDS patient with protracted diarrhea caused by *Enterocytozoon bieneusi*. *Rev. Inst. Med. Trop. Sao Paulo* **40**, 215–218.

Brasil, P., de Lima, D.B., de Paiva, D.D., Lobo, M.S., Sodre, F.C., Silva, S.P., Villela, E.V., Silva, E.J., Peralta, J.M., Morgado, M., and Moura, H. (2000) Clinical and diagnostic aspects of intestinal microsporidiosis in HIV-infected patients with chronic diarrhea in Rio de Janeiro, Brazil. *Rev. Inst. Med. Trop. Sao Paulo* **42**, 299–304.

Brennhovd, O., Kapperud, G. and Langeland, G. (1992) Survey of thermotolerant *Campylobacter* spp. and *Yersinia* spp. in three surface water sources in Norway. *Int. J. Food Microbiol.* **15**, 327–338.

Brugha, R., Vipond, I.B., Evans, M.R., Sandifer, Q.D., Roberts, R.J., Salmon, R.L., Caul, E.O. and Mukerjee, A.K. (1999) A community outbreak of food-borne small round-structured virus gastroenteritis caused by a contaminated water supply. *Epidemiol. Infect.* **122**, 145–154.

Buck, A.C. and Cook, E.M. (1969) The fate of ingested *Pseudomonas aeruginosa* in normal persons. *J. Med. Microbiol.* **2**, 521–525.

Buckley, J.T. and Howard, S.P. (1999) The cytotoxic enterotoxin of *Aeromonas hydrophila* is aerolysin. *Infect. Immunol.* **67**, 466–467.

Bunn, J.E., MacKay, W.G., Thomas, J.E., Reid, D.C. and Weaver, L.T. (2002) Detection of *Helicobacter pylori* DNA in drinking water biofilms: implications for transmission in early life. *Lett. Appl. Microbiol.* **34**(6), 450–454.

Burke, V., Robinson, J., Gracey, M., Peterson, D. and Partridge, K. (1984) Isolation of *Aeromonas hydrophila* from a Metropolitan water supply: seasonal correlation with clinical isolates. *Appl. Environ. Microbiol.* **48**, 361–366.

Buswell, C.M., Herlihy, Y.M. Lawrence L.M., McGuiggan, J.T., Marsh, P.D., Keevil, C.W. and Leach, S.A. (1998) Extended survival and persistence of *Campylobacter* spp. in water and aquatic biofilms and their detection by immunofluorescent-antibody and –rRNA staining. *Appl. Environ. Microbiol.* **64**, 733–741.

Byleveld, P.M., Hunt, A. and McAnulty, J.M. (1999) Cryptosporidiosis in the immunocompromised: weighing up the risk. *Med. J. Aust.* **171**, 426–428.

Caccio, S., Homan, W., van Dijk, K. and Pozio, E. (1999) Genetic polymorphism at the beta-tubulin locus among human and animal isolates of *Cryptosporidium parvum. FEMS Microbiol. Lett.* **170**,173–179.

Cappelier, J.M., Minet, J., Magras, C., Colwell, R.R. and Federighi, M. (1999) Recovery in embryonated eggs of viable but nonculturable *Campylobacter jejuni* cells and maintenance of ability to adhere to HeLa cells after resuscitation. *Appl. Environ. Microbiol.* **65**, 5154–5157.

Casadevall, A. and Pirofski, L.A. (1999) Host-pathogen interactions redefining the basic concepts of virulence and pathogenicity. *Infect. Immun.* **67**, 3703–3713.

Casadevall, A. and Pirofski, L.A. (2000) Host-pathogen interactions: basic concepts of microbial commensalism colonization, infection, and disease. *Infect. Immun.* **68**, 6511–6518.

Casemore, D.P. (1990) Epidemiological aspects of human cryptosporidiosis. *Epidemiol. Infect.* **104**, 1–28.

Casemore, D.P. (1995) Disinfection options. In *Proceedings of Workshop on Treatment Optimisation for Cryptosporidium Removal from Water Supplies* (ed. P.A. West and M.S. Smith), pp. 19–24. HMSO, London.

Cavalieri d'oro L., Merlo, E., Ariano, E., Silvestri, M.G., Ceraminiello, A., Negri, E. and La Vecchia, C. (1999) *Vibrio cholerae* outbreak in Italy. *Emerg. Infect. Dis.* **5**, 300–301.

Cellini, L., Dainelli, B., Angelucci, D., Grossi, L., Di Bartolomeo, S., Di Campli, E. and Marzio, L. (1999) Evidence for an oral–faecal transmission of *Helicobacter pylori* infection in an experimental murine model. *APMIS* **107**, 477–484.

Center for Disease Control and Prevention (1993) Surveillance for waterborne disease outbreaks – United States, 1991–1992. *Morb. Mortal. Wkly Rep.* **42**, 1–22.

Center for Disease Control and Prevention (1996) Surveillance for waterborne disease outbreaks – United States, 1993–1994. *Morb. Mortal. Wkly Rep.* **45**, 1–33.

Chakraborty, S., Mukhopadhyay, A.K., Bhadra, R.K., Ghosh, A.N., Mitra, R., Shimada, T., Yamasaki, S., Faruque, S.M., Takeda, Y., Colwell, R.R. and Nair, G.B. (2000) Virulence genes in environmental strains of *Vibrio cholerae. Appl. Environ. Microbiol.* **66**, 4022–4028.

Chalmers, R.M., Aird, H. and Bolton, F.J. (2000) Waterborne *Escherichia coli* O157. *J. Appl. Microbiol. Symp. Suppl.* **88**, 124S–132S.

Chany, C., Moscovici, O., Lebon, P. and Rousset, S. (1982) Association of Coronavirus infection with neonatal necrotizing enterocolitis. *Pediatrics*, **69**, 209–214.

Chappell, C.L., Okhuysen, P.C., Sterling C.R. and DuPont, H.L. (1996) *Cryptosporidium parvum*: intensity of infection and oocysts excretion patterns in healthy volunteers. *J. Infect. Dis.* **173**, 232–236.

Chappell, C.L., and Okhuysen P.C. (2000) Infectivity and virulence of *Cryptosporidium* genotype 1/H oocysts in healthy adults volunteers. *Report* EPA Grant Number R828035. University of Texas Health Science Center.

Chart, H. (2000) VTEC enteropathogenicity. *J. Appl. Microbiol.* **SS**. **88**, 12S–23S.

Chauret, C., Nolan, K., Chen, P., Springthorpe, S. and Sattar, S. (1998) Aging of *Cryptosporidium parvum* oocysts in river water and their susceptibility to disinfection by chlorine and monochloramine. *Can. J. Microbiol.* **44**, 1154–1160.

Cirillo, J.D., Falkow, S., Tompkins, L.S., Bermudez, L.E. (1997) Interaction of *Mycobacterium avium* with environmental amoebae enhances virulence. *Infect. Immun.* **65**, 3759–3767.

Clark, C.G., Kravetz, A.N., Alekseenko, V.V., Krendelev, Y. and Johnson, W.M. (1998) Microbiological and epidemiological investigation of cholera epidemic in Ukraine during 1994 and 1995. *Epidemiol. Infect.* **121**, 1–13.

Colford, J.M., Rees, J.R., Wade, T.J., Khalakdina, A., Hilton, J.F., Ergas, I.J., Burns, S., Benker, A., Ma, C., Bowen, C., Mills, D.C., Vugia, D.J., Juranek, D.D. and Levy, D.A. (2002) Participant blinding and gastrointestinal illness in a randomized, controlled trial of an in-home drinking water intervention. *Emerg. Infect. Dis.* **8**(1), 29.

Collins, C.H., Grange, J.M. and Yates, M.D. (1984) Mycobacteria in water. *J. Appl. Bacteriol.* **57**, 193–211.

Colwell, R.R. and Huq, A. (1994) Vibrios in the environment: viable but non-culturable *Vibrio cholerae*. In *Vibrio cholerae and Cholera: Molecular to Global Perspectives* (ed. I.K. Wachsmuth, P.A. Blake, and O. Olsvik), pp. 117–133, ASM Press, Washington, D.C.

Costerton, J.W., Lewandowski, Z., Caldwell, D.E., Korber, D.R. and Lappin-Scott, H.M. (1995) Microbial biofilms. *Annu. Rev. Microbiol.* **49**, 711–745.

Cotte, L., Rabodonirina, M., Chapuis, F., Bailly, F., Bissuel, F., Raynal, C., Gelas, P., Persat, F., Piens, M.A. and Trepo, C. (1999) Waterborne outbreak of intestinal microsporidiosis in persons with and without human immunodeficiency virus infection. *J. Infect. Dis.* **180**(6), 2003–2008.

Craun, G.F. (1986) *Waterborne Diseases in the United States*. CRC Press, Boca Raton, FL.

Craun, G.F. (1992) Waterborne disease outbreaks in the United States of America: causes and prevention. *Rapport trimestriel de statistiques sanitaires mondiales* **45**, 192.

Craun, G.F., Swerdlow, D., Tauxe, R., Clark, R., Fox, K., Geldreich, E., Reasoner, D. and Rice, E. (1991) Prevention of waterborne cholera in the United States. *JAWWA* **83**, 40–47.

Cubitt, W.D. (1989) Diagnosis, occurrence and clinical significance of the human "candidate" caliciviruses. *Prog. Med. Virol.* **36**, 103–119.

Current, W.L. and Garcia, L.S. (1991) Cryptosporidiosis. *Clin. Microbiol. Rev.* **4**, 325–358.

Curry, A. and Smith, H.V. (1998) Emerging pathogens: Isospora, Cyclospora and microsporidia. *Parasitology* **117S**, 143–159.

D'Antonio, R.G., Winn, R.E., Taylor, J.P. *et al.* (1985) A waterborne outbreak of cryptosporidiosis in normal hosts. *Ann. Intern. Med.* **103**, 886–888.

Davey, M.E. and O' Toole, G.A. (2000) Microbial biofilms: from ecology to molecular genetics. *Microbiol. Mol. Biol. Rev.* **64**(4), 847–867.

De Serres, G., Cromeans, T.L., Levesque, B., Brassard, N., Barthe, C., Dionne, M., Prud'homme, H., Paradis, D., Shapiro, C.N., Nainan, O.V. and Margolis, H.S. (1999) Molecular confirmation of hepatitis A virus from well water: epidemiology and public health implications. *J. Infect. Dis.* **179**(1), 37–43.

Delhaes L., Bourel, B., Scala, L., Muanza, B., Dutoit, E., Wattel, F., Gosset, D., Camus, D. and Dei-Cas, E. (2001) Case report: recovery of *Calliphora vicina* first-instar larvae from a human traumatic wound associated with a progressive necrotizing bacterial infection. *Am. J. Trop. Med. Hyg.* **64**, 159–161.

Deltenre, M. and de Koster, E. (2000) How come I've got it? (A review of *Helicobacter pylori* transmission). *Eur. J. Gastroenterol. Hepatol.* **12**, 479–482.

Dennis, D.T., Smith, R.P., Welch, J.J., Chute, C.G., Anderson, B., Herndon, J.L. and Fordham von Reyn, F. (1993) Endemic giardiasis in New Hampshire: a case–control study of environmental risks. *J. Infect. Dis.* **167**, 1391–1395.

Deodhar, L.P., Saraswathi, K. and Varudkar, A. (1991) *Aeromonas* spp. and their association with human diarrheal disease. *J. Clin. Microbiol.* **29**, 853–856.

Dold, H. (1918) On pyocyaneus sepsis and intestinal infections in Shanghai due to *Bacillus pyocyaneus. Chin. Med. J.* **32**, 435.

Dowd, S.E., Gerba, C.P. and Pepper, I.L. (1998) Confirmation of the human-pathogenic microsporidia *Enterocytozoon bieneusi*, *Encephalitozoon intestinalis*, and *Vittaforma corneae* in water. *Appl. Environ. Microbiol.* **64**, 3332–3335.

du Moulin, G.C. and Stottmeier, K.D. (1986) Waterborne mycobacteria: an increasing threat to health. *Am. Soc. Microbiol. News* **52**, 525–529.

du Moulin, G.C., Stottmeier, K.D., Pelletier, P.A., Tsang, A.Y. and Hedley-Whyte, J. (1988) Concentration of *Mycobacterium avium* by hospital hot water systems. *J. Am. Med. Ass.* **260**, 1599–1601.

Dumoulin, A., Guyot, K., Lelievre, E., Dei-Cas, E. and Cailliez. J.C. (2000) *Cryptosporidium* et faune sauvage: un risque pour l'homme? *Parasitology* 167–172.

Duncan, H.E. and Edberg, S.C. (1995) Host-microbe interaction in the gastrointestinal tract. *Crit. Rev. Microbiol.* **21**(2), 85-100.

DuPont, H.L., Chappell, C.L., Sterling, C.R., Okhuysen, P.C., Rose, J.B. and Jakubowski, W. (1995). Infectivity of *Cryptosporidium parvum* for adult humans. *N. Engl. J. Med.* **332**, 855–859.

DuPont, H.L., Levine, M.M., Hornick, R.B. and Formal, S.B. (1989) Inoculum size in shigellosis and implications for expected mode of transmission. *J. Infect. Dis.* **159**, 1126–1128.

Edberg, S.C. (1996) Assessing health risk in drinking water from naturally occurring microbes. *Environ. Health* January/February, 15–24.

Edberg, S.C. and Allen, M.J. (2003) Issues for microbial regulations: *Aeromonas* as a model. Accepté pour publication dans. *Crit. Rev. Microbiol.*

Edberg, S.C., Kops, S., Kontnick, C. and Escarzaga, M. (1997) Analysis of cytotoxicity and invasiveness of heterotrophic plate count bacteria (HPC) isolated from drinking water on blood media. *J. Appl. Microbiol.* **82**, 455–461.

Eden, K.V., Rosenberg, M.L., Stoopler, M., Wood, B.T., Highsmith, A.K., Skaliy, P., Wells, J.G., and Feeley, J.C. (1977) Waterborne gastrointestinal illness at a ski resort. *Publ. Health Rep.* **92**, 245–250.

Engel, H.W.B., Berwald, L.G. and Havelaar, A.H. (1980) The occurrence of *Mycobacterium kansasii* in tap water. *Tuberculosis* **61**, 21–26.

Ernst, P.B. and Gold, B.D. (2000) The disease spectrum of *Helicobacter pylori*: the immunopathogenesis of gastroduodenal ulcer and gastric cancer. *Annu. Rev. Microbiol.* **54**, 615–640.

Falkinham, J.O. (1996) Epidemiology of infection by nontuberculous mycobacteria. *Clin. Microbiol. Rev.* 177–215.

Falkinham, J.O., Norton, C.D. and LeChevallier, M.W. (2001) Factors influencing numbers of *Mycobacterium avium*, *Mycobacterium intracellulare*, and other mycobacteria in drinking water distribution systems. *Appl. Environ. Microbiol.* **67**(3), 1225–1231.

Farmer, J.J., Arduino, M.J. and Hickman-Brenner, F.W. (1992) The genera *Aeromonas* and *Plesiomonas*. In *The Prokaryotes* (ed. A. Balows, G. Trüper, M. Dworkin, W. Harder, and K.H. Schleifer), 2nd edn, Springer-Verlag, New York.

Faruque, S.M., Albert, M.J. and Mekalanos, J.J. (1998) Epidemiology, genetics, and ecology of toxigenic *Vibrio cholerae*. *Microbiol. Mol. Biol. Rev.* **62**,1301–1314.

Faruque, S.M., Saha, M.N., Asadulghani, Bag, P.K., Bhadra, R.K., Bhattacharya, S.K., Sacl, R.B., Takeda, Y. and Nair, G.B. (2000) Genomic diversity among *Vibrio cholerae* O139 strains isolated in Bangladesh and India between 1992 and 1998. *FEMS Microbiol. Lett.* **184**, 279–284.

Ferley, J.P., Zmirou, D., Collin, J.F. and Charrel, M. (1986) Etude longitudinale des risques liés à la consommation d'eaux non conformes aux normes bactériologiques. *Rev. Epidemiol. Santé Publique* **34**, 89–99.

Fields, B.S., Benson, R.F. and Besser, R.E. (2002) *Legionella* and Legionnaires' disease: 25 years of investigation. *Clin. Microbiol. Rev.* **15**(3), 506–526.

Fliermans, C.B. (1996) Ecology of *Legionella*: from data to knowledge with a little Wisdom. *Microbiol. Ecol.* **32**, 203–228.
Florman, A.L. and Schifrin, N. (1950) Observations on a small outbreak of infantile diarrhea associated with *Pseudomonas aeruginosa. J. Pediatr.* **36**, 758–766.
Foca, M., Jakob, K., Whittier, S., Della Latta, Ph.D., Factor, S., Rubenstein, D. and Saiman, L. (2000) Endemic *Pseudomonas aeruginosa* infection in a neonatal intensive care unit. *N. Engl. J. Med.* **343**, 695–700.
Ford, T.E. 1999. Microbiological safety of drinking water: United States and global perspectives. *Environ. Health Perspect.* **107**, 191.
Fournier, S., Liguory, O., Santillana-Hayat, M., Guillot, E., Sarfati, C., Dumoutier, N., Molina, J. and Derouin, F. (2000) Detection of microsporidia in surface water: a one-year follow-up study. *FEMS Immunol. Med. Microbiol.* **29**, 95–100.
Franco, A.A., Fix, A.D., Prada, A., Paredes, E., Palomino, J.C., Wright, A.C., Johnson, J.A., McCarter, R., Guerra, H. and Morris, J.G., Jr. (1997) Cholera in Lima, Peru, correlates with prior isolation of *Vibrio cholerae* from the environment. *Am. J. Epidemiol.* **146**, 1067–1075.
Fraser, G.G. and Cooke, K.R. (1991) Endemic giardiasis and municipal water supply. *Am. J. Publ. Health* **81**, 760.
Fricker, C.R. (2002) The presence of bacteria in water after regrowth. Presented at the *NSF International/World Health Organization Symposium on HPC Bacteria in Drinking Water*, Geneva, 22–24 April.
Fritsche, T.R., Sobek, D. and Gautom, R.K. (1998) Enhancement of *in vitro* cytopathogenicity by *Acanthamoeba* spp. following acquisition of bacterial endosymbionts. *FEMS Microbiol. Lett.* **166**, 231–236.
Furtado, C., Adak, G.K., Stuart, J.M., Wall, P.G., Evans, H.S. and Casemore, D.P. (1998) Outbreaks of waterborne infectious intestinal disease in England and Wales, 1992–1995. *Epidemiol. Infect.* **121**, 109–119.
Galbraith, N.S. (1994) A historical review of microbial disease spread by water in England and Wales. In *Water and Public Health* (ed. N. Noah and R. Stanwell-Smith), pp. 15–37, Smith-Gordon & Co., Golding AMB, London.
Gale, P. (1996) Developments in microbiological risk assessment models for drinking water-a short review. *J. Appl. Bacteriol.* **81**, 403–410.
Gerba, C.P. and Rose, J.B. (1990) Viruses in source and drinking water. In *Drinking Water Microbiology* (ed. G.A. McFeters), pp. 380–396, Springer-Verlag, New York.
Gibson, C.J., III, Haas, C.N. and Rose, J.B. (1998) Risk assessment of waterborne protozoa: current status and future trends. *Parasitology* **117**, S205–S212.
Giordano, M.O., Martinez, L.C., Rinaldi, D., Espu, I C., Martinez, N., Isa, M.B., Depetris, A.R., Medeot, S.I. and Nates, S.V. (1999) Diarrhea and enteric emerging viruses in HIV-infected patients. *AIDS Res. Hum. Retroviruses* **15**, 1427–1432.
Glass, R.I. (1995) Other viral agents of gastroenteritis. In *Infections of the Gastrointestinal Tract* (ed. M.J. Blaser, P.D. Smith, J.I. Ravdin, H.B. Greenberg and R.L. Guerrant), pp. 1055–1064, Raven Press, New York.
Glass, R.I., Noel, J., Ando, T., Fankhauser, R., Belliot, G., Mounts, A., Parashar, U.D., Brese, J.S. and Monroe, S.S. (2000) The epidemiology of enteric caliciviruses from humans: a reassessment using new diagnostics. *J. Infect. Dis.* **181**, S254–261.
Glover, N., Holtzman, A., Aronson, T., Froman, S., Berlin, O.G.W., Dominguez, P., Kunkel, K.A., Overturf, G., Stelma, G. Jr, Smith, C. and Yakrus, M. (1994) The isolation and identification of *Mycobacterium avium* complex (MAC) recovered from Los Angeles potable water, a possible source of infection in AIDS patients. *Int. J. Environ. Health Res.* **1**, 63–72.
Godoy, P., Artigues, A., Nuin, C., Aramburu, J., Perez, M., Dominguez, A. and Salleras, L. (2002) Outbreak of gastroenteritis caused by *Campylobacter jejuni* transmitted through drinking water. *Med. Clin.* **119**(18), 695–698.

Goldstein, S.T., Juranek, D.D., Ravanholt, O., Hightower, A.W., Martin, D.G., Mesnik, J.L., Griffiths, S.D., Bryant, A.J., Reich, R.R. and Herwald, B.L. (1996) Cryptosporidiosis: an outbreak associated with drinking water despite state-of-the-art water treatment. *Ann. Intern. Med.* **124**, 459.

Gonzalez, P., Sanches, A., Rivera, P., Jimenez, C. and Hernandez, F. (1997) Rotavirus and coronavirus outbreak: etiology of annual diarrhea in Costa Rican children. *Rev. Biol. Trop.* **45**, 989–991.

Gosling, P.J. (1996) Pathogenic mechanisms. In *The Genus Aeromonas* (ed. B. Austin *et al.*), pp. 24–265, Wiley, London.

Goyal, S.M. (1983) Indicators of viruses. In *Viral Pollution of the Environment.* (ed. G. Berg), pp. 211–227, CRC Press, Boca Raton, FL.

Grabow, W.O.K, Taylor, M.B., Clay, C.G. and De Villiers, J.C. (1999) Molecular detection of viruses in drinking water implications for safety and disinfection. In *The Safety of Water Disinfection: Balancing Chemical and Microbial Risks*, ILSI Press, Miami Beach, FL.

Graham, D.Y., Dufour, G.R. and Estes, M.K. (1987) Minimal infective dose of rotavirus. *Arch. Virol.* **92**, 261–271.

Gramley, W.A., Asghar, A., Frierson, H.R. Jr. and Powell, S.M. (1999) Detection of *Helicobacter pylori* DNA in faecal samples from infected individuals. *J. Clin. Microbiol.* **37**, 2236–2240.

Grange, J.M., Yates, M.D. and Boughton, E. (1990) Review: the avian tubercle bacillus and its relatives. *J. Appl. Bacteriol.* **68**, 411–431.

Green, K.Y., Ando, T., Balayan, M.S., Berke, T., Clarke, I.N., Estes, M.K., Matson, J.D., Studdert, M.J. and Thiel, H.J. (2000) Taxonomy of the caliciviruses. *J. Infect. Dis.* **2**, S322–S330.

Griffin, P.M. and Tauxe, R.V. (1991) The epidemiology of infections caused by *Escherichia coli* O157:H7, other enterohemorrhagic *E. coli*, and the associated hemolytic uremic syndrome. *Epidemiol. Rev.* **13**, 60–98.

Grimwood, K., Carzino, R., Barnes, G.L. and Bishop, R.F. (1995) Patients with enteric adenovirus gastroenteritis admitted to an pediatric teaching hospital from 1981 to 1992. *J. Clin. Microbiol.* **33**(1), 131–136.

Gumbo, T., Sarbah, S., Gangaidzo, I.T., Ortega, Y., Sterling, C.R., Carbille, A., Tzipori, S. and Wiest, P.M. (1999) Intestinal parasites in patients with diarrhea and human immunodeficiency virus infection in Zimbabwe. *AIDS* **13**, 819–821.

Guyot, K., Follet-Dumoulin, A., Recourt, C., Lelièvre, E., Cailliez, J.C. and Dei-Cas, E. (2002) PCR-RFLP analysis of a diagnostic 452 bp DNA fragment discriminates between *C. parvum* and *C. meleagridis* and between *C. parvum* isolates of human and animal origin. *Appl. Env. Microbiol.* **68**, 2071–2076.

Guyot, K., Follet-Dumoulin, A., Lelièvre, E., Sarfati, C., Rabodonirina, M., Nevez, G., Cailliez, J.C., Camus, D. and Dei-Cas, E. (2001) Molecular characterization of *Cryptosporidium* obtained from humans in France. *J. Clin. Microbiol.* **39**, 3472–3480.

Guyot, K., Gireaudot-Liepmann, M.F., Cabon, A., Riveau-Ricard, I., Lange, M., Delattre, J.M. and Dei-Cas, E. (2000) Influence of US EPA 1622 method successive steps on the viability of *Cryptosporidium* oocysts. *Water Sci. Technol.* **41**(7), 189–196.

Haas, C.N. (2000) Epidemiology, microbiology, and risk assessment of waterborne pathogens including *Cryptosporidium*. *J. Food Protect.* **63**, 827–831.

Haas, C.N. and Rose, J.B. (1994) Reconciliation of microbial risk models and outbreak epidemiology: the case of the Milwaukee outbreak. In *Proceedings of the American Water Works Association*, pp. 517–523, American Water Works Association, Denver, CO.

Hafliger, D., Hubner, P. and Luthy, J. (2000) Outbreak of viral gastroenteritis due to sewage-contaminated drinking water. *Int. J. Food Microbiol.* **54**, 123–126.

Halliday, M.L., Kang, L.Y., Zhou, T.K., Hu, M.D., Pan, Q.C., Fu, T.Y. and Huang, Y.S. (1991) An epidemic of hepatitis A attributable to the ingestion of raw China. *J. Infect. Dis.* **164**(5), 852–859.

Hänninen, M.L. (1994) Phenotypic characteristics of the three hybridization groups of *Aeromonas hydrophila* complex isolated from different sources. *J. Appl. Bacteriol.* **76**, 455–462.

Hanninen, M.L., Haajanen, H., Pummi, T., Wermundsen, K., Katila, M.L., Sarkkinen, H., Miettinen, I. and Rautelin, H. (2003) Detection and typing of *Campylobacter jejuni* and *Campylobacter coli* and analysis of indicator organisms in three waterborne outbreaks in Finland. *Appl. Environ. Microbiol.* **69**(3), 1391–1396.

Hardalo, C. and Edberg, S.C. (1997) *Pseudomonas aeruginosa*: assessment of risk from drinking water. *Crit. Rev. Microbiol.* **23**, 47–75.

Hardy, M.E. (1999) Norwalk and "Norwalk-like viruses" in epidemic gastroenteritis. *Clin. Lab. Med.* **19**, 675–690.

Havelaar, A.H., Schets, F.M., Van Silfhout, A., Jansen, W.H., Wieten, G. and Van Der Kooij, D. (1992) Typing of *Aeromonas* strains from patients with diarrhoea and from drinking water. *J. Appl. Bacteriol.* **72**, 435–344.

Havelaar, A.H., Versteegh, J.F.M. and During, M. (1990) The presence of *Aeromonas* in drinking water supplies in the Netherlands. *Zentralbl. Hyg.* **190**, 236–256.

Hazen, T.C., Fliermans, C.B., Hirsch, R.P. and Esch, G.W. (1978) Prevalence and distribution of *Aeromonas hydrophila* in the United States. *Appl. Environ. Microbiol.* **36**, 731–738.

Hegarty, J.P., Dowd, M.T. and Baker, K.H. (1999) Occurrence of *Helicobacter pylori* in surface water in the United States. *J. Appl. Microbiol.* **87**, 697–701.

Hellard, M.E., Sinclair, M.I., Forbes, A.B. and Fairley, C.K. (2001) A randomized, blinded, controlled trial investigating the gastrointestinal health effects of drinking water quality. *Environ. Health Perspect.* **109**(8), 773.

Henderson, A., MacLaurin, J. and Scott, J.M. 1969. *Pseudomonas* in a Glasgow baby unit. *Lancet* **i**, 316–317.

Herman-Taylor, J., Bull, T.J., Sheridan, J.M, Cheng, J., Stellakis, M.L. and Sumar, N. (2000) Causation of Crohn's disease by *Mycobacterium avium* subspecies paratuberculosis. *Can. J. Gastroenterol.* **14**, 521–539.

Herrmann, J.E. and Blacklow, N.R. (1995) Enteric adenoviruses. In *Infections of the Gastrointestinal Tract* (ed. M.J. Blaser, P.D. Smith, J.I. Ravdin, H.B. Greenberg, and R.L. Guerrant), pp. 1047–1053, Raven Press Ltd, New York.

Herwaldt, B., Ackers, M.L. and the Cyclospora working group. (1997) An outbreak in 1996 of cyclosporiasis associated with imported raspberries. *N. Engl. J. Med.* **336**, 1548–1556.

Herwaldt, B.L. (2000) Cyclospora cayetanensis: a review, focusing on the outbreaks of cyclosporiasis in the 1990s. *Clin. Infect. Dis.* **31**, 1040–1057.

Holmberg, S.D., Schell, W.I. and Fanning, G.R. (1986) *Aeromonas* intestinal infections in the United States. *Ann. Intern. Med.* **105**, 683–689.

Horsburgh, C.R. (1992) Epidemiology of mycobacterial diseases in AIDS. *Res. Microbiol.* **143**, 373–377.

Horsburgh, C.R., Chin, D.P., Yajko, D.M., Hopewell, P. C., Nassos, P. S., Elkin, E.P., Hadley, W.K., Stone, E.N., Simon, E.M., Gonzalez, P., Ostroff, S. and Reingold, A.L. (1994) Environmental risk factors for acquisition of *Mycobacterium avium* complex in persons with human immunodeficiency virus infection. *J. Infect. Dis.* **170**, 362–367.

Huang, P., Weber, J.T., Sosin, D.M., Griffin, P.M., Long, E.G., Murphy, J.J., Kocka, F., Peters, C. and Kallick, C. (1995) The first reported outbreak of diarrheal illness associated with Cyclospora in the United States. *Ann. Intern. Med.* **123**, 409–414.

Hulten, K., Enroth, H., Nyström, T. and Engstrand, L. (1998) Presence of *Helicobacter* species DNA in Swedish water. *J. Appl. Microbiol.* **85**, 282–286.

Hulten, K., Han, S.W., Enroth, H., Klein, P.D., Opekun, A.R., Gilman, R.H., Evans, D.G., Engstrand, L., Graham, D.Y. and El-Zaatari, F.A. (1996) *Helicobacter pylori* in the drinking water in Peru. *Gastroenterology* **110**, 1031–1035.

Hunter, C.A. and Ensign, P.R. (1947) An epidemic of diarrhea in a new-born nursery caused by *Pseudomonas aeruginosa. Am. J. Publ. Health* **37**, 1166–1169.

Hunter, P.R. (1997) *Waterborne Disease. Epidemiology and Ecology*, 372 pp., John Wiley & Sons, Chichester.

Hunter, P.R. (2002) *Epidemiological Evidence of Disease Linked to HPC bacteria*. Presented at the *NSF International/World Health Organization Symposium on HPC Bacteria in Drinking Water*, Geneva, 22–24 April.

Huys, G., Kampfer, P., Albert, M.J., Kuhn, I., Denys, R. and Swings, J. (2002) *Aeromonas hydrophila* subsp. *dhakensis* subsp. nov., isolated from children with diarrhoea in Bangladesh, and extended description of *Aeromonas hydrophila* subsp. *hydrophila* (Chester 1901) Stanier 1943 (approved lists 1980). *Int. J. Syst. Evol. Microbiol.* **52**, 705–712.

Imanishi, Y., Ogata, T., Matsuzuka, A., Tasaki, T., Fujioka, T., Akashi, M., Makino and Y., Nishizono, A. (2003) Possibility for the presence of *Helicobacter pylori* in drinking well water. *Kansenshogaku Zasshi* **77**(1), 18–23.

Isenberg, H.D. (1988) Pathogenicity and virulence: another view. *Clin. Microbiol. Rev.* **1**(1), 40–53.

Isonhood, J.H. and Drake, M. (2002) *Aeromonas* species in foods. *J. Food Protect.* **65**(3), 575–582.

Jacob, J., Woodward, D., Feuerpfeil, I. and Johnson, W.M. (1998) Isolation of *Arcobacter butzleri* in raw water and drinking water treatment plants in Germany. *Zentralbl. Hyg. Umwelmed.* **201**, 189–198.

Janda, J.M. (1991) Recent advances in the study of the taxonomy, pathogenicity, and infectious syndromes associated with the genus *Aeromonas. Clin. Microbiol. Rev.* **4**, 397–410.

Janda, J.M. and Abbott, S.L. (1998) Evolving concepts regarding the genus *Aeromonas*: an expanding panorama of species, disease presentations, and unanswered questions. *Clin. Infect. Dis.* **27**, 332–344.

Jenkins, P.A. (1991) Mycobacteria in the environment. *J. Appl. Bacteriol.* **70**, 137S–141S.

Jesudason, M.V., Balaji, V., Mukindan, U. and Thomson, C.J. (2000) Ecological study of *Vibrio cholerae* in Vellore. *Epidemiol. Infect.* **124**, 201–206.

Jiang, S.C., Matte, M., Matte, G., Huq, A. and Colwell, R.R. (2000) Genetic diversity of clinical and environmental isolates of *Vibrio cholerae* determined by amplified fragment length polymorphism fingerprinting. *Appl. Environ. Microbiol.* **66**, 148–153.

Jiang, X., Cubitt, W.D., Berke, T., Zhong, W., Dai, X., Nakata, S., Pickering, L.K. and Matson, D.O. (1997) Sapporo-like human caliciviruses are genetically and antigenically diverse. *Arch. Virol.* **142**, 1813–1827.

Jiang, X. and Doyle, M.P. (1998) Effect of environmental and substrate factors on survival and growth of *Helicobacter pylori. J. Food Protect.* **61**, 929–933.

Jiang, X., Matson, D.O., Cubitt, W.D. and Estes, M.K. (1996) Genetic and antigenic diversity of human caliciviruses (HuCVs) using RT-PCR and new EIAs. *Arch. Virol. Suppl.* **12**, 251–262.

Johnson, C.H., Rice, E.W. and Reasoner, D.J. (1997) Inactivation of *Helicobacter pylori* by chlorination. *Appl. Environ. Microbiol.* **63**, 4969–4970.

Jones, D.M., Sutcliffe, E.M. and Curry, A. (1991) Recovery of viable but non-culturable *Campylobacter jejuni. J. Gen. Microbiol.* **137**, 2477–2482.

Kaminski, J.C. (1994) *Cryptosporidium* and the public water supply (letter). *N. Engl. J. Med.* **31**, 1529–1530.

Kapikian, A.Z. and Chanock, R.M. (1990) Norwalk group of viruses. In *Virology* (ed. B.N. Fields, D.M. Knipe, R.M. Chanock *et al.*), 2nd edn, pp. 61–93, Raven Press, New York.

Kapikian, A.Z., Estes, M.K. and Chanock, R.M. (1996) Norwalk group of viruses. In *Fields Virology* (ed. B.N. Fields, D.M. Knipe, P.M. Howley, R.M. Chanock, J.L. Melnick, T.P. Monath, B. Roizman and S.E. Straus), 3rd edn, Lippincott-Raven, Philadelphia, PA.

Kaplan, J.E., Gary, G.W., Baron, R.C., Singh, N., Schonberger, L.B., Feldman, R. and Greenberg, H.B. (1982) Epidemiology of Norwalk gastroenteritis and the role of

Norwalk virus in outbreaks of acute nonbacterial gastroenteritis. *Ann. Int. Med.* **96**, 756.

Kapperud, G. (1977) *Yersinia enterocolitica* and *Yersinia*-like microbes isolated from mammals and water in Norway and Denmark. *Acta Pathol. Microbiol. Scan.* **85**, 129–135.

Karanis, P., Schoenen, D., Maier, W.A. and Seitz, H.M. (1993) Drinking water and parasites. *Immun. Infekt.* **21**, 132–136.

Kaustova, J., Olsovsky, Z., Kubin, M., Zatloukai, O., Pelikan, M. and Hradil, Y. (1981) Endemic occurrence of *Mycobacterium kansasii* in water supply systems. *J. Hyg. Epidemiol. Microbiol. Immunol.* **25**, 24–30.

Kersters, I., Huys, G., Van Duffel, H., Vancanneyt, M., Kersters, K. and Verstraete, W. (1996) Survival potential of *Aeromonas hydrophila* in freshwaters and nutrient-poor waters in comparison with other bacteria. *J. Appl. Bacteriol.* **80**, 266–276.

Kersters, I., Van Vooren, L., Huys, G., Janssen, P., Kersters, K. and Verstraete, W. (1995) Influence of temperature and process technology on the occurrence of *Aeromonas* species and hygienic indicator organisms in drinking water production plants. *Microbiol. Ecol.* **30**, 203–218.

Kirov, S.M. (1993) The public health significance of *Aeromonas* spp. in foods. *Intern. J. Food Microbiol.* **20**, 179–198.

Kirov, S.M. (1997) *Aeromonas* and *Plesiomonas* species. In *Food Microbiology, Fundamentals and Frontiers* (ed. M. Doyle, L. Beuchat, and T. Montville), pp. 265–287, SM Press, Washington, D.C.

Kirschner, R.A., Parker, B.C. and Falkinham, J.O. (1999) Humic and fulvic acids stimulate the growth of *Mycobacterium avium*. *FEMS Microbiol. Ecol.* **30**, 327–332.

Kjelleberg, S. (1993) *Starvation in Bacteria*. Plenum Press, New York and London.

Kogure, K. and Ikemoto, E. (1997) Wide occurrence of enterohemorragic *Escherichia coli* O157 in natural freshwater environment. *Nippon Saikingaku. Zasshi.* **52**(3), 601–607.

Kool, J.L., Bergmire-Sweat, D., Butler, J.C., Brown, E.W., Peabody, D.J., Massi, D.S., Carpenter, J.C., Pruckler, J.M., Benson, R.F. and Fields, B.S. (1999) Hospital characteristics associated with colonization of water systems by *Legionella* and risk of nosocomial Legionnaires' disease: a cohort study of 15 hospitals. *Infect. Control. Hosp. Epidemiol.* **20**,798–805.

Koopmans, M., Von Bonsdorff, C.H., Vinje, J., De Medici, D. and Monroe, S. (2002) Foodborne viruses. *FEMS Microbiol. Rev.* **26**, 187–205.

Koopmans, M.P.G., Bijen, M.H., Monroe, S.S. and Vinje, J. (1998) Age-stratified seroprevalence of neutralizing antibodies to astrovirus types 1 to 7 in humans in the Netherlands. *Clin. Diagn. Lab. Immunol.* **5**, 33–37.

Koopmans, M.P.G., Goosen, E.S., Lima, A., McAuliffe, I.T., Nataro, J.P., Barrett, L.J., Glass, R.I. and Guerrant, R.L. 1997. Association of torovirus with acute and persistent diarrhea in children. *Pediatr. Infect. Dis. J.* **16**, 504–507.

Koumans, E.H.A., Katz, D.J., Malecki, J.M., Kumar, S., Wahlquist, S.P., Arrowood, M.J. *et al.* (1998) An outbreak of cyclosporiasis in Florida 1995: a harbinger of multistate outbreaks in 1996 and 1997. *Am. J. Trop. Med. Hyg.* **59**, 235–242.

Krovacek, K., Faris, A., Baloda, S.B., Lindberg, T., Peterz, M. and Mansson, I. (1992) Isolation and virulence profiles of *Aeromonas* spp. from different municipal drinking water supplies in Sweden. *Food Microbiol.* **9**, 215–222.

Kühn, I., Allestram, G., Huys, G., Janssen, P., Kersters, K., Krovacek, K. and Stenstrom, T.A. (1997) Diversity, persistence, and virulence of *Aeromonas* strains isolated from drinking water distribution systems in Sweden. *Appl. Environ. Microbiol.* **63**, 2708–2715.

Kuijper, E.J., Bol, P., Peeters, M.F., Steigerwalt, A.G., Zanen, H.C. and Brenner, D.J. (1989) Clinical and epidemiologic aspects of members of *Aeromonas* groups isolated from human feces. *J. Clin. Microbiol.* **27**(7), 1531–1537.

Kukkula, M., Maunula, L., Silvennoinen, E. and von Bonsdorff, C.H. (1999) Outbreak of viral gastroenteritis due to drinking water contaminated by Norwalk-like viruses. *J. Infect. Dis.* **180**, 1771–1776.

Kurtz, B. and Lee, T.W. (1987) Astroviruses: human animal. In *Novel Diarrhoea Viruses* (ed. G. Bock, and J. Whelan), *Foundation Symposium 128,* pp. 92–107, John Wiley & Sons, Chichester, UK.

Kwa, B.H., Moyad, M., Pentella, M.A. and Rose, J.B. (1993) A nude mouse model as an *in vivo* infectivity assay for cryptosporidiosis. *Water. Sci. Technol.* **27**(3–4), 65–68.

Langeland, G. (1983) *Yersinia enterocolitica* and *Yersinia enterocolitica*-like bacteria in drinking water and sewage sludge. *Acta Pathol. Microbiol. Scand.* **91**, 179–185.

Le Dantec, C., Duguet, J.P., Montiel, A., Dumoutier, N., Dubrou, S. and Vincent, V. (2002) Occurrence of mycobacteria in water treatment lines and in water distribution systems. *Appl. Environ. Microbiol.* **68**(11), 5318–5325.

LeChevallier, M.W., Abbaszadegan, M., Camper, A.K., Hurst, C.J., Izaguirre, G., Marshall, M.M., Naumovitz, D., Payment, P., Rice, E.W., Rose, J., Schaub, S., Slifko, T.R., Smith, D.B., Smith, H.V., Sterling, C.R. and Stewardt, M. (1999a) Committee report: Emerging pathogens - viruses, protozoa, and algal toxins. *J. Am. Wat. Wks. Ass.* **91**, 110–121.

LeChevallier, M.W., Abbaszadegan, M., Camper, A.K., Hurst,C.J., Izaguirre, G., Marshall, M.M., Naumovitz, D., Payment, P., Rice, E.W., Rose, J., Schaub, S., Slifko, T.R., Smith, D.B., Smith, H.V., Sterling, C.R. and Stewardt, M. (1999b) Committee report: Emerging pathogens - bacteria. *J. Am. Wat. Wks. Ass.* **91**, 101–109.

LeChevallier, M.W., Abbazaszadegan, M., Camper, A.K., Izaguire, G., Stewart, M., Naumovitz, D., Marshall, M., Sterling, C.R., Payment, P., Rice, E.W., Hurst, C.J., Schaub, S. Slifko, T.R., Rose, J.B., Smith, H.V. and Smith, D.B. (1999c) Emerging pathogens: names to know and bugs to watch out for. In *Identifying Future Drinking Water Contaminants*, pp. 139–172, National Research Council, Washington, D.C.

LeChevallier, M.W., Norton, C.D., Falkinham, J.O., Willimas, M.D., Taylor, R.H. and Cowan, H.E. (2001) Occurrence and control of *Mycobacterium avium* complex, pp. 1–115, AWWA Research Foundation and American Water Works Association, Denver, CO.

Le Guyader, H., Lecointre, G. (2001) *Classification Phylogénétique du Vivant*, Belin, Paris.

Leclerc, H. (2003) Y-a-t-il des infections bactériennes opportunistes transmises par les eaux d'alimentation? *J. Eur. Hydrol.* **34**(1).

Leclerc, H. and Buttiaux, R. (1962) Fréquence des *Aeromonas* dans les eaux d'alimentation. *Ann. Inst. Pasteur* **103**, 97–100.

Leclerc, H. and Oger, C. (1974) Les eaux usées des hôpitaux et leur importance épidémiologique. *Rev. Epidémiol. Santé Publique* **22**, 185–198.

Leclerc, H. and Oger, C. (1975) Les eaux usées des abattoirs et leur importance épidémiologique. *Rev. Epidémiol. Santé Publique* **23**, 429–444.

Leder, K., Ryan, N., Spelman, D. and Crowe, S.M. (1998) Microsporidial disease in HIV-infected patients: a report of 42 patients and review of the literature. *Scand. J. Infect. Dis.* **30**, 331–338.

Lee, J.V. and West A.A. (1991) Survival and growth of *Legionella* species in the environment. *J. Appl. Bacteriol.* **70**, 121S–129S.

Lepow, M.L. (1994) *Pseudomonas aeruginosa* colonization and infection of the gastrointestinal tract. In *Pseudomonas aeruginosa Infections and Treatment* (ed. A.L. Baltch and R.P. Smith), pp. 421–440, Marcel Dekker, New York.

Lerner, J., Brumberger, V. and Preac-Mursic, V. (1994) Severe diarrhea associated with *Arcobacter butzleri*. *Eur. J. Clin. Microbiol. Infect. Dis.* **13**, 660–662.

Leung, W.K., Siu, K.L., Kwok, C.K., Chan, S.Y., Sung, R. and Sung, J.J. (1999) Isolation of *Helicobacter pylori* from vomitus in children and its implication in gastro-oral transmission. *Am. J. Gastroenterol.* **94**, 2881–2884.

Lindsay, D.S., Dubey, J.P. and Blagburn, B.L. (1997) Biology of *Isospora* spp. from humans, nonhuman primates, and domestic animals. *Clin. Microbiol. Rev.* **10**, 19–34.

Lisle, J.T. and Rose, J.B. (1995) *Cryptosporidium* contamination of water in the USA and UK: a mini-review. *J. Water. Supply: Res. Technol. Aqua.* **44**, 103–117.

Lodder, W.J., Vinje, J., Van de Heide, R., Roda Husman, A.M., Leenen, E.J.T. and Koopmans, M.P.G. (1999) Molecular detection of Norwalk-like caliciviruses in sewage. *Appl. Environ. Microbiol.* **65**, 5624–5627.

Lopez-Velez, R., Turrientes, M.C., Garron, C., Montilla, P., Navajas, R., Fenoy, S. and del Aguila, C. (1999) *Microsporidiosis* in travelers with diarrhea from the tropics. *J. Travel. Med.* **6**, 223–227.

Lopmans, B.A., Brown, D.W. and Koopmans, M. (2002) Human caliciviruses in Europe. *J. Clin. Virol.* **24**(3), 137–160.

Mac Kenzie, W.R., Hoxie, N.J., Proctor, M.E., Gradus, M.S., Blair, K.A., Peterson, D.E. *et al.* (1994) A massive outbreak in Milwaukee of cryptosporidium infection transmitted through the public water supply. *N. Engl. J. Med.* **331**, 161–167.

Maggi, P., Carbonara, S., Fico, C., Santantonio, T., Romanelli, C., Sforza, E. and Pastore, G. (1997) Epidemiological, clinical and therapeutic evaluation of the Italian cholera epidemic in 1994. *Eur. J. Epidemiol.* **13**, 95–97.

Makristathis, A., Pasching, E., Schutze, K., Wimmer, M., Rotter, M.L. and Hirschl, A.M. (1998) Detection of *Helicobacter pylori* in stool specimens by PCR and antigen enzyme immunoassay. *J. Clin. Microbiol.* **36**, 2772–2774.

Marshall, M.M., Naumovitz, D., Ortega, Y. and Sterling, C.R. (1997) Waterborne protozoan pathogens. *Clin. Microbiol. Rev.* **10**, 67–85.

Martinelli, F., Caruso, A., Moschini, L., Turano, A., Scarcella, C. and Speziani, F. (2000) A comparison of *Legionella pneumophila* occurrence in hot water tanks and instantaneous devices in domestic, nosocomial, and community environments. *Curr. Microbiol.* **41**, 374–376.

Matsui, M., Ushijima, H., Hachiya, M., Kakizawa, J., Wen, L., Oseto, M., Morooka, K. and Kurtz, J.B. (1998) Determination of serotypes of astroviruses by reverse transcription–polymerase chain reaction and homologies of the types by the sequencing of Japanese isolates. *Microbiol. Immunol.* **42**, 539–547.

Maurer, A.M. and Sturchler, D. (2000) A waterborne outbreak of small round structured virus, campylobacter and shigella co-infections in La Neuveville, Switzerland, 1998. *Epidemiol. Infect.* **125**(2), 325–332.

McFadden, J.J., Collins, J., Beaman, B., Arthur, M. and Gitnick, G. (1992) Mycobacteria in Crohn's disease: DNA probes identify the wood pigeon strain of *Mycobacterium avium* and *Mycobacterium paratuberculosis* from human tissue. *J. Clin. Microbiol.* **30**, 3070–3073.

McKeown, I., Orr, P., Macdonald, S., Kabani, A., Brown, R., Coghlan, G., Dawood, M., Embil, J., Sargent, M., Smart, G. and Bernstein, C.N. (1999) *Helicobacter pylori* in the Canadian arctic: seroprevalence and detection in community water samples. *Am. J. Gastroenterol.* **94**, 1823–1829.

Medema, G.J., Schets, F.M., van de Giessen, A.W. and Havelaar, A.H. (1992) Lack of colonization of 1 day old chicks by viable, non-culturable *Campylobacter jejuni*. *J. Appl. Bacteriol.* **72**, 512–516.

Megraud, F. and Serceau, R. (1990) Search for *Campylobacter* species in the public water supply of a large urban community. *Zentralbl. Hyg. Umweltmed.* **189**, 536–542.

Meinhardt, P.L., Casemore, D.P. and Miller, K.B. (1996) Epidemiologic aspects of human cryptosporidiosis and the role of waterborne transmission. *Epidemiol. Rev.* **18**, 118.

Melby, K., Gondrosen, B., Gregusson, S., Ribe, H. and Dahl, O.P. (1991) Waterborne campylobacteriosis in northern Norway. *Inter. J. Food Microbiol.* **12**, 151–156.

Melby, K.K., Svently, J.G., Eggebo, T., Holmen, L.A., Andersen, B.M., Lind Sjogren, E. and Kaijser, B. (2000) Outbreak of *Campylobacter* infection in a subArctic community. *Eur. J. Clin. Microbiol. Infect. Dis.* **19**(7), 542–544.

Merritt, A., Miles, R. and Bates, J. (1999) An outbreak of *Campylobacter enteritis* on an island resort north Queensland. *Commun. Dis. Intell.* **23**(8), 215–219.

Metcalf, T.G., Melnick, J.L. and Estes, M.K. (1995) Environmental virology: from detection of virus in sewage and water by isolation to identification by molecular biology - A trip of over 50 years. *Annu. Rev. Microbiol.* **49**, 461–487.

Miller, N.M. and van den Ende, J. (1986) Seasonal prevalence of *Cryptosporidium* associated diarrhoea in young children. *S. Afr. Med. J.* **70**, 636–637.

Millson, M., Bokhout, M., Carlson, J., Spielberg, L., Aldis, R., Borczyk, A. and Lior, H. (1991) An outbreak of *Campylobacter jejuni* gastroenteritis linked to meltwater contamination of a municipal well. *Can. J. Publ. Health* **82**, 27–31.

Mizoguchi, H., Fujioka, T. and Nasu, N. (1999) Evidence for viability of coccoid forms of *Helicobacter pylori. J. Gastroenterol.* **34**, 32–36.

Moe, C.L. (1997) Waterborne transmission of infectious agents. In *Manual of Environmental Microbiology* (ed. C.J. Hurst, G.R. Knudsen, M.J. McInerney, L.D. Stetzenbach, and M.V. Walter), pp. 136–153. ASM Press, Washington, D.C.

Montecalvo, M.A., Forester, G., Tsang, A.Y., Du Moulin, G. and Wormser, G.P. (1994) Colonisation of potable water with *Mycobacterium avium* complex in homes of HIV-infected patients. *Lancet* **343**, 1639.

Morgan, D.R., Johnson, P.C., Dupont, H.L., Satterwhite, T.K. and Wood, L.V. (1985) Lack of correlation between known virulence properties of *Aeromonas hydrophila* and enteropathogenicity for humans. *Infect. Immun.* **50**, 62–65.

Morita, R.Y. (1997) *Bacteria in Oligotrophic Environments. Starvation-Survival Lifestyle*, 529 p., Chapman and Hall, New York.

Morris, R.D. and Levin, R. (1995) Estimating the incidence of waterborne infectious disease related to drinking water in the United States. In *Assessing and Managing Health Risks from Drinking Water Contamination: Approaches and Applications* (ed. E.G. Reichard, and G.A. Zapponi), pp. 75–88, International Association of Hydrological Sciences, Wallingford, UK.

Morrissey, A.B., Aisu, T.O., Falkinham, J.O., Eriki, P.P., Eliner, J.J. and Daniel, T.M. (1992) Absence of *Mycobacterium avium* complex disease in patients with AIDS. *J. Acquir. Immune Defic. Syndr.* **5**, 477–478.

Mortensen, M.L., Ray, C.G., Payne, C.M., Friedman, A.D., Minnich, L.L. and Rousseau, C. (1985) Coronavirus-like particles in human gastrointestinal disease: epidemiologic, clinical and laboratory observations. *Am. J. Dis. Child.* **139**, 928–934.

Mosley, J.W. (1967) Transmission of viral diseases by drinking water. In *Transmission of Viruses by the Water Route* (ed G. Berg), pp. 5, Interscience, New York.

Mota, P., Rauch, C.A. and Edberg, S.C. (2000) Microsporidia and Cyclospora: epidemiology and assessment of risk from the environment. *Crit. Rev. Microbiol.* **26**, 69–90.

Muder, R.R. and Yu, V.L. (2002) Infection due to *Legionella* species other than *L. pneumophila. Clin. Infect. Dis.* **35**(8), 990–998.

Murga, R., Forster, T.S., Brown, E., Pruckler, J.M., Fields, B.S. and Donlan, R.M. (2001) Role of biofilms in the survival of *Legionella pneumophila* in a model potable-water system. *Microbiology* **147**, 3121–3126.

Naik, S.R., Aggarwal, R., Salunke, P.N. and Mehrotra, N.N. (1992) A large waterborne viral hepatitis E epidemic in Kanpur, India. *Bull. World Health Organ.* **70**, 597–604.

Nichols, G. and Lee, J. (2002) Crohn's disease, Johne's disease, *Mycobacterium avium* subsp. *paratuberculosis* and water. In *International Symposium on Waterborne Pathogens*, American Water Works Association.

Nilsson, H.O., Blom, J., Al-Soud, W.A., Ljungh, A., Leif, Andersen, L.P. and Wadström T. (2002) Effect of cold starvation, acid stress, and nutrients on metabolic activity of *Helicobacter pylori. Appl. Environ. Microbiol.* **68**(1), 11–19.

Noel, J.S., Liu, B.L., Humphrey, C.D., Rodriguez, E.M., Lambden, P.R., Clarke, I.N., Dwyer, D.M., Ando, T., Glass, R.I. and Monroe, S.S. (1997) Parkville virus: a novel genetic variant of human calicivirus in the Sapporo virus clade, associated with an outbreak of gastroenteritis in adults. *J. Med. Virol.* **52**, 173–178.

O'Brien, S. and Bhopal, R.S. (1993) Legionnaires' disease: the infective dose paradox. *Lancet* **342**, 5.

Obiri-Danso, K. and Jones, K. (1999) Distribution and seasonality of microbial indicators and thermophilic campylobacters in two freshwater bathing sites on the River Lune in northwest England. *J. Appl. Microbiol.* **87**, 822–832.

Okhuysen, P. C., Chappell, C.L., Crabb, J.H., Sterling, C.R. and DuPont, H.L. (1999) Virulence of three distinct *Cryptosporidium parvum* isolates for healthy adults. *J. Infect. Dis.* **180**, 1275–1281.

Oliver, J.D. (1993) Formation of viable but nonculturable cells. In *Starvation in Bacteria* (ed. S. Kjelleberg), pp. 239–268, Plenum Press, New York.

Onishchenko, G.G., Toporkov, V.P., Prometnoi, V.I. and Naumov, A.V. (1995) The cholera epidemic in mountainous regions of Dagestan in relation to the probable role of the water factor in its spread. *Zh. Mikrobiol. Epidemiol. Immunobiol.* **2**, 64–69.

Ostroff, S.M., Kapperud, G., Hutwagner, L.C., Nesbakken, T. and Bean, N.H. (1994) Sources of sporadic *Yersinia enterocolitica* infections in Norway: a prospective case–control study. *Epidemiol. Infect.* **112**, 133–141.

Park, S.R., Mackay, W.G. and Reid, D.C. (2001) *Helicobacter* sp. recovered from drinking water biofilm sampled from a water distribution system. *Water Res.* **35**(6), 1624–1626.

Parks, C.G., Moe, C.L., Rhodes, D., Lima, A., Barrett, L., Tseng, F., Baric, R., Talal, A. and Guerrant, R. (1999) Genomic diversity of "Norwalk like viruses" (NLVs): pediatric infections in a Brazilian shantytown. *J. Med. Virol.* **58**, 426–434.

Parsonnet, J., Shmuely, H. and Haggerty, T. (1999) Faecal and oral shedding of *Helicobacter pylori* from healthy infected adults. *JAMA* **282**, 2240–2245.

Payment, P. (1991) Antibody levels to selected enteric viruses in a French-Canadian population in the province of Quebec (Canada). *Immunol. Infect. Dis.* **1**, 317–322.

Payment, P. (1993) Viruses: prevalence of disease, levels and sources. In *Safety of Water Disinfection: Balancing Chemical and Microbial Risks* (ed. G.F. Craun), pp. 99–115, ILSI Press, Washington, D.C.

Payment, P. and Armon, R. (1989) Virus removal by drinking water treatment processes. *Crit. Rev. Environ. Cont.* **19**, 15–31.

Payment, P. and Morin, E. (1989) Minimal infective dose of the OSU strain of porcine rotavirus. *Arch. Virol.* **112**, 277–282.

Payment, P., Berte, A., Prevost, M., Menard, B. and Barbeau, B. (2000) Occurrence of pathogenic micro-organisms in the Saint Lawrence river (Canada) and comparison of health risks for populations using it as their source of drinking water. *Can. J. Microbiol.* **46**, 565–576.

Payment, P., Coffin, E. and Paquette, G. (1994) Blood agar to detect virulence factors in tap water heterotrophic bacteria. *Appl. Environ. Microbiol.* **60**, 1179–1183.

Payment, P., Franco, E., Richardson, L. and Siemiatychi, J. (1991a) Gastrointestinal health effects associated with the consumption of drinking water produced by point-of-use domestic reverse-osmosis filtration units. *Appl. Environ. Microbiol.* **57**, 945–948.

Payment, P., Richardson, L., Siemiatycki, J., Dewar, R., Edwardes, M. and Franco, E. (1991b) A randomized trial to evaluate the risk of gastrointestinal disease due to consumption of drinking water meeting current microbiological standards. *Am. J. Publ. Health* **81**, 703.

Pedalino, B., Feely, E., McKeown, P., Foley, B., Smyth, B. and Moren, A. (2003). An outbreak of Norwalk-like viral gastroenteritis in holidaymakers travelling to Andorra, January-February 2002. *Eurosurveillance* **8**(1), 1–8.

Peng, M.M., Xiao, L., Freeman, A.R., Arrowood, M.J., Escalante, A.A., Weltman, A.C., Ong, C.S., Mac Kenzie, W.R., Lal, A.A. and Beard, C.B. (1997) Genetic polymorphism among *Cryptosporidium parvum* isolates: evidence of two distinct human transmission cycles. *Emerg. Infect. Dis.* **3**, 567–573.

Pereira, S.J., Ramirez, N.E., Xiao, L. and Ward, L.A. (2002) Pathogenesis of human and bovine *Cryptosporidium parvum* in gnotobiotic pigs. *J. Infect. Dis.* **186**, 715–718.

Perz, J.F., Ennever, F.K. and Le Blancq, S.M. (1998) *Cryptosporidium* in tap water. *Am. J. Epidemiol.* **147**, 289.

Pettibone, G.W. (1998) Population dynamics of *Aeromonas* spp. in an urban river watershed. *J. Appl. Microbiol.* **85**, 723–730.

Popowski, J., Lekowska-Kochaniak, A. and Korsak, D. (1997) The incidence of heat tolerant *Campylobacter* in rivers and lakes of the Warsaw region. *Rocz. Panstw. Zakl. Hig.* **48**, 253–262.

Pringle, C.R. (1998) Virus taxonomy-San Diego 1998. *Arch. Virol.* **143**, 1449–1459.

Rabold, J.G., Hoge, C.W., Shlim, D.R., Kefford, C., Rajah, R. and Echeverria, P. (1994) *Cyclospora* outbreak associated with chlorinated drinking water. *Lancet* **344**, 1360–1361.

Rangel-Frausto, M.S., Rhomberg, P., Hollis, R.J., Pfaller, M.A., Wenzel, R.P., Helms, C.M. and Herwaldt, L.A. (1999) Persistence of *Legionella pneumophila* in a hospital's water system: a 13-year survey. *Infect. Control. Hosp. Epidemiol.* **20**, 793–797.

Rautelin, H. and Hanninen, M.L. (2000) Campylobacters: the most common bacterial enteropathogens in the Nordic countries. *Ann. Med.* **32**, 440–445.

Rautelin, H., Koota, K., von Essen, R., Jahkola, M., Siitonen, A. and Kosunen, T.U. (1990) Waterborne *Campylobacter jejuni* epidemic in a Finnish hospital for rheumatic diseases. *Scand. J. Infect. Dis.* **22**, 321–326.

Reeves, P.R. and Lan, R. (1998) Cholera in the 1990s. *Br. Med. Bull.* **54**, 611–623.

Rice, E.W., Johnson, C.H., Wild, D.K. and Reasoner, D.J. (1992) Survival of *Escherichia coli* O157:H7 in drinking water associated with a waterborne disease outbreak of hemorrhagic colitis. *Lett. Appl. Microbiol.* **15**, 38–40.

Rice, E.W., Rodgers, M.R., Wesley, I.V., Johnson, C.H. and Tanner, S.A. (1999) Isolation of *Arcobacter butzleri* from ground water. *Lett. Appl. Microbiol.* **28**, 31–35.

Riley, L.W., Remis, R.S., Helgerson, S.D., McGee, H.B., Wells, J., Davis, B.R., Hebert, R.J., Olcott, E.S., Johnson, L.M., Hargrett, N.T., Blake, P.A. and Cohen, M.L. (1983) Hemorrhagic colitis associated with a rare *Escherichia coli* serotype. *N. Engl. J. Med.* **308**, 681–685.

Ristola, M.A., Von Reyn, C.F., Arberit, R.D., Soini, H. Lumio, J., Ranki, A., Buhler, S., Waddell, R., Tosteson, A.N., Falkinham, J.O. and Sox, C.H. (1999) High rates of disseminated infection due to non-tuberculous mycobacteria among AIDS patients in Finland. *J. Infect.* **39**(1), 61–67.

Robinson, D.A. 1981. Infective dose of *Campylobacter jejuni* in milk. *Br. Med. J.* **282**, 1584.

Roefer, P.A., Monscvitz, J.T. and Rexing, D.J. (1996) The Las Vegas cryptosporidiosis outbreak. *J. Am. Wat. Wks. Ass.* **88**, 95–106.

Rogers, J. and Keevil, C.W. (1992) Immunogold and fluorescein immuno-labelling of *Legionella pneumophila* within an aquatic biofilm visualized by using episcopic differential interference contrast microscopy. *Appl. Environ. Microbiol.* **58**, 2326–2330.

Rogers, J., Dowsett, A.B. and Keevil C.W. (1995) A paint incorporating silver to control mixed biofilms containing *Legionella pneumophila. J. Ind. Microbiol.* **15**, 377–383.

Rogers, J., Dowsett, A.B., Dennis, P.J., Lee, J.V. and Keevil, C.W. (1994) Influence of temperature and plumbing material selection on biofilm formation and growth of *Legionella pneumophila* in a model potable water system containing complex microbial flora. *Appl. Environ. Microbiol.* **60**, 1585–1592.

Rollins, D.M. and Colwell, R.R. (1986) Viable but nonculturable stage of *Campylobacter jejuni* and its role in survival in natural, aquatic environment. *Appl. Environ. Microbiol.* **58**, 531–538.

Römling, U., Wingender, J., Müller, H. and Tümmler, B. (1994) A major *Pseudomonas aeruginosa* clone common to patients and aquatic habitats. *Appl. Environ. Microbiol.* **60**, 1734–1738.

Rose, J.B. (1990) Occurrence and control of *Cryptosporidium* in drinking water. In *Drinking Water Microbiology. Progress and Recent Developments* (ed. G.A. McFeters), pp. 294–322, Springer-Verlag, New York.

Rose, J.B. (1993) Enteric waterborne protozoa: hazard and exposure assessment. In *Safety of Water Disinfection: Balancing Chemical and Microbial Risks* (ed. G.F. Craun), pp. 115–126, ILSI Press, Washington, D.C.

Roszak, D.B. and Colwell, R.R. (1987) Survival strategies of bacteria in the natural environment. *Microbiol. Rev.* **51**, 365–379.

Rowbotham, T.J. (1980) Preliminary report on the pathogenicity of *Legionella pneumophila* for freshwater and soil amoebae. *J. Clin. Pathol.* **33**, 1179–1183.

Rusin, P.A., Rose, J.B., Haas, C.N. and Gerba, C.P. (1997) Risk assessment of opportunistic bacterial pathogens in drinking. *Rev. Environ. Contam. Toxicol.* **152**, 57–83.

Russo, F., Notarnicola, M., Di Matteo, G., Leoci, C., Caruso, M.L., Pirrelli, M., Caradonna, M., Morandi, L. and Di Leo, A. (1999) Detection of *Helicobacter pylori* cagA gene by polymerase chain reaction in faecal samples. *Eur. J. Gastroenterol. Hepatol.* **11**, 251–256.

Saebo, A., Kapperud, G., Lassen, J. and Waage, J. (1994) Prevalence of antibodies to *Yersinia enterocolitica* O:3 among Norwegian military recruits: association with risk factors and clinical manifestations. *Eur. J. Epidemiol.* **10**, 749–755.

Sasaki, K., Tajiri, Y., Sata, M., Fujii, Y., Matsubara, F., Zhao, M., Shimizu, S., Toyonaga, A. and Tanikawa, K. (1999) *Helicobacter pylori* in the natural environment. *Scand. J. Infect. Dis.* **31**, 275–279.

Satterthwaite, P., Pritchard, K., Floyd, D. and Law, B. (1999) A case–control study of *Yersinia enterocolitica* infections in Auckland. *Austr. N. Z. J. Publ. Health* **23**, 482–485.

Schieman, D.A. (1990) *Yersinia enterocolitica* in drinking water. In *Drinking Water Microbiology* (ed. G.A. McFeters), pp. 322–340, Springer-Verlag, New York.

Schubert, R.H. (1991) Aeromonads and their significance as potential pathogens in water. *J. Appl. Bacteriol. Symp. Suppl.* **70**, 131S–135S.

Schubert, R.H. (2000) Intestinal cell adhesion and maximum growth temperature of psychrotrophic aeromonads from surface water. *Int. J. Hyg. Environ. Health* **203**, 83–85.

Schwartz, J., Levin, R. and Goldstein, R. (2000) Drinking water turbidity and gastrointestinal illness in the elderly of Philadelphia. *J. Epidemiol. Community Health* **54**(1), 45–51.

Schwartzbrod, L. (1991) *Virologie des milieux hydriques*, Tec. & Doc – Lavoisier, Paris.

Seidler, R.J., Allen, D.A., Lockman, H., Colwell, R.R., Joseph, S.W. and Daily, O.P. (1980) Isolation, enumeration and characterization of *Aeromonas* from polluted waters encountered in diving operations. *Appl. Environ. Microbiol.* **39**, 1010–1018.

Shinnick,T. and Good, R.C. (1994) Mycobacterial taxonomy. *Eur. J. Clin. Microbiol. Infect. Dis.* **13**, 884–901.

Simon, L., Di Majo, P., Pichon, M. and Hartemann, Ph. (2001) *Hygienes* vol. IX.

Singh, N. and Yu, V.L. (1994) Potable water and *Mycobacterium avium* complex in HIV patients is prevention possible? *Lancet* **343**, 1110.

Sobottka, I., Schwartz, D.A., Schottelius, J., Visvesvara, G.S., Pieniazed, N.J., Schmetz, C., Kock, N.P., Laufs, R. and Albrecht, H. (1998) Prevalence and clinical significance of intestinal microsporidiosis in human immunodeficiency virus-infected patients with and without diarrhea in Germany: a prospective coprodiagnostic study. *Clin. Infect. Dis.* **26**, 475–480.

Spano, F., Putignani, L., Guida, S. and Crisanti, A. (1998) *Cryptosporidium parvum*: PCR-RFLP analysis of the TRAP-C1 (thrombospondin-related adhesive protein of Cryptosporidium-1) gene discriminates between two alleles differentially associated with parasite isolates of animal and human origin. *Exp. Parasitol.* **90**, 195–198.

Sparfel, J.M., Sarfati, C., Liguory, O., Caroff, B., Dumoutier, N., Gueglio, B., Billaud, E., Raffi, F., Molina, J.M., Miegeville, M. *et al.* (1997) Detection of microsporidia and identification of *Enterocytozoon bienensi* in surface water by filtration followed by specific PCR. *J. Eukar. Microbiol.* **44**, 78S.

Squier, C., Yu, V. L. and Stout, J.E. (2000) Waterborne nosocomial infections. *Curr. Infect. Dis. Rep.* **2**(6), 490–496.

Stanley, K., Cunningham, R. and Jones, K. (1998) Isolation of *Campylobacter jejuni* from groundwater. *J. Appl. Microbiol.* **85**, 187–191.

States, S.J., Wadowsky, R.M., Kuchta, J.M., Wolford, R.S., Conley, L.F. and Yee, R.B. (1990) *Legionella* in drinking water. In *Drinking Water Microbiology* (ed. G.A. McFeters), pp. 340–368, Springer-Verlag, New York.

Stehr-Green, J.K., Nicholls, C., McEwan, S., Payne, A. and Mitchell, P. (1991) Waterborne outbreak of *Campylobacter jejuni* in Christchurch: the importance of a combined epidemiologic and microbiologic investigation. *N. Z. Med. J.* **104**, 356–358.

Steinert, M., Birkness, K., White, E., Fields, B. and Quinn, F. (1998) *Mycobacterium avium* bacilli grow saprozoically in coculture with *Acanthamoeba polyphaga* and survive within cyst walls. *Appl. Environ. Microbiol.* **64**, 2256–2261.

Steinert, M., Emödy, L., Amann, R. and Hacker, J. (1997) Resuscitation of viable but nonculturable *Legionella pneumophila* Philadelphia JR32 by *Acanthamoeba castellanii*. *Appl. Environ. Microbiol* **63**, 2047–2053.

Stelma, G.N., Johson, C.H. and Spaulding, P. (1986) Evidence for the direct involvement of β-hemolysin in *Aeromonas hydrophila* enteropathogenicity. *Curr. Microbiol.* **14**, 71–77.

Stenström, T.A. (1994) A review of waterborne outbreaks of gastroenteritis in Scandinavia. In *Water and Public Health* (ed. N. Noah and R. Stanwell Smith), pp. 137–143, Smith Gordon & Co. Golding AMB, London.

Sterling, C.R. and Ortega, Y.R. (1999) *Cyclospora*: an enigma worth unraveling. *Emerg. Infect. Dis.* **5**, 48–53.

Stiles, M.E. (1989) Less recognized or presumptive foodborne pathogenic bacteria. In *Foodborne Bacterial Pathogens* (ed. M.P. Doyle), pp. 673–733, Marcel Dekker, New York.

Swanson, M.S. and Hammer, B.K. (2000) *Legionella pneumophila* pathogenesis: a fateful journey from amoebae to macrophages. *Annu. Rev. Microbiol.* **54**, 567–613.

Swerdlow, D.L., Woodruff, B.A., Brady, R.C., Griffin, P.M., Tippen, S., Donnell, H.D.J, Geldreich, E., Payne, B.J., Meyer, A.J. and Wells, J.G. (1992) A waterborne outbreak in Missouri of *Escherichia coli* O157:H7 associated with bloody diarrhea and death. *Ann. Intern. Med.* **117**, 812–819.

Szewzyk, R., Bäz, G. and Bartocha, W. (2000a) Survival and growth of *Legionella pneumophila* in biofilms of drinking water bacteria and amoebae. In *American Society for Microbiology Conference on Biofilms 2000*, Big Sky, MT, USA, poster no 145, American Society for Microbiology, Washington, D.C.

Szewzyk, U., Szewzyk, R., Manz, W. and Schleifer, K. (2000b) Microbiological safety of drinking water. *Annu. Rev. Microbiol.* **54**, 81–127.

Talibart, R., Denis, M., Castillo, A., Cappelier, J.M. and Erme, G. (2000) Survival and recovery of viable but noncultivable forms of *Campylobacter* in aqueous microcosm. *Int. J. Food Microbiol.* **55**, 263–267.

Tauxe, R.V., Hargrett-Bean, N., Patton, C.M. and Wachsmuth, I.K. (1988) *Campylobacter* isolates in the United States, 1982–1986. *Morb. Mortal. Wkly Rep.* **37**, 1–13.

Taylor, D.N., McDermott, K.T., Little, J.R., Wells, J.G. and Blaser, M.J. (1983) *Campylobacter enteritis* from untreated water in the Rocky Mountains. *Ann. Intern. Med.* **99**, 38–40.

Tholozan, J.L., Cappelier, J.M., Tissier, J.P., Delattre, G. and Federighi, M. (1999) Physiological characterization of viable-but-nonculturable *Campylobacter jejuni* cells. *Appl. Environ. Microbiol.* **65**(3), 1110–1116.

Thomas, C., Hill, D.J. and Mabey, M. (1999) Evaluation of the effect of temperature and nutrients on the survival of *Campylobacter* spp. in water microcosms. *J. Appl. Microbiol.* **86**, 1024–1032.

Thomson, C.J., Jesudason, M.V., Balaji, V., Malathi, B., Mukundan, U. and Amyes, S.G. (1998) The prevalence of *Vibrio* spp. in drinking water and environmental samples in Vellore South India. *Epidemiol. Infect.* **121**, 67–76.

Thorel, M.F., Krichevsky, M.I. and Levy-Frebault, V. (1990) Numerical taxonomy of mycobactin-dependent mycobacteria, emended description *of Mycobacterium avium*, and description of *Mycobacterium avium* subsp. *avium* subsp. nov., *Mycobacterium avium* subsp. *paratuberculosis* subsp. nov., *Mycobacterium avium* subsp. *silvaticum* subsp. nov. *Intern. J. Syst. Bacteriol.* **40**, 254–260.

Tison, D.L., Pope, D.H., Cherry, W. B. and Fliermans, C.B. (1980) Growth of *Legionella pneumophila* in association with blue green algae (cyanobacteria). *Appl. Environ. Microbiol.* **39**, 456-459.

Tursi, A., Cammarota, G., Papa, A., Cuoco, L., Gentiloni, N., Fedeli, P., Fedeli, G. and Gasbarrini, G. (1997) The modes of transmission of *Helicobacter pylori* infection. *Recenti Prog. Med.* **88**, 232–236.

Uchiyama, H. (1998) A number of *Vibrio cholerae* non-O1 isolated from aquatic environments. *Kansenshogaku Zasshi* **72**, 720–726.

Vaira, D., Malfertheiner, P., Megraud, F., Axon, A.T., Deltenre, M., Hirschl, A.M., Gasbarrini, G., O'Morain, C., Garcia, J.M., Quina, M. and Tytgat, G.N. (1999) Diagnosis of *Helicobacter pylori* infection with a new non-invasive antigen-based assay. HpSA European study group. *Lancet* **354**, 30–33.

Van der Poel, W.H.M, Vinje, J., Van der Heide, R., Herrera, M.I., Vivo, A. and Koopmans, M.P.G. (2000) Norwalk-like calicivirus genes in farm animals. *Emerg. Infect. Dis.* **6**, 36.

Vincent, P. (1995) Transmission and acquisition of *Helicobacter pylori* infection: evidences and hypothesis. *Bioméd. Pharmacol.* **49**, 11–18.

Vinje, J., Deijl, H., van der Heide, R., Lewis, D., Hedlund, K.O., Svensson, L. and Koopmans, M.P. (2000) Molecular detection and epidemiology of sapporo-like viruses. *J. Clin. Microbiol.* **38**, 530–536.

Vishwanathan, R. (1957) Infectious hepatitis in Delhi (1955–1956): a critical study; epidemiology. *Indian J. Med. Res.* **45**, 1–30.

Von Graevenitz, A. (1977) The role of opportunistic bacteria in human disease. *Annu. Rev. Microbiol.* **31**, 447–471.

Von Reyn, C.F., Maslow, J.N., Barber, T.W., Falkinham, J.O. and Arbeit, R.D. (1994) Persistent colonisation of potable water as a source of *Mycobacterium avium* infection in AIDS. *Lancet* **343**, 1137.

Von Reyn, C.F., Waddell, R.D., Eaton, T., Arbeit, R.D., Maslow, J.N., Barber, T.W., Brindle, R.J., Gilks, C.F., Lumio, J. and Lahdevirta, J. (1993) Isolation of *Mycobacterium avium* complex from water in the United States, Finland, Zaire, and Kenya. *J. Clin. Microbiol.* **31**, 3227–3230.

Wadström, T. and Ljungh, A. (1991) *Aeromonas* and *Plesiomonas* as food and waterborne pathogens. *Int. J. Food Microbiol.* **12**, 303–312.

Wang, G., Doyle, M.P. (1998) Survival of enterohemorrhagic *Escherichia coli* O157:H7 in water. *J. Food Protect.* **61**, 662–667.

Ward, R.L., Bernstein, D.I., Ylung, E.C., Sherwood, J.R., Knowlton, D.R. and Schiff, G.M. (1986) Human rotavirus studies in volunteers: determination of infectious dose and serological response to infection. *J. Infect. Dis.*154, 871–880.

Watanabe, T., Tomita, S., Kudo, M., Kurokawa, M., Orino, A., Todo, A. and Chiba, T. (1998) Detection of *Helicobacter pylori* gene by means of immunomagnetic separation-based polymerase chain reaction in feces. *Scand. J. Gastroenterol.* **33**, 1140–1143.

Wayne, L.G. and Kubica, G.P. (1986) The mycobacteria. In *Bergey's Manual of Systematic Bacteriology* (ed. P.H.A. Sneath, N.S. Mair, M.E. Sharpe, and J.G. Holt), pp. 1436–1457, Williams & Wilkins, Baltimore, MD.

Weber, R. Bryan, R.T., Schwartz, D.A. and Owen, R.L. (1994) Human microsporidial infections. *Clin. Microbiol. Rev.* **7**, 426–461.

WHO (1998) *Conquering Suffering-Enriching Humanity*. World Health Organization, Geneva.

WHO (2002a) *Guidelines for Drinking Water Quality*, 2nd edn, Addendum microbiological agents in drinking water, World Health Organization, Geneva.

WHO (2002b) *Heterotrophic Plae Count Measurement in Drinking Water Safety Management.* WHO/SDE/WSH/02. Report of an expert meeting, Geneva, 24–25 April.

Widdowson, M.A., Jaspers, W.J., van der Poel, W.H., Verschoor, F., de Roda, H.L., Zaaijer, H.L. and Koopmans, M. (2003) Cluster of cases of acute hepatitis associated with hepatitis E v acquired in the Netherlands. *Clin. Infect. Dis.* **36**(1), 29–33.

Widmer, G., Tchack, L., Chappell, C.L. and S. Tzipori. (1998) Sequence polymorphism in the beta-tubulin gene reveals heterogeneous and variable population structures in *Cryptosporidium parvum*. *Appl. Environ. Microbiol.* **64**, 4477–4481.

Williams, E.P. and Cameron, B.A. (1897) Upon general infection by *Bacillus pyocyaneus* in children. *J. Pathol. Bacteriol.* **3**, 344–351.

Wimpenny, J., Manz, W. and Szewzyk, U. (2000) Heterogeneity in biolfilms. *FEMS Microbiol. Rev.*, **24**, 661–671.

Winiecka-Krusnell, J. and Linder, E. (1999) Free-living amoebae protecting *Legionella* in water: the tip of an iceberg? *Scand. J. Infect. Dis.* **31**, 383–385.

Winiecka-Krusnell, J., Wreiber, K., von Euler, A., Engstrand, L. and Linder, E. (2002) Free-living amoebae promote growth and survival of *Helicobacter pylori*. *Scand. J. Infect. Dis.* **34**(4), 253–256.

Winn, W.C. (1999) Enterically transmitted hepatitis. Hepatitis A and E viruses. *Clin. Lab. Med.* 19, 661–673.

Wolinsky, E. (1979) Nontuberculous mycobacteria and associated diseases. *Am. Rev. Respir. Dis.* **119**, 107–159.

Wolk, D.M., Johnson, C.H., Rice, E.W., Marshall, M.M., Grahn, K.F., Plummer, C.B. and Sterlin, C.R. (2000) A spore counting method and cell culture model for chlorine disinfection studies of *Encephalitozoon* syn. *Septata intestinalis*. *Appl. Environ. Microbiol.* **66**, 1266–1273.

Wright, P.J., Gunesekere, I.C., Doultree, J.C. and Marshall, J.A. (1998) Small round-structured (Norwalk-like) viruses and classical human caliciviruses in southeastern Australia, 1980–1996. *J. Med. Virol.* **55**, 312–320.

Xu, X.J., Ferguson, M.R., Popov, V. L., Houston, C.W., Peterson, J.W. and Chopra, A.K. (1998) Role of a cytotoxic enterotoxin in *Aeromonas*-mediated infections: development of transposon and isogenic mutants. *Infect. Immunol.* **66**, 3501–3509.

Yajko, D.M., Chin, D.P., Gonzalez, P.C., Nassos, P.S., Hopewell, P.C., Reingold, A.L., Horsburgh, C.R., Yakrus, M.A., Ostroff, S.M. and Hadley, W.K. (1995) *Mycobacterium avium* complex in water, food, and soil samples collected from the environment of HIV-infected individuals. *J. Acquir. Immune. Defic. Syndr. Hum. Retrovirol.* **9**, 176–182.

Yamamoto, N., Urabe, K., Takaoka, M., Nakazawa, K., Gotoh, A., Haga, M., Fuchigami, H., Kimata, I. and Iseki, M. (2000) Outbreak of cryptosporidiosis after contamination of the public water supply in Saitama Prefecture, Japan, in 1996. *Kansenshogaku Zasshi* **74**, 518–526.

Yates, M.D., Pozniak, A. and Grange, J.M. (1993) Isolation of mycobacteria from patients seropositive for the human immunodeficiency virus (HIV) in south east England: 1884–1922. *Thorax* **48**, 990–995.

Yates, M.V., Citek, R.W., Kamper, M.F. and Salazar, A.M. (2000) Detecting enteroviruses in water: comparing infectivity, molecular, and combination methods.

Yu, V.L. (1993) Could aspiration be the major mode of transmission of Legionaires' disease. *J. Infect. Dis.* **147**, 362.

Yu, V. L. and Stout, J.E. (2000) Hospital characteristics associated with colonization of water systems by *Legionella* and risk of nosocomial legionnaires' disease: a cohort study of 15 hospitals. *Infect. Control. Hosp. Epidemiol.* **21**, 434–435.

Zaaijer, H.L., Kok, M., Lelie, P.N., Timmerman, R.J., Chau, K. and van der Pal, H.J. (1993) Hepatitis E in the Netherlands: imported and endemic. *Lancet* **341**, 826.

Zmirou, D., Ferley, J.P., Collin, J.F., Charrel, M. and Berlin, J. (1987). A follow-up study of gastro-intestinal diseases related to bacteriologically substandard drinking water. *Am. J. Publ. Health* **77**, 582–584.

Zmirou, D., Reys, S., Courtois, X. *et al.* (1995) Residual microbiological risk after simple chlorine treatment of drinking groundwater in small community systems. *Eur. J. Publ. Health* **5**, 75–81.

2

Emerging infectious waterborne diseases: bacterial agents

L.H. Nel and W. Markotter

2.1 INTRODUCTION

Bacterial pathogens, together with protozoa, fungi, viruses and infectious proteins (prions), are the cause of infectious disease, which plague all the higher forms of life, including humanity. Of all these diseases, it is the bacterial diseases that, in general, have become most controllable due to the discovery and continued development of antibiotics, a class of anti-microbials that specifically act against bacteria. Additional factors, such as the development of vaccines, education, improved diagnostics, improved public health systems and improvements in the quality of water and sanitation have also contributed significantly to the control of infectious diseases. However, despite these spectacular sociological and scientific advances, infectious diseases have become increasingly less manageable in recent years. Today, infectious diseases that are resurging, emerging or re-emerging, or are being recognized as a new disease, contribute largely to the infectious disease problems of the world. An appropriate and popular definition for emerging infections has been put forward in a recent report of the Institute of Medicine (http://fas.org/promed): "Emerging infections are those whose incidence in humans has increased within the past two decades or threatens to increase in the near future. Emergence may be

 Microbial Waterborne Pathogens edited by T.E. Cloete, J. Rose, L.H. Nel and T. Ford. ISBN: 1 84339 055 8. Published by IWA Publishing, London, UK.

due to the spread of a new agent, recognition of an infection that has been present in the population but has gone undetected or realization that an established disease has an infectious origin. Emergence may also be used to describe the reappearance (or 're-emergence') of a known infection after a decline in incidence." Thus, emerging diseases may be either new diseases or diseases re-appearing. New diseases would usually be caused by organisms that jump species or by organisms that were already present in a non-pathogenic form and now become pathogenic through specific modification, mutation or recombination. Many bacterial diseases, however, fall in the category of re-emerging diseases due to resistance to antibiotics. For many decades these anti-microbial agents have been used with great success in controlling bacterial diseases and the continued and serious emergence of drug resistance has now become a major health concern. Apart from changes in the genotype of the pathogen that causes increased drug resistance, virulence and disease incidence, the past few decades have seen unprecedented increases in global travel, population explosions (in concert with poverty, breakdown of public health systems) and incursions into virgin ecosystems (with species displacement and increased opportunities for zoonoses and related events which lead to broader host ranges and wider distribution of pathogens). Taken together, all of these factors play some role in the present-time emergence and re-emergence of infectious diseases around the globe. Crucially, the availability and quality of potable water are also directly compromised by the global events described above. Indeed, here we want to focus on waterborne infection, with pathogen entry through the mouth and alimentary track or, in some instances, through epithelial cells of the respiratory system.

Waterborne diseases are classified as those that are transmitted through the ingestion of contaminated water, with water acting as the passive carrier of the infectious agent (Leclerc *et al.*, 2002). Most commonly, waterborne infections are caused by human and animal faecal pollution of water – going hand in hand with scarce water supplies and poor sanitation. Although some emerging waterborne bacterial infections do not enter through the oral route, but indeed through the respiratory route, it is the faecal–oral route of infection that contributes to the three to five billion cases of diarrhoea that are reported every year – resulting in 3 million deaths per year. Waterborne bacterial infections may account for half of these episodes (CDC disease information). These diseases will be discussed in more detail in the following sections. In general though, waterborne bacterial pathogens include enteric and aquatic bacteria. Enteric bacteria can persist in water, e.g. *E. coli*, *Vibrio cholerae*, *Campylobacter* and *Shigella*. Other emerging bacterial agents are natural inhabitants of water, e.g. *Legionella* and *Mycobacterium avium*. Their primary water-related transmission is thought to be through the inhalation of aerosols. The most prominent factor in the resurgence of common bacterial infectious diseases and epidemic bacterial diarrhoea such as those caused by waterborne *Escherichia coli*, *Shigella* and *Vibrio*, is antibiotic resistance. The use (and uncontrolled use) of anti-microbials in agriculture has significantly contributed to the development of drug resistance. Apart from resistance, some new zoonotic diseases and the conversion of non-pathogens into pathogens by the addition of virulence factors will be discussed. Table 2.1 presents a concise and comparative summary of the emerging waterborne bacterial diseases discussed in the following sections.

Table 2.1 Waterborne new and emerging diseases of bacterial origin.

Pathogen and disease	Geographical incidence	Factors involved in disease emergence	Clinical features	Treatment and detection
E. coli O157:H7 Haemorrhagic colitis (Reviewed in Besser, 1999; Chalmers *et al*., 2000; Su and Brandt, 1995)	Occurs worldwide	Emergence of new serotypes AIDS epidemic Antibiotic resistance Resistance to chlorine Increased screening for the disease	Acute bloody diarrhoea and abdominal cramps with little or no fever Vomiting may occur	Laboratories do not test routinely Selective culturing Serological tests and molecular methods Only supportive treatment
Campylobacter Campylobacteriosis (Reviewed in Allos, 2001; Coker *et al*., 2002; Friedman *et al*., 2000)	Occurs worldwide	Improved diagnostics Acquisition of drug resistance Emergence of toxin producing strains	Diarrhoea, which may be watery, sticky or bloody Fever, abdominal pain, nausea, headache and muscle pain	Selective culturing and molecular methods Treatment with erythromycin and ciprofloxacin
Helicobacter pylori Gastritis and ulcer disease (Reviewed in Velazquez and Feirtag, 1999; Petersen and Krogfelt, 2003; Bani-Hani, 2002)	Occurs worldwide	Adaptation leads to new genetic diversity Poor sanitation Antibiotic resistance	Abdominal pain, sometimes include nausea, vomiting and loss of appetite and bleeding (may lead to stomach ulcers and stomach cancer)	Selective culturing, urea breath test, serology and molecular methods Antibiotic resistance complicates treatment

(*continued*)

Table 2.1 (*continued*)

Pathogen and disease	Geographical incidence	Factors involved in disease emergence	Clinical features	Treatment and detection
Legionella pneumophila Legionnaires' disease and Pontiac fever. (Reviewed in Fields *et al*., 2002; Steinert *et al*., 2002; Swanson and Hammer, 2000)	Occurs worldwide	Modern practices involving aerosolization Improved diagnostic techniques	Legionnaires' disease: fever, chills, and a cough Other symptoms may include muscle aches, headache, myalgia, loss of appetite, diarrhoea Pontiac fever: Fever and muscle aches, not pneumonia	Culturing on selective media Serological assays, molecular methods Treatment with antibiotics
Non–tuberculosis mycobacteria Pulmonary infections (Reviewed in Collins *et al*., 1984)	Occurs worldwide	Development of drug resistant strains AIDS epidemic Modern practices involving aerosolization Resistance to chlorine	Mostly asymptomatic, but symptoms may be a bad cough lasting longer than 2 weeks, chest pain, weakness, fatigue, weight loss, chills and fever	Staining techniques and selective culturing Multiple drug resistance complicates therapy
Pseudomonas aeruginosa Dermatitis	Occurs worldwide	Antibiotic resistance	Symptoms primarily skin rashes, redness, pain and itching	Selective culturing Treatment with aminoglycoside and β-lactam antibiotics

Shigella Shigellosis (Reviewed in Wittenberg, 1999)	Occurs worldwide	Antibiotic resistance Emergence of toxin-producing strains	Diarrhoea, often bloody, fever and stomach cramps	Detection by culturing and serological assays Multiple drug resistance complicates therapy, but the disease is usually self-limiting
Vibrio cholera Serotypes O1 and O139 Cholera (Reviewed in Faruque and Nair, 2002; Reidl and Klose, 2002; Colwell, 1996)	Serotype O1 occurs worldwide Serotype O139 occurs only in South East Asia	Emergence of new serotypes AIDS epidemic Emergence of new toxin producing strains	Symptoms include diarrhoea, vomiting without nausea and muscle cramps	Selective culturing Microscopy Serological assays, e.g. slide agglutination and fluorescent antibody test Molecular methods The disease is usually self-limiting and treatment supportive with re-hydration Vaccines, for travellers

2.2 SPECIFIC EMERGING WATERBORNE INFECTIOUS DISEASES CAUSED BY BACTERIAL INFECTION

2.2.1 Campylobacteriosis

Campylobacter jejuni is the aetiological agent for campylobacteriosis. Like most of the waterborne diseases this is a diarrhoeal disease and it occurs worldwide (Allos, 2001; Coker *et al.*, 2002; Friedman *et al.*, 2000). Although *Campylobacter* species have been known to infect animals and livestock for many decades, the recognition of these bacteria as an important pathogen of man only came in the middle to late 1970s – an achievement greatly facilitated by the development of appropriate selective culture media (Szewzyk *et al.*, 2000; Allos, 2001). Clearly it is poor hygiene, poor sanitation and close proximity to animals that contribute to easy and frequent acquisition of this pathogen, and today it is one of the most frequently isolated bacteria from the stools of infants with diarrhoea in developing countries (Taylor, 1992; Oberhelman and Taylor, 2000). The disease is usually self-limiting, but treatment with antibiotics may be useful. It comes as no surprise that the recent emergence of toxin-producing strains together with anti-microbial resistance has been met with some alarm.

Campylobacter infections are thought to be more common than infections with other enteric bacteria such as *Shigella*, *Salmonella* or *Escherichia coli* O157:H7 (Altekruse *et al.*, 1999; Slutsker *et al.*, 1997). Whereas other *Campylobacter* species (e.g., *C. laridis*, *C. hyointestinalis* (Van den Berg *et al.*, 2001)) may also be involved in disease, *C. jejuni* and *C. coli* are of major public health importance (Nachamkin and Blaser, 2000). The bacterium is Gram-negative, has a curved rod-shaped morphology and is motile. It is highly infectious and the symptoms are diarrhoea, fever, abdominal pain, nausea, headache and myalgia (Blaser *et al.*, 1983). Complications are rare but the infection may progress towards reactive arthritis, haemolytic uraemic syndrome and septicaemia. Guillain–Barre syndrome (Nachamkin *et al.*, 1998; Allos, 1997) is also recognized as a rare (1 per 1000 infected individuals) complication leading to disease of the peripheral nervous system. The incubation time for campylobacteriosis is 2–5 days. As mentioned above, treatment with antibiotics may reduce bacterial numbers and the duration of the infection, and may be of particular value in cases involving immuno-compromised individuals (Quinn, 1997) or where complications are evident. From this perspective the emergence of antibiotic-resistant strains of *Campylobacter* all over the world is disconcerting. In the past, flouroquinolones were the antibiotic of choice, but in recent years *Campylobacter* strains have become resistant to these drugs. This resistance coincided with the use of antibiotics in animal foods (Endtz *et al.*, 1991) and the use of these drugs in the treatment of other infections. Currently, the drugs of choice are erythromycin and ciprofloxacin (Engberg *et al.*, 2001). The organism can be isolated using antibiotic-containing media (cephalothin) and microaerophilic conditions, since most other stool flora are susceptible to cephalothin (Nachamkin, 1995). The use of PCR analysis and PCR-enzyme-linked immunosorbent assays has also been described (Lawson *et al.*, 1999).

2.2.2 Cholera

Vibrio cholerae is a Gram-negative, motile, rod-shaped bacterium and is responsible for the serious cholera disease (Faruque and Nair, 2002; Reidl and Klose, 2002). Both pathogenic and non-pathogenic strains of *V. cholerae* exist as natural inhabitants of aquatic systems (Islam *et al.*, 1994) and by acquisition of virulence factor-encoding genes, they evolve into a dangerous human pathogen by adapting to the human intestinal environment (Reidl and Klose, 2002). About 200 serogroups of *V. cholerae* are known; however, only serogroup O1 and the newly emerged O139 have been associated with cholera pandemics (Faruque *et al.*, 1998). Cholera is an acute, severe and serious infection of the small intestine with diarrhoea often accompanied by vomiting with dehydration and shock (Bennish, 1994). Epidemics occur in developing countries (Faruque *et al.*, 1998; Kaper *et al.*, 1995) but the organism is implicated in disease worldwide. The symptoms associated with cholera are caused by the enterotoxins produced by *V. cholerae* during the infection (Kaper *et al.*, 1995). The infection starts with the oral ingestion of contaminated water or food. The bacteria must then survive passage through the gastric acid barrier of the stomach and then adhere to the intestinal epithelial cells where it produces the cholera toxin (CT) (Bennish *et al.*, 1994). This pathogenic effect is complex and involves a number of virulence-encoding genes that aid the bacteria in producing the toxin (Hacker *et al.*, 1997), e.g. the integrated genome of a bacteriophage CTXϕ coding for CT (cholera toxin) (Waldor and Mekalanos, 1996) and a TCP pathogenic island (Herrington *et al.*, 1988). These genes aid the bacteria in reaching the epithelium of the intestine and colonizing it, producing the CT. Accurate records of cholera date back to almost a century, but cholera may have been known in India and elsewhere for thousands of years (Barua, 1991). In 1817 an epidemic outbreak, called the first pandemic, was reported in India and spread across the whole Indian subcontinent (Colwell, 1996). During the nineteenth century, six cholera pandemics caused by the classic biotype of *V. cholerae* O1 took place, primarily affecting mostly continents in the southern hemisphere, North America and Europe (Barua, 1991; Colwell, 1996). Many believed that cholera would not return because water supplies had been improved worldwide but in 1961 *V. cholerae* O1, biotype El Tor, not previously implicated in widespread epidemics, emerged and caused a pandemic spread over six continents during the past 40 years (Blake, 1994; Swerdlow and Isaäcson, 1994; Faruque *et al.*, 1998; Colwell, 1996). This seventh pandemic spread from Indonesia to Eastern Asia, India, the former USSR, Iran and Iraq. Thereafter, in 1970 serotype O1 took hold in West Africa and then became endemic in most of the African continent. In 1991, it spread throughout South and Central America, involving more than a million known cases with a mortality rate of about 1%. Much higher mortality rates (33%) were observed in 1994 in a large epidemic among Rwandan refugees in Zaire. In recent times, Africa continued to host large cholera epidemics. These included a 1997 outbreak in Kenya, Tanzania and Mozambique, a 1998 outbreak in Congo and Zaire and in 2000–2001, a large epidemic in South Africa.

Until 1992, serogroup O1 caused endemic cholera; however, the emerging serotype O139 appeared in 1992 (Ramamurthy *et al.*, 1993; Swerdlow and Ries, 1993) and has as yet only been reported from South East Asia where it causes epidemics in populations that were previously exposed to other serotypes of the same pathogen. This epidemic has been described as the eighth pandemic (Faruque *et al.*, 1998). Cholera is endemic in Southern Asia, parts of Africa and Latin America, and the pattern of cholera epidemics depends largely on environmental factors, the immune status or antigenic exposure of the population and the properties of the *V. cholerae* organism. A virulence-associated gene pool exists in the environment (Mukhopadhay *et al.*, 2001) and virulence genes can be transferred to *V. cholerae*, transforming a benign strain into a pathogen (Berche *et al.*, 1994; Bik *et al.*, 1995). Gene transfer is also responsible for acquiring antibiotic resistance (Dalsgaard *et al.*, 2000). *V. cholerae* can survive and persist in aquatic environments and it has been postulated that the bacteria are converted to a dormant uncultivable life form, representing a spore-like stage under stress conditions (Gauthier, 2000). Cholera occurs in endemic areas with seasonal patterns and it has been suggested that the bacteria exist in association with aquatic organisms during endemic periods (Islam *et al.*, 1994). This association can prolong the survival of the pathogen.

A wet mount of liquid stool is examined microscopically for rapid diagnosis (finding motile bacteria). Samples can also be cultured on selective agar. Other diagnostic tests, e.g. slide agglutination tests with specific antiserum, fermentation tests, fluorescent antibody tests and PCR (Miyagi *et al.*, 1999) can also be used after a strain has been identified as *V. cholerae* (Reidl and Klose, 2002). Giving oral re-hydration salts can treat cholera, but patients who are severely dehydrated must be given intravenous fluids (Shears, 2001). In severe cases antibiotics can be used but resistance is increasing (Garg *et al.*, 2000). People with damaged immunity, reduced gastric acidity or malnutrition may suffer more severe forms of the disease.

Besides hygienic and sanitary control measures, vaccine strategies are one of the main efforts to combat disease (Tacket *et al.*, 1992). Two oral vaccines for immunization against *V. cholerae* O1 have become available in limited quantities. These are at present mostly used for vaccination of travellers to endemic cholera regions and are only partially effective.

2.2.3 Gastritis and ulcer diseases

Helicobacter pylori, formerly known as *Campylobacter pylori*, a spiral Gram-negative bacterium, was discovered in 1983 as a cause of gastritis (Velazquez and Feirtag, 1999; Marshall and Warren, 1984). In 1989 this pathogen was transferred to a new genus, *Helicobacter pylori* (Goodwin *et al.*, 1989). It is still unclear whether this bacterium is a *bona fide* waterborne pathogen since its mode of transmission remains under investigation (Bani-Hani, 2002). It appears that the human stomach is the only recognized reservoir and infection is common in the general population (Parsonnet, 1998). Although most infected individuals are asymptomatic and unlikely to develop a serious medical problem from the infection, approximately 16% will develop peptic ulcer disease (Qureshi and Graham, 1999). *H. pylori* causes chronic gastritis and 90% of duodenal ulcers and up to 80% of gastric ulcers (Dubois *et al.*, 1994; Hopkins *et al.*, 1996). Infected persons have a two- to sixfold increased risk of developing gastric cancer (Recavarren-Arce *et al.*, 1991; Dunn *et al.*, 1997; Forman, 1996). Acute infections, when they do occur, are characterized by abdominal pain, nausea and vomiting. Human populations throughout the world are affected (Graham *et al.*, 1991; Parsonnet, 1998) and the rate increases in developing countries where up to 90% of some populations may be infected (Hulten *et al.*, 1996; Goodman and Correa, 1995). The transmission of *H. pylori* remains unclear although the bacteria are most likely spread from person to person through faecal–oral or oral–oral routes (Deltenre and De Koster, 2000; Enroth and Engstrand, 1995; Hulten *et al.*, 1998; Nilsson *et al.*, 1996). The main risk factors for acquiring *H. pyloris* are childhood, low socio-economic status and poor sanitation (Deltenre and De Koster, 2000). Significant genetic diversity identified amongst *H. pylori* strains is likely to be attributable to adaptation and genomic rearrangement when the bacteria infect new hosts. The disease-causing mechanism of *H. pylori* is unknown but the bacteria produce a toxin (Vac A) that may be involved (Figura *et al.*, 1989), as well as other virulence factors, enabling it to survive the gastric acids (CagA pathogenic island) (Perez-Perez *et al.*, 1999; Konturek *et al.*, 1999).

Methods for identifying *H. pylori* are invasive and there is a need for new cost-effective and accurate diagnostic techniques. *H. pylori* can be diagnosed by culturing on selective media (VNC forms will not be detected) (Poms, 1997), the urea breath test and immuno-separation followed by molecular techniques such as PCR or hybridization (Cohen, 1996; Nilsson *et al.*, 1996; Hulten *et al.*,1996; Hulten *et al.*, 1998). Autoradiographic detection (Shahamat *et al.*, 1993) or ATP bioluminescence can be used to determine viability. Recently, an antigen test to diagnose *H. pylori* in stool samples was approved (Van Doorn *et al.*, 2001). Isolation of *H. pylori* from samples (e.g., water) other than gastric tissue is very difficult (Goodman and Correa, 1995) and only sophisticated and expensive techniques are generally successful. The lack of success is due to the fact that adverse environmental conditions trigger the development of a viable but non-culturable state (VNC) (Bode *et al.*, 1993). This phenomenon is also observed in *V. cholerae* (Duncan *et al.*, 1994) and *E. coli* O157:H7 (Rigsbee *et al.*, 1997), making it difficult to detect the presence of these pathogens in water. These VNC forms can reintroduce symptoms when conditions again become favourable, e.g. if treatment is withdrawn (Aleljung *et al.*, 1996). Failure in eradicating *H. pylori* may also be due to resistance of the bacteria to antibiotics used for treatment (Labenz and O'Morain, 1995; Tytgat *et al.*, 2000).

2.2.4 Haemorrhagic colitis and related complications

Enterohaemorrhagic *Escherichia coli* (EHEC), a Gram-negative rod-shaped bacterium (Besser, 1999), has only recently emerged as a major waterborne enteric pathogen (Levine, 1987). The pathogen was first recognized in 1982 (Riley *et al.*, 1983) and the first waterborne outbreak associated with this pathogen was in

Missouri in 1989 (Swerdlow *et al.*, 1992). The main reservoir of *E. coli* O157:H7 implicated in water outbreaks is animals grazing in water and surrounding areas, with resultant faecal contamination of the potable water resources. Raw sewage and contaminated water from abattoirs have also been implicated (Chalmers *et al.*, 2000). Once in water the bacteria are capable of surviving at cold temperatures (Rice *et al.*, 1992). Whereas *E. coli* bacteria are usually symbiotic as part of the normal intestinal flora of animals and humans, some strains of the organism are capable of causing disease by producing large quantities of Vero toxins (Johnson *et al.*, 1983; Karmali, 1989), also called the Shiga-like toxin because of its close relation to the Shiga toxin of *Shigella* (Noda *et al.*, 1987). Bacteriophages infecting the bacterial cells integrate into the bacterial genomes and from there provide the toxin-producing capability. Today, a large number of *E. coli* serotypes that can produce Vero toxins are recognized (Pai *et al.*, 1988; Von Wulffen *et al.*, 1989), but the serotype O157:H7 is the predominant pathogen that is most frequently associated with haemorrhagic colitis in humans. Other virulence factors may also be involved in causing disease, e.g. a virulence plasmid involved in the organism's ability to attach to mucosal surfaces (Karch *et al.*, 1987; McKee and O'Brien, 1995). The disease is serious (Su and Brandt, 1995), usually characterized by bloody diarrhoea (can also present non-bloody diarrhoea (Griffin and Tauxe, 1991)), abdominal cramps and nausea (Griffin *et al.*, 1988), but more severe complications known as haemolytic uraemic syndrome (HUS) (Pickering *et al.*, 1994; Karmali *et al.*, 1985; Riley, 1987), leading to complete renal and kidney failure, may follow in 2–7% of the people infected (Griffin and Tauxe, 1991; Slutsker *et al.*, 1997). Thrombotic thrombo-cytopaenic purpura (Kovacs *et al.*, 1990) and neurological impairment can also be a severe complication especially in the elderly (Bale *et al.*, 1980; Hamano *et al.*, 1993). Case fatality rates for *E. coli* O157:H7 range from 3% to 36%, the fate of patients being dependent on age and immuno-competence (Ryan *et al.*, 1986; Pavia *et al.*, 1990; Carter *et al.*, 1987).

Waterborne transmissibility through both drinking and recreational waters together with antibiotic and chlorine resistance and an ability to survive in water for long periods of time, are major factors in the widespread emergence of EHEC. In addition, these organisms are highly infectious. The infectious dose for *E. coli* O157:H7 is less than 100 organisms (Griffin and Tauxe, 1991). Outbreaks associated with water continue to be recognized worldwide (Griffin, 1995; Chalmers *et al.*, 2000) with inclusion of the United States (Swerdlow *et al.*, 1992; Keene *et al.*, 1994; Ackman *et al.*, 1997; Warner *et al.*, 1996), Japan (Akashi *et al.*, 1994), United Kingdom (Cransberg *et al.*, 1996; Hildebrand *et al.*, 1996), Europe (Pebody *et al.*, 1999), Scotland (Dev *et al.*, 1991; Brewster *et al.*, 1994; Jones and Roworth, 1996), Finland (Paunio *et al.*, 1999) and throughout Africa (Paquet *et al.*, 1993; Isaacson *et al.*, 1993; Effler *et al.*, 2001). The rate of infection follows a seasonal pattern, with the peak incidence from June to September (Ostroff *et al.*, 1989; Waters *et al.*, 1994).

Numerous *E. coli* O157:H7 infections may go unrecognized (Su and Brandt, 1995) because laboratories do not routinely test for this pathogen and outbreaks are sometimes attributed to *Shigella* infection because of the overlap in the spectrum of clinical illness (Wittenberg, 1999). Increased screening and an increase in reporting infections may have accounted for the increase in the disease (Boyce *et al.*, 1995). Most microbiological assays rely on a single phenotype, the inability to ferment sorbitol, to detect and isolate *E. coli* O157:H7 on selective media (March and Ratnam, 1986; Chapman *et al.*, 1991). This method is 86% accurate in detecting *E. coli* O157:H7 provided that isolation of samples occurs quickly after the onset of illness (Wells *et al.*, 1983). Positive samples must be serologically confirmed by further testing with commercially available antisera in agglutination tests. These techniques are specific and sensitive and can detect phenotypic variants of *E. coli* O157:H7. Molecular diagnostic techniques may facilitate detection of these different variants and some techniques (e.g., hybridization, ribotyping, pulsed field electrophoresis (Bender *et al.*, 1997) and lambda restriction fragment length polymorphism (Fields *et al.*, 1997; Samadpour *et al.*, 1993)) are commercially available at some reference laboratories. Isolates can also be examined by more sophisticated techniques for the presence of toxins (Cebula *et al.*, 1995) and the *uidA* allele specific for *E. coli* O157:H7 (Feng, 1993), using the polymerase chain reaction. Antibodies specific to the O157 antigen can also be used in serological assays to identify O157:H7 and its variants. For the detection of *E. coli* in water, methods need to be sensitive since the number of target organisms may be low. Before sub-culturing the sample on selective media, the bacteria must be concentrated by membrane filtration, incubated in enrichment broth and then isolated by immuno-magnetic separation with appropriate antibodies (Chalmers *et al.*, 2000).

E. coli is susceptible to chlorine and adjustments in the chlorination of water may prevent outbreaks (Mintz *et al.*, 1995; Kaneko, 1998). However, it has been indicated that in nutrient-limiting conditions *E. coli* O157:H7 develops a chlorine-resistant phenotype (Lisle *et al.*, 1998). Private water supplies are not treated with chlorine and reports of *E. coli* O157:H7 will continue. These organisms are already resistant to many antibiotics, e.g. erythromycin, metronidazole, vancomycin, tetracycline and sulfisoxole (Kim *et al.*, 1994; Slutsker *et al.*, 1997). Therefore, treatment is as a rule supportive by managing fluids and electrolytes, in the absence of complications. Treating infections with antibiotics may predispose the patient to develop serious complications (Wong *et al.*, 2000).

2.2.5 Legionellosis

Legionnaires' disease, a severe form of legionellosis, is an acute pneumonia caused in 90% of cases by a Gram-negative bacterium, *Legionella pneumophila* (Winn, 1988; Fields *et al.*, 2002). Seventeen additional *Legionella* species have also been associated with disease (Steinert *et al.*, 2002). This is a flu-like illness, typically occurring 3 days after exposure and followed within a week by high fever, chills, dry cough, muscle aches, headache and occasionally diarrhoea (Cunha, 1998). The disease is usually self-limited, but can lead to higher mortalities in the presence of various risk factors such as immuno-deficiency, cigarette smoking, chronic lung diseases, lung malignancies, heart disease and old age (Marston *et al.*, 1994). A milder flu-like illness with no pneumonia, Pontiac fever, can also occur (Fields *et al.*, 1990). The mortality rate of Legionnaires' disease is about 7–24% depending on the age and immune status of the patient. Many cases may go unreported because of the difficulty in distinguishing these diseases from other pneumonia and flu-like diseases (Paszko-Kolva *et al.*, 1993). When this disease was first recognized in the late 1970s, the associated aetiological agent was hard to isolate, given that *Legionella* is a particularly fastidious organism that does not grow on typical culture media. The importance of these waterborne Gram-negative, rod-shaped bacilli has since become clear, following the development of suitable diagnostic techniques and occurrence worldwide. Of significant importance is the fact that these bacteria are natural parasites of protozoa (e.g., Amoeba sp.), which commonly occur in rivers, lakes and streams (Abu *et al.*, 1998). Within these hosts *Legionella* proliferates and is protected from environmental hazards, including chlorination (Barker *et al.*, 1992; Kilvington and Price, 1990; Rowbotham, 1986). However, these natural sources are rarely connected to disease because the conditions do not allow *Legionella* to grow to high numbers (Fliermanns *et al.*, 1981). Legionella can also survive within biofilms in building water systems, which may also play a role in their persistence (Rogers *et al.*, 1994). Their optimum growth temperature is 25–42°C (Katz and Hammel, 1987) and the only natural waters that may be the source of these diseases are naturally occurring warm water spas. *Legionella* species are introduced into drinking water from natural water sources and can then multiply if nutrients are available and higher temperatures are maintained (Sanden *et al.*, 1989). Bacterial proliferation is further enhanced by water stagnation, sediment build-up and the intracellular growth of *Legionella* in protozoa (e.g., amoeba) (Fields, 1996). Following intracellular replication in protozoal cells, *L. pneumophila* converts to a virulent form and expresses a number of traits (including enhanced motility) that assist in extracellular survival and transmission to new cells (Swanson and Hammer, 2000). This phenomenon is thought to explain the virulence of *L. pneumophila* for the macrophages in the human lung, given the similarities in the cell biology of these and protozoal cells. The virulent form of the bacteria can enter the lungs via aspiration (foreign particles escape the gag reflex and fall directly into the respiratory tract) or inhalation. Replication may occur within the alveolar macrophages and disease will ensue. Transmission occurs easily in the presence of airborne water droplets and legionellosis has become particularly important and common following the various practices leading to aerosolization of water (e.g., showers, spraying of produce in large markets, air-conditioning systems, cooling towers, whirlpools and spas (Atlas, 1999)). Similar to *E. coli* O157:H7 and *V. cholerae*, *Legionella* is also able to enter a viable but non-culturable state (Hussong *et al.*, 1987).

The bacteria can be identified by selective media (Fields, 1992; Feeley *et al.*, 1979; Edelstein, 1982). Serological methods can also be used but cross reactivity with new species limits the specificity of these tests (Maiwald *et al.*, 1998). Other methods used are pulsed field gel electrophoresis, amplified fragment length polymorphism, PCR (Benson and Fields, 1998), and fluorescence *in situ* hybridization (Grimm *et al.*, 1998). The disease can be treated with antibiotics (Breiman and Butler, 1998), but the patient can be non-responsive due to resistance.

2.2.6 Shigellosis

Shigella is a highly infectious Gram-negative rod-shaped bacterium, of which most species are implicated in waterborne infections. Although not new, it is the virulence of these organisms and their ability to develop drug resistance through plasmid transfer, conferring resistance to a large spectrum of antibiotics, which characterize a resurgence of shigellosis colitis. The disease symptoms include abdominal pains, fever and rectal pain while complications may include sepsis, seizure, renal failure and haemolytic uraemia syndrome depending on the age and immune status of the patient (Perdomo *et al.*, 1994). The bacterium infects epithelial cells of the intestinal mucosa where it replicates and causes tissue destruction and some strains can produce enterotoxins (Shiga toxin), which contribute to disease (Fasano *et al.*, 1995). *Shigella* is a highly infective organism and as few as 10 cells can initiate infection.

Shigella was identified as the causative agent in a large dysentery epidemic in Japan more than a hundred years ago (Wittenberg, 1999) and has since then been implicated in large epidemics in developing countries worldwide. *Shigella dysenteriae* type 1 (Sd1) is the cause of epidemic dysentery that usually originates from water polluted with human faeces. This type is a major cause of dysentery in Asia and Central and South America and has only recently arrived in Africa. After major pandemics in southern Africa in the early and middle 1990s, *S. dysenteriae* type 1 has now become endemic in many countries in Africa (Wittenberg, 1998) and can easily become epidemic due to climatic changes, immune status of the population and/or poor water sanitation. Other species of *Shigella* that are frequently responsible for waterborne infections are *Shigella sonnei* and *Shigella flexneri*.

Despite scientific advances over the past century, the epidemiological characteristics, virulence and ability to develop drug resistance has led to renewed and increased incidence of *Shigella*. Shigellosis can usually be treated with antibiotics, e.g. ampicillin, trimethoprim/sulfamethoxazole, nalidixic acid or ciprofloxacin (Bennish *et al.*, 1992) but plasmids that encode for resistance against antibiotics are known in these bacterial species (Wharton *et al.*, 1990). *Shigella* can be diagnosed by selective culturing of stool samples. Sensitive and rapid methodology for identification of *Shigella* species utilizes DNA probes that hybridize with common virulence plasmid genes or DNA primers that amplify plasmid genes by polymerase chain reaction (PCR). Enzyme-linked immuno-sorbent assays (ELISA) can also be used to screen stool samples for *Shigella* species.

2.2.7 Pulmonary diseases

The genus *Mycobacterium* includes the *Mycobacterium tuberculosis* complex (that is responsible for tuberculosis) and the *Mycobacterium avium* complex (MAC) (compromising *M. avium* and *M. intracellulare*) (Shinnick and Good, 1994). MAC causes non-tuberculosis diseases, with water being the suspected vehicle of transmission (Collins *et al.*, 1984). These organisms live in the environment as free-living saprophytes and can survive and grow in water at an optimum temperature of

45°C (Collins *et al.*, 1984). They are resistant to chlorine, contributing to their persistence in drinking water (Falkinham, 1996). Protozoa may also serve as a host (as in the case of Legionella), favouring their survival (Cirillo *et al.*, 1997). MAC causes several different infections. In HIV-infected persons, manifestations include night sweats, weight loss, abdominal pain, fatigue, diarrhoea and anaemia. Pulmonary diseases are associated with disease in non–immuno-compromised persons. This disease is similar to tuberculosis with chronic cough, weight loss and malaise. Exposure usually occurs through inhalation of infected droplets and thus aerosolization of water contributes to the transmission of disease. With the emergence of AIDS, the concern for diseases caused by non-tuberculosis bacteria has increased (Von Reyn *et al.*, 1994).

These organisms can be identified microscopically and with biochemical tests (Collins *et al.*, 1984). Molecular methods, e.g. hybridization and PCR, are able to distinguish between *M. tuberculosis* and non-tuberculosis mycobacteria. Treatment of non-tuberculosis bacteria can be problematic due to the resistance of these bacteria to antibiotics, but drugs such as clarithromycin or azithromycin plus ethambutol can be used in combinations, while other useful cocktails can be composed of clofazimine, rifabutin, fluoroquinolones and amikacin (Pierce *et al.*, 1996).

2.2.8 Dermatitis/folliculitis

This is an inflammation of the skin marked by redness, pain and itching (skin rashes) and/or pus-filled pimples around a hair follicle and is caused by the Gram-negative and rod-shaped bacterium, *Pseudomonas aeruginosa* (Pollack, 1995). Until the recent recognition of waterborne dermatitis caused by *P. aeruginosa*, this organism was rarely implicated in disease. With the increased popularity of whirlpools and hot tubs, sometimes not chlorinated, the occurrence of this organism is on the increase. The bacterium typically survives in biofilms and antibiotic-resistant strains are also appearing, but effective control should be possible through proper water treatment. The true incidence of *P. aeruginosa* is difficult to determine because symptoms are often self-limited and therefore patients do not report it. Most outbreaks have occurred in high-temperature water, e.g. whirlpools (Gustafson *et al.*, 1983; Ratnam *et al.*, 1986), but conventional swimming pools have also been implicated (Hopkins *et al.*, 1981). Crowding is a high risk factor (Washburn *et al.*, 1976). Outbreaks usually occur because of low water disinfectant levels (Gustafson *et al.*, 1983). Selective culturing can be used to diagnoze the presence of the bacteria. *Pseudomonas* are resistant to a variety of antibiotics and symptoms normally resolve without treatment.

REFERENCES

Abu, K.Y., Gao, L.-Y. Stone, B.J., Venkataraman, C. and Harb, O.S. (1998) Invasion of protozoa by *Legionella pneumophila* and its role in bacterial ecology and pathogenesis. *Appl. Environ. Microbiol.* **64**, 3127–3133.

Ackman, D., Marks, S., Mack, P., Caldwell, M., Roodt, T. and Birkhead, G. (1997) Swimming-associated haemorrhagic colitis due to *Escherichia coli* O157:H7 infection: evidence of prolonged contamination of a fresh water lake. *Epidemiol. Infect.* **119**, 1–8.

Akashi, S., Joh, K., Tsuji, A., Ito, H., Hawakawa, T., Ihara, J., Abe, T., Hanori, M. and Mori, N.T. (1994) A severe outbreak of haemorrhagic colitis and hemolytic uremic syndrome associated with *Escherichia coli* O157:H7 in Japan. *Eur. J. Pediatr.* **153**, 650–655.

Allos, B.M. (1997) *Campylobacter jejuni* infection and Guillain-Barré syndrome. *J. Infect. Dis.* **2**(Suppl), S125–S128.

Allos, B.M. (2001) *Campylobacter jejuni* infections: update on emerging issues and trends. *Food Safety* **32**, 1201–1206.

Altekruse, S.F., Stern, N.J., Fields, P.I. and Swerdlow, D.L. (1999) *Campylobacter jejuni*-an emerging foodborne pathogen. *Emerg. Infect. Dis.* **5**, 28–35.

Atlas, R.M. (1999) *Legionella*: from environmental habitats to disease, pathology, detection and control. *Environ. Microbiol.* **4**, 283–293.

Bale, J.F. Jr., Brasher, C. and Siegler, R.L. (1980) CNS manifestations of the hemolytic-uremic syndrome. Relationship to metabolic alterations and prognosis. *Am. J. Dis. Child.* **134**, 869–872.

Bani-Hani, K.E. (2002) The current status of *Helicobacter pylori*. *Saudi Med. J.* **23**(4), 379–383.

Barker, J., Brown, M.R., Collier, P.J., Farrell, I. and Gisbert, P. (1992) Relationship between *Legionella pneumophila* and *Acanthamoeba polyphaga*: physiological status and susceptibility to chemical inactivation. *Appl. Environ. Microbiol.* **58**, 2420–2425.

Barua, D. (1991) History of cholera. In *Cholera* (ed. D. Barua and W.B. Greenough III), pp. 1–35, Plenum, New York.

Bender, J.B., Hedberg, C.W., Besser, J.M., Boxrud, D.J., MacDonald, K.L. and Osterholm, M.T. (1997) Surveillance for *Escherichia coli* O157:H7 infections in Minnesota by molecular subtyping. *N. Engl. J. Med.* **337**, 388–394.

Bennish, M.L. (1994) Cholera: pathophysiology, clinical features, and treatment. In *Vibrio cholerae and Cholera, Molecular to Global Perspectives* (ed. K.I. Wachsmuth, P.A. Blake and O. Olsik), pp. 229–255, American Society for Microbiology, Washington, D.C.

Bennish, M.L., Salam, M.A., Khan W.A. and Khan, A.M. (1992). Treatment of shigellosis. 3. Comparison of one-dose or 2-dose ciprofloxacin with standard 5-day therapy – a randomized, blinded trial. *Ann. Int. Med.* **117**, 727.

Benson, R.F. and Fields, B.S. (1998) Classification of the genus *Legionella*. *Semin. Respir. Infect.* **13**, 90–99.

Berche, P., Poyart, C., Abachin, E., Lelievre, H., Vandepitte, J., Dodin, A. and Fournier, J.M. (1994) The novel epidemic strain O139 is closely related to the pandemic strain O1 of *Vibrio cholerae*. *J. Infect. Dis.* **170**, 701–704.

Besser, R.E. (1999) *Escherichia Coli* O157:H7 gastroenteritis and the hemolytic uremic syndrome: an emerging infectious disease. *Annu. Rev. Med.* **50**, 355–367.

Bik, E.M., Bunschoten, A.E., Gouw, R.D. and Mooi, F.R. (1995) Genesis of the novel epidemic *Vibrio cholerae* O139 strain: evidence for horizontal transfer of genes involved in polysaccharide synthesis. *EMBO J.* **14**, 209–216.

Blake, P.A. (1994) Historical perspectives on pandemic cholera. In *Vibrio cholerae and Cholera: Molecular to Global Perspectives* (ed. K.I. Wachsmuth, P.A. Blake and O. Olsik), pp. 293–295, American Society for Microbiology Press, Washington, D.C.

Blaser, M.J., Wells, J.G., Feldman, R.A., Pollard, R.A. and Allen, J.R. (1983) *Campylobacter enteritis* in the United States: a multicenter study. *Ann. Intern. Med.* **98**, 360–365.

Bode, G., Mauch, F. and Malfertheiner, P. (1993) The coccoid forms of *Helicobacter pylori*. Criteria for their viability. *Epidemiol. Infect.* **111**, 483–490.

Boyce, T.G., Pemberton, A.G., Wells, J.G. and Griffin, P.M. (1995) Screening for *Escherichia coli* O157:H7-a nationwide survey of clinical laboratories. *J. Clin. Microbiol.* **33**, 3275–3277.

Breiman, R.F. and Butler, J.C. (1998) Legionnaires' disease: clinical, epidemiological, and public health perspectives. *Semin. Respir. Infect.* **13**, 84–89.

Brewster, D.H., Brown, M.I., Robertson, D., Houghton, G.L., Bomson, J. and Sharp, J.C.M. (1994) An outbreak of *Escherichia coli* O157 associated with a children's paddling pool. *Epidemiol. Infect.* **112**, 441–447.

Carter, A.O., Borczyk, A.A., Carlson, J.A., Harvey, B., Hockin, J.C., Karmali, M.A., Krishnan, C., Korn, D.A. and Lior, H. (1987) A severe outbreak of *Escherichia coli* O157:H7-associated hemorrhagic colitis in a nursing home. *N. Engl. J. Med.* **316**, 1496–1500.

Cebula, T.A., Payne, W.L. and Feng, P. (1995) Simultaneous identification of *Escherichia coli* of the O157:H7 serotype and their Shiga-like toxin type by MAMA/multiplex PCR. *J. Clin. Microbiol.* **33**, 248–250.

Chalmers, R.M., Aird, H. and Bolton, F.J. (2000) Waterborne *Escherichia coli* O157. *J. Appl. Microbiol. Symp. Suppl.* **88**, 124S–132S.

Chapman, P.A., Siddon, C.A., Zadik, P.M. and Jewes, L. (1991) An improved selective medium for the isolation of *Escherichia coli* O157. *J. Med. Microbiol.* **35**, 107–110.

Cirillo, J.D., Falkow, S., Tompkins, L.S. and Bermudes, L.E. (1997) Interaction of *Mycobacterium avium* with environmental amoebae enhances virulence. *Infect. Immun.* **65**, 3759–3767.

Cohen, A.E. (1996) A PCR method for the detection of *Helicobacter pylori* in food products: US Food and Drug Administration (FDA) study. In *16th Food Microbiology Symposium. Current Concepts in Foodborne Pathogens and Rapid and Automated Methods in Food Microbiology*, University of Wisconsin, River Falls, WI.

Coker, A.O., Isokpehi, R.D., Thomas, B.N., Amisu, K.O. and Obi, C.L. (2002) Human Campylobacteriosis in developing countries. *Emerg. Infect. Dis.* **8**, 237–243.

Collins ,C.H., Grange, J.M. and Yates, M.D. (1984). *Mycobacteria* in water. *J. Appl. Bacteriol.* **57**, 193–211.

Colwell, R.R. (1996) Global climate and infectious disease: the cholera paradigm. *Science* **274**, 2025–2031.

Cransberg, K., Van den Kerkhof, J.H.C.T., Banffer, J.R.J., Stijnen, C., Wernars, K., van de Kar, N.C.A.J., Nauta, J. and Wolff, E.D. (1996) Four cases of hemolytic uremic syndrome-source contaminated swimming water? *Clin. Nephrol.* **46**, 45–49.

Cunha, B.A. (1998) Clinical features of Legionnaires' disease. *Semin. Resp. Infect.* **33**, 116–127.

Current, W.L., Reese, N.C., Ernst, J.V., Bailey, W.S., Heyman, M.B. and Weinstein, W.N. (1983) Human cryptosporidiosis in immunocompetent and immunodeficient persons:studies of an outbreak and experimental transmission, *N. Engl. J. Med.* **308**, 1252–1257.

Dalsgaard, A., Forslund, A., Petersen, A., Broun, D.J., Dias, F., Monteiro, S., Molbak, K., Aaby, P., Rodrigues, A. and Sandström, A. (2000) Class 1 integron-borne, multiple antibiotic resistance encoded by a 150-kilobase conjugative plasmid in epidemic *Vibrio cholerae* O1 strains isolated in Guinea-Bissau. *J. Clin. Microbiol.* **38**, 3744–3779.

Deltenre, M. and De Koster, E. (2000) How come I've got it? (A review of *Helicobacter pylori* transmission). *Eur. J. Gastroenterol. Hepatol.* **12**, 479–482.

Dev, V.J., Main, M. and Gould, I. (1991) Waterborne outbreak of *Escherichia coli* O157. *Lancet* **337**, 1412.

Dubois, A., Fiala, N., Heman-Ackah, L.M., Drazek, E.S., Tarnawski, A., Fishbein, W.N., Perez-Perez, G.I. and Blaser, M.J. (1994) Natural gastric infection with *Helicobacter pylori* in monkeys: a model for spiral bacteria infection in humans. *Gastroenterology* **106**, 1405–1417.

Duncan, S., Glover, L.A., Killham, K. and Prosser, J.I. (1994) Luminescence-based detection of activity of starved and viable but nonculturable bacteria. *Appl. Environ. Microbiol.* **60**, 1308–1316.

Dunn. B.E., Cohen, H. and Blaser, M.J. (1997) *Helicobacter pylori. Clin. Microbiol. Rev.* **10**, 720–741.

Edelstein, P.H., (1982) Comparative study of selective media for isolation of *Legionella pneumophila* from potable water. *J. Clin. Microbiol.* **16**, 697–699.

Effler, P., Isaäcson,M., Arntzen, L., Heenan, R., Canter, P. Barrett, T., Lee, L., Mambo, C., Levine, W., Zaidi, A. and Griffin, P.M. (2001) Factors contributing to the emergence of *Escherichia coli* O157 in Africa. *Emerg. Infect. Dis.* **7**(5), 812–819.

Endtz, H.P., Ruijs, G.J., Van Klingeren, B., Jansen, W.H., Van der Reyden, T. and Mouton, R.P. (1991) Quinolone resistance in *Campylobacter* isolated from man and poultry following the introduction of fluoroquinolones in veterinary medicine. *J. Antimicrob. Chemother.* **27**, 199–208.

Engberg, J., Aarestrup, F.M., Taylor, D.E., Gerner-Smidt, P. and Nachamkin, I. (2001) Quinolone and macrolide resistance in *Campylobacter jejuni* and *C. coli*: resistance and trends in human isolates. *Emerg. Infect. Dis.* **7**, 24–34.

Enroth, H. and Engstrand, L. (1995) Immunomagnetic separation and PCR for detection of *Helicobacter pylori* in water and stool specimens. *J. Clin. Microbiol.* **33**, 2162–2165.

Falkinham, J.O. (1996) Epidemiology of infection by nontuberculous *mycobacteria. Clin. Microbiol. Rev.* **9**, 177–215.

Faruque, S.M. and Nair, G.B. (2002) Molecular ecology of toxigenic *Vibrio cholerae. Microbiol. Immunol.* **46**(2), 59–66.

Faruque, S.M., Albert, M.J. and Mekalanos, J.J. (1998) Epidemiology, genetics, and ecology of toxigenic *Vibrio cholerae. Microbiol. Mol. Biol. Rev.* **62**, 1301–1314.

Fasano, A, Noriega, F.R. and Maneval, D.R. (1995) *Shigella* enterotoxin 1: an enterotoxin of *Shigella flexneri* 2a active in rabbit small intestine *in vivo* and *in vitro*. *J. Clin. Invest.* **95**, 2853–2861.

Feeley, J.C., Gibson, R.J., Gorman, G.W., Langford, N.C., Rasheed, J.K., Mackel, D.C. and Baine, W.B. (1979) Charcoal-yeast extract agar: primary isolation medium for *Legionella pneumophila*. *J. Clin. Microbiol.* **10**, 437–441.

Feng, P. (1993) Identification of *Escherichia coli* O157:H7 by DNA probe specific for an allele of *uidA* gene. *Mol. Cell. Probes* **7**, 151–154.

Fields, B.S. (1992) Procedures for the recovery of *Legionella* from the environment vol 2. U.S. Department of Health and Human Services, Atlanta, GA.

Fields, B.S. (1996) The molecular ecology of *legionellae*. *Trends Microbiol.* **4**, 286–290.

Fields, B., Barbaree, J.M., Sanden, G.N. and Morrill, W.E. (1990) Virulence of a *Legionella anisa* strain associated with Pontiac fever: an evaluation using protozoan, cell culture, and guinea pig models. *Infect. Immun.* **58**, 3139–3142.

Fields, B.S., Benson, R.F. and Besser, R.E. (2002). *Legionella* and Legionnaires' Disease: 25 years of investigation. *Clin. Microbiol. Rev.* **15**, 506–526.

Fields, P.I., Blom, K., Hughes, H.J., Helsel, L.O., Feng, P. and Swaminathan, B. (1997) Molecular characterization of the gene encoding H antigen in *Escherichia coli* and development of a PCR-restriction fragment length polymorphism test for identification of *E. coli* O157:H7 and O157:NM. *J. Clin. Microbiol.* **35**, 1066–1070.

Figura, N., Guglielmetti, P., Rossolini, A., Barberi, A., Cusi, G., Musmanno, R.A., Russi, M. and Quaranta, S. (1989) Cytotoxin production by *Campylobacter pylori* strains isolated from patients with peptic ulcers and from patients with chronic gastritis only. *J. Clin. Microbiol.* **27**, 225–226.

Fliermanns, C.B., Chery, W.B., Orrison, L.H., Smith, S.J., Tison, D.L. and Pope, D.H. (1981) Ecological distribution of *Legionella pneumophila*. *Appl. Environ. Microbiol.* **41**, 9–16.

Forman, D. (1996) *Helicobacter pylori* and gastric cancer. *Scand. J. Gastroenterol.* **31**, 48–51.

Friedman, C.R., Neimann, J., Wegener, H.C. and Tauxe, R.V. (2000) Epidemiology of *Campylobacter jejuni* infections in the United States and other industrialized nations. In *Campylobacter* (ed. I. Nachamkin and M.J. Blaser), 2nd edn, pp. 121–138, ASM Press, Washington, D.C.

Garg, P., Chakraborty, S., Basu, I., Datta, S., Rajendran, K., Bhattacharya, T., Yamasami, S., Bhattacharya, S.K., Takeda, Y., Nair, G.B. and Ramamurthy, T. (2000). Expanding muliple antibiotic resistance among clinical strains of *Vibrio cholerae* isolated from 1992–1997 in Calcutta, India. *Epidemiol. Infect.* **124**, 393–399.

Gauthier, M.J. (2000) Environmental parameters associated with the viable but nonculturable state. In *Nonculturable Micro-organisms in the Environment* (ed. R.R. Colwell and D.J. Grimes), pp. 87–112, ASM Press, Washington, D.C.

Goodman, K.J. and Correa, P. (1995) The transmission of *Helicobacter pylori*. A critical review of the evidence. *Int. J. Epidemiol.* **24**, 875–887.

Goodwin, C.S., Armstrong, J.A., Chilvers, T., Peters, M., Collins, M.D., Sly, L., McConnell, W. and Harper, W.E.S. (1989) Transfer of *Campylobacter pylori* and *Campylobacter mustelae* to *Helicobacter* gen. nov. as *Helicobacter pylori* comb. nov. and *Helicobacter mustelae* comb. nov., respectively. *Int. J. Syst. Bacteriol.* **39**, 397–405.

Graham, D.Y. Malaty, H.M. Evans, D.G., Evans, D.J. Jr., Klein, P.D. and Adam, E. (1991) Epidemiology of *Helicobacter pylori* in an asymptomatic population in the United States. Effect of age, race and socioeconomic status. *Gastroenterology* **100**, 1495–1501.

Griffin, P.M. (1995) *Escherichia coli* O157:H7 and other enterohemorrhagic *Escherichia coli*. In *Infections of the Gastrointestinal Tract* (ed. M.J. Blaser, P.D. Smith, J.I. Ravdin, H.B. Greenberg and R.L. Guerrant), Vol. 1, pp. 739–761, Raven, New York.

Griffin, P.M. and Tauxe, R.V. (1991) The epidemiology of infections caused by *Escherichia coli* O157:H7, other enterohemorrhagic *Escherichia coli*, and the associated hemolytic-uremic syndrome. *Epidemiol. Rev.* **13**, 60–98.

Griffin, P.M., Ostroff, S.M., Tauxe, R.V., Greene, K.D., Wells, J.G., Lewis, J.H. and Blake, P.A. (1988) Illnesses associated with *Escherichia coli* O157:H7 infections. A broad clinical spectrum. *Ann. Intern. Med.* **109**, 705–712.

Grimm, D., Merkert, H., Ludwig, W., Schleifer, K.H., Hacker, J. and Brand, B.C. (1998) Specific detection of *Legionella pneumophila*: construction of a new 16S rRNA-targeted oligonucleotide probe. *Appl. Environ. Microbiol.* **64**, 2686–2690.

Gustafson, T.L. Band, J.D. Hutcheson, R.H. and Schaffner, W. (1983) *Pseudomonas folliculitis*: an outbreak and review. *Rev. Infect. Dis.* **5**(1), 1–8.

Hacker, J., Blum-Oehler, G., Muhldorfer, I. and Tschape, H. (1997) Pathogenicity islands of virulent bacteria: structure, function and impact on microbial evolution. *Mol. Microbiol.* **23**, 1089–1097.

Hamano, S., Nakanishi, Y., Nara, T., Seki, T., Ohtani, T., Oishi, T., Joh, K., Oikawa, T., Muramatsu, Y., Ogawa, Y. and Akashi, S. (1993) Neurological manifestations of hemorrhagic colitis in the outbreak of *Escherichia coli* O157:H7 infection in Japan. *Acta Paediatr.* **82**, 454–458.

Herrington, D.A., Hall, R.H., Losonsky, G., Mekalanos, J.J., Taylor, R.K. and Levine, M.M. (1988) Toxin, toxin-coregulated pili and the toxR regulation are essential for *Vibrio cholerae* pathogenesis in humans. *J. Exp. Med.* **168**, 1487–1492.

Hildebrand, J.M., Maguire, H.C., Holliman, R.E. and Kangesu, E. (1996) An outbreak of *Escherichia coli* O157 infection linked to paddling pools. *Commun. Dis. Rep.* **6**, R33–R36.

Hopkins, R.S., Abbott, D.O. and Wallace, L.E. (1981). Follicular dermatitis outbreak caused by *Pseudomonas aeruginosa* associated with a motel's indoor swimming pool. *Public Health Rep.* **96**, 246–249.

Hopkins, R.J., Girardi, L.S. and Turney, E.A. (1996) Relationship between *Helicobacter pylori* eradication and reduced duodenal and gastric ulcer recurrence: a review. *Gastroenterology* **110**, 1244–1252.

Hulten, K., Enroth, H., Nyström, T. and Engstrand, L. (1998) Presence of *Helicobacter* species DNA in Swedish water. *J. Appl. Microbiol.* **85**, 282–286.

Hulten, K., Han, S.W., Enroth, H., Klein, P.D., Opekun, A.R., Gilman, R.H., Evans, D.G., Engstrand, L., Graham, D.Y. and Elzaatari, F.A.K. (1996) *Helicobacter pylori* in the drinking water in Peru. *Gastroenterology* **110**, 1031–1035.

Hussong, D., Colwell, R.R., O'Brien, M., Weiss, E., Pearson, A.D., Weiner, R.M. and Burge, W.D. (1987) Viable *Legionella pneumophila* not culturable by culture on agar medium. *BioTechnology* **5**, 947–950.

Isaäcson, M., Canter, P.H., Effler, P., Arntzen, L., Bomans, P. and Heenan, R. (1993) Haemorrhagic colitis epidemic in Africa. *Lancet* **341**, 961.

Islam M.S., Drasar, B.S. and Sack, R.B. (1994) The aquatic flora and fauna as reservoirs of *Vibrio cholerae*: a review. *J. Diarrh. Dis* **12**, 87–96.

Johnson, W.M., Lior, H. and Bezanson, G.S. (1983) Cytotoxic *Escherichia coli* O157:H7 associated with haemorrhagic colitis in Canada. *Lancet* **1**, 76.

Jones, I.G. and Roworth, M. (1996) An outbreak of *Escherichia coli* O157 and Campylobacteriosis associated with contamination of a drinking water supply. *Public Health* **110**, 277–282.

Kaneko, M. (1998). Chlorination of pathogenic *E. coli* O157. *Water Sci. Technol.* **38**(12), 141–144.

Kaper, J.B., Morris, J.G. and Levine, M.M. (1995) Cholera. *Clin. Microbiol. Rev.* **8**, 48–86.

Karch, H., Heesemann, J., Laufs, R., O'Brien, A.D., Tacket, C.O. and Levine, M.M. (1987) A plasmid of enterohemorrhagic *Escherichia coli* O157:H7 is required for expression of a new fimbrial antigen and for adhesion to epithelial cells. *Infect. Immun.* **55**, 455–461.

Karmali, M.A. (1989) Infection by verocytotoxin-producing *Escherichia coli*. *Clin. Microbiol. Rev.* **2**, 15–38.

Karmali, M.A., Petric, M., Lim, C., Fleming, P.C., Arbus, G.S. and Lior, H. (1985) The association between idiopathic hemolytic-uremic syndrome and infection by verotoxin-producing *Escherichia coli*. *J. Infect. Dis.* **151**, 775–782.

Katz, S.M. and Hammel, J.M. (1987). The effect of drying, heat and pH on the survival of *Legionella pneumophila*. *Ann. Clin. Lab. Sci.* **17**, 150–156.

Keene, E.E., McAnulty, J.M., Hoesly, F.C., Williams, L.P., Hedberg, K., Oxman, G.L., Barret, T.J., Pfaller, M.A. and Fleming, D.W. (1994) A swimming associated outbreak of hemorrhagic colitis caused by *Escherichia coli* O157: H7 and *Shigella sonnei*. *New Engl. J. Med.* **331**, 579–584.

Kilvington, S. and Price, J. (1990) Survival of *Legionella pneumophila* within cysts of *Acanthamoeba polyphaga* following chlorine exposure. *J. Appl. Bacteriol.* **68**, 519–525.

Kim, H.H., Samadpour, M., Grimm, L., Clausen, C.R., Besser, T.E., Baylor, M., Kobayashi, J.M., Neill, M.A., Schaenknecht, F.D. and Tarr, P.L. (1994) Characteristics of antibiotic resistant *Escherichia coli* O157:H7 in Washington state, 1984–91. *J. Infect. Dis.* **170**, 1606–1609.

Konturek, P.C., Bielanski, W., Konturek, S.J. and Hahn, E.G. (1999) *Helicobacter pylori* associated gastric pathology. *J. Physiol. Pharmacol.* **50**, 695–710.

Kovacs, M.J., Roddy, J., Gregoire, S., Cameron, W., Eidus, L. and Drouin, J. (1990) Thrombotic thrombocytopenic purpura following hemorrhagic colitis due to *Escherichia coli* O157:H7. *Am. J. Med.* **88**, 177–179.

Labenz, J. and O'Morain, C. (1995) Eradication. In *The Year in Helicobacter pylori 1995* (ed. P. Malfertheiner, F.M. Megraud, P. Michetti, and A. Price), pp. 47–51, Current Science, Philadelphia, PA.

Lawson, A.J., Logan, A.M., O'Neill, G.L., Desai, M. and Stanley, J. (1999) Large-scale survey of *Campylobacter* species in human gastroenteritis by PCR and PCR-enzyme-linked immunosorbent assay. *J. Clin. Microbiol.* **37**, 3860–3864.

Leclerc, H., Schwartzbrod, L. and Dei-Cas, E. (2002) Microbial agents associated with waterborne diseases. *Crit. Rev. Microbiol.* **28**(4), 371–409.

Levine, M.M. (1987) *Escherichia coli* that cause diarrhea: enterotoxigenic, enteropathogenic, enteroinvasive, enterohemorrhagic, and enteroadherent. *J. Infect. Dis.* **155**, 377–389.

Lisle, J.T., Broadway, S.C., Prescott, A.M., Pyle, B.H., Fricker, C. and McFeters, G.A. (1998) Effects of starvation on physiological activity and chlorine disinfection resistance of *E. coli* O157:H7. *Appl. Environ. Microbiol.* **64**, 4658–4662.

Maiwald, M., Helbig, J.H. and Lück, P.C. (1998) Laboratory methods for diagnosis of *Legionella* infections. *J. Microbiol. Methods* **33**, 59–79.

March, S.B. and Ratnam, S. (1986) Sorbitol-MacConkey medium for detection of *Escherichia coli* O157:H7 associated with hemorrhagic colitis. *J. Clin. Microbiol.* **23**, 869–872.

Marshall, B.J. and Warren, J.R. (1984) Unidentified curved bacilli in the stomach of patients with gastritis and peptic ulceration. *Lancet* **1**, 1311–1315.

Marston, B.J., Lipman, H.B. and Breiman, R.F. (1994). Surveillance for Legionnaires' disease. Risk factors for morbidity and mortality. *Arch. Intern. Med.* **154**, 2417–2411.

McKee, M.L. and O'Brien, A.D. (1995) Investigation of enterohemorrhagic *Escherichia coli* O157:H7 adherence characteristics and invasion potential reveals a new attachment pattern shared by intestinal *E. coli. Infect. Immun.* **63**, 2070–2074.

Mintz, E.D., Reiff, F.M. and Tauxe, R.V. (1995) Safe water treatment and storage in the home: a practical new strategy to prevent waterborne disease. *JAMA* **273**, 948–953.

Miyagi, K., Sano, K., Morita, C., Imura, S., Morimatsu, S., Goto, T., Nakano, Y., Omura, K., Matsumoto, Y., Maeda, K., Hashimoto, S. and Honda, T. (1999). An improved method for detecting faecal *Vibrio cholerae* by PCR of the toxin A gene. *J. Med. Microbiol.* **48**, 883–889.

Mukhopadhay, A.K., Chakraborty, S., Takeda, Y., Nair, G.B. and Berg, D.E. (2001) Characterization of VPI pathogenicity islands and CTXϕ prophage in environmental strains of *Vibrio cholerae. J. Bacteriol.* **183**, 4737–4746.

Nachamkin, I. (1995) *Campylobacter* and *Acrobacter*. In *Manual of Clinical Microbiology* (ed. P.R. Murray, E.J. Baron, M.A. Pfaller, F.C. Tenover, R.H. Yolken), pp. 483–491, American Society for Microbiology, Washington, D.C.

Nachamkin, I. and Blaser, M.J. (2000) *Campylobacter*, 2nd edn, American Society for Microbiology, Washington, D.C.

Nachamkin, I., Allos, B.M. and Ho, T. (1998) *Campylobacter* species and Guillain-Barré syndrome. *Clin. Microbiol. Rev.* **11**, 555–567.

Nilsson, H.O., Aleljung, P., Nilsson, I., Tyszkiewicz, T. and Wadström, T. (1996) Immunomagnetic bead enrichment and PCR for detection of *Helicobacter pylori* in human stools. *J. Microbiol. Methods* **27**, 73–79.

Noda, M., Yutsudo, T., Nakabayashi, N., Hirayama, T. and Takeda, Y. (1987) Purification and some properties of Shiga-like toxin from *Escherichia coli* O157:H7 that is immunologically identical to Shiga toxin. *Microb. Pathog* **2**, 339–349.

Oberhelman, R.A. and Taylor, D.N. (2000) *Campylobacter* infections in developing countries. In *Campylobacter* (ed. I. Nachamkin and M.J. Blaser), 2nd edn, pp. 139–153, American Society for Microbiology, Washington, D.C.

Ostroff, S.M., Kobayashi, J.M. and Lewis, J.H. (1989) Infections with *Escherichia coli* O157:H7 in Washington state: the first year of statewide disease surveillance. *J. Am. Med. Ass.* **262**, 355–359.

Pai, C.H., Ahmed, N., Lior, H., Johnson, W.M., Sims, H.V. and Woods, D.E. (1988) Epidemiology of sporadic diarrhea due to verocytotoxin-producing *Escherichia coli*: a two-year prospective study. *J. Infect. Dis.* **157**, 1054–1057.

Paquet, C., Perea, W., Grimont, F., Collin, M. and Guillod, M. (1993) Aetiology of haemorrhagic colitis epidemic in Africa. *Lancet* **342**, 175.

Parsonnet, J. (1998) *Helicobacter pylori. Infect. Dis. Clin. N. Am.* **12**, 185–197.

Paszko-Kolva, C., Shahmat, M., Keiser, J. and Colwell, R.R. (1993) Prevalence of antibodies against *Legionella* species in healthy and patient populations. In *Legionella Current*

Status and Emerging Perspectives (ed. J.M. Barbaree *et al.*), pp. 24–25, American Society for Microbiology, Washington, D.C.

Paunio, M., Pebody, R., Keskimaki, M., Kokki, M., Ruutu, P., Oininen, S., Vuotari, V., Siitonen, A., Lahti, E. and Leinikki, P. (1999). Swimming associated outbreak *Escherichia coli* O157:H7. *Epidemiol. Infect.* **122**, 1–5.

Pavia, A.T., Nichols, C.R., Green, D.P., Tauxe, R.V., Mottice, S. Greene. K.D., Wells, J.G., Siegler, R.L., Brewer, E.D. and Hannon, D. (1990) Hemolytic–uremic syndrome during an outbreak of *Escherichia coli* O157:H7 infections in institutions for mentally retarded persons: clinical and epidemiologic observations. *J. Pediatr.* **116**, 544–551.

Pebody, R.G., Furtado, C., Rojas, A., McCarthy, N., Nylen, G., Ruutu, P., Donnelly, M., Fisher, I., Gilham, C., Graverson, L., Cheasty, T., Navarro, M., Leinikki, P., Wall, P. and Bartlett, C. (1999) An international travel associated outbreak of *Escherichia coli* O157 infection; a challenge for European outbreak investigation. *Epidemiol. Infect.* **123**, 217–223.

Perdomo, O.J.J., Cavaillon, J.M., Huerre, M., Ohayon, H., Gounon, P. and Sansonetti. P.J. (1994) Acute inflammation causes epithelial invasion and mucosal destruction in experimental shigellosis. *J. Exp. Med.* **180**, 1307.

Perez-Perez, G.I., Peek, R.M., Legath, A.J., Heine, P.R. and Graff, L.B. (1999) The role of CagA status in gastric and extragastric complications of *Helicobacter pylori*. *J. Physiol. Pharmacol.* **50**, 833–845.

Petersen, A.M. and Krogfelt, K.A. (2003). *Helicobacter pylori*: an invading microorganism? A review. *FEMS Immunol. Med. Microbiol.* **1498**, 1–10.

Pickering, L.K., Obrig, T.G. and Stapleton, F.B. (1994) Hemolytic–uremic syndrome and enterohemorrhagic *Escherichia coli*. *Pediatr. Infect. Dis. J.* **13**, 459–475.

Pierce, M., Crampton, S. Henry, D., Heifets, L., LaMarca, A., Montecalvo, M., Wormser, G.P., Jablonowski, H., Jemsek, J., Cynamon, M., Yangco, B.G., Notario, G. and Craft, J.C. (1996) A randomized trial of clarithromycin as prophylaxis against disseminated *Mycobacterium avium* complex infection in patients with advanced acquired immunodeficiency syndrome. *N. Engl. J. Med.* **335**, 384–391.

Pollack, M. (1995) *Pseudomonas aeruginosa*. In *Principles and Practice of Infectious Disease* (ed. D. Mandell), pp. 1980–2003, Churchill Livingstone, New York.

Poms, R.E. (1997) Evaluation of conditions for recovery of *Helicobacter pylori* from foods and their survival in ready to eat foods, M.S. Thesis, University of Minnesota.

Quinn, T.C. (1997) Diversity of *Campylobacter* species and its impact on patients infected with human immunodeficiency virus. *Clin. Infect. Dis.* **24**, 1114–1117.

Qureshi, W.A. and Graham, D.Y. (1999) Diagnosis and management of *Helicobacter pylori* infection. *Clin. Cornerstone* **1**, 18–28.

Ramamurthy, T., Grag, S. and Sharma, R. (1993) Emergence of novel strain of *Vibrio cholerae* with epidemic potential in southern and eastern India. *Lancet* **341**, 703–704.

Ratnam, S., Hogan, K., March, S.B. and Butler, R.W. (1986). Whirlpool-associated folliculitis caused by *Pseudomonas aeruginosa*: Report of an outbreak and review. *J. Clin. Microbiol.* **23**(3), 655–659.

Recavarren-Arce, S., Leon-Barua, R., Cok, J., Berendson, R., Gilman, R.H., Ramirez-Ramos, A., Rodriquez, C. and Spira, W.M. (1991) *Helicobacter pylori* and progressive gastric pathology that predisposes to gastric cancer. *Scand. J. Gastroenterol.* **26**, 51–57.

Reidl, J. and Klose, K.E. (2002) *Vibrio cholerae* and cholera: out of the water and into the host. *FEMS Microbiol. Rev.* **26**, 125–139.

Rice, E.W., Johnson, C.H., Wild, D.K. and Reasoner, D.J. (1992) Survival of *Escherichia coli* O157:H7 in drinking water associated with a waterborne disease outbreak of hemorrhagic colitis. *Lett. Appl. Microbiol.* **15**, 38–40.

Rigsbee, W., Simpson, L.M. and Oliver, J.D. (1997) Detection of the viable but nonculturable state in *Escherichia coli* O157:H7. *J. Food Safety* **16**, 255–262.

Riley, L.W. (1987) The epidemiologic, clinical, and microbiologic features of hemorrhagic colitis. *Annu. Rev. Microbiol.* **41**, 383–407.

Riley, L.W., Remis, R.S., Helgerson, S.D., McGee, H.B., Wells, J.G., Davis, B.R. *et al.* (1983) Hemorrhagic colitis associated with a rare *Escherichia coli* serotype. *N. Engl. J. Med.* **308**, 681–685.

Rogers, J., Dowsett, A.B., Dennis, P.J., Lee, J.V. and Keevil, C.W. (1994). Influence of temperature and plumbing material selection on biofilm formation and growth of *Legionella pneumophila* in a model potable water system containing complex microbial flora. *Appl. Environ. Microbiol.* **60**, 1585–1592.

Rowbotham, T.J. (1986) Current views on the relationships between *amoebae*, *legionellae* and man. *Isr. J. Med. Sci.* **22**, 678–689.

Ryan, C.A., Tauxe, R.V., Hosek, G.W., Wells, J.G., Stoesz, P.A., McFadden, H.W. Jr, Smith, P.W., Wright, G.F. and Blake, P.A. (1986) *Escherichia coli* O157:H7 diarrhea in a nursing home: clinical, epidemiological, and pathological findings. *J. Infect. Dis.* **154**, 631–638.

Samadpour, M., Grimm, L.M., Desai, B., Alfi, D., Ongerth, J.E. and Tarr, P.I. (1993) Molecular epidemiology of *Escherichia coli* O157:H7 strains by bacteriophage lambda restriction fragment length polymorphism analysis: application to a multistate foodborne outbreak and a day-care center cluster. *J. Clin. Microbiol.* **31**, 3179–3183.

Sanden, G., Fields, B.S., Barbaree, J.M. and Feeley, J.C. (1989) Viability of *Legionella pneumophila* in chlorine-free waters at elevated temperatures. *Curr. Microbiol.* **18**, 61–65.

Shahamat, M., Mai, U., Paszko-Kolva, C.P., Kessel, M. and Colwell, R.R., (1993). Use of autoradiography to assess viability of *Helicobacter pylori* in water. *Appl. Environ. Microbiol.* **59**, 1231–1235.

Shears, P. (2001). Recent developments in cholera. *Curr. Opin. Infect. Dis.* **14**, 553–558.

Shinnick, T. and Good, R.C. (1994). Mycobaterial taxonomy. *Eur. J. Clin. Infect. Dis.* **13**, 884–901.

Slutsker, L.A., Ries, A.A., Greene, K.D., Wells, J.G., Hutwagner, L. and Griffin, P.M. (1997) *Escherichia coli* O157:H7 diarrhea in the United States: clinical and epidemiological features. *Ann. Intern. Med.* **126**, 505–13.

Steinert, M., Hentschel, U. and Hacker, J. (2002). *Legionella pneumophila*:an aquatic microbe goes astray. *FEMS Microbiol. Rev.* **26**, 149–162.

Su, C. and Brandt, L.J. (1995). *Escherichia coli* O157:H7 infection in humans. *Ann. Intern. Med.* **123**(9), 698–707.

Swanson, M.S. and Hammer, B.K. (2000). *Legionella pneumophila* pathogenesis: a fateful journey from Amoebae to macrophages. *Annu. Rev. Microbiol.* **54**, 567–613.

Swerdlow, D.L. and Isaäcson, M. (1994) The epidemiology of cholera in Africa. In *Vibrio cholerae and Cholera: Molecular to Global Perspectives* (ed. K.I. Wachsmuth, P.A. Blake and O. Olsik), pp. 297–307, American Society for Microbiology Press, Washington, D.C.

Swerdlow, D.L. and Ries, A.A. (1993) *Vibrio cholerae* non-O1 the 8th pandemic? *Lancet* **342**, 382–383.

Swerdlow, D.L., Woodruff, B.A., Brady, R.C., Griffin, P.M., Tippen, S., Donnell, H.D. Jr, Geldreich, E., Payne, B.J., Meyer, A., Wells, J.G., Greene, K.D., Bright, M., Bean, N.H. and Blake, P.A. (1992) A waterborne outbreak in Missouri of *Escherichia coli* O157:H7 associated with bloody diarrhea and death. *Ann. Intern. Med.* **116**, 812–819.

Szewzyk, U., Szewzyk, R., Manz, W. and Schleifer, K.H. (2000) Microbiological safety of drinking water. *Annu. Rev. Microbiol.* **54**, 81–127.

Tacket, C.O., Clemens, J. and Kaper, J.B. (1992) Cholera vaccines. In *Vaccines* (ed. R.W. Ellis), pp. 53–68, Butterworth and Heinemann, Reading, MA.

Taylor, D.N. (1992) *Campylobacter* infections in developing countries. In *Campylobacter jejuni: Current Status and Future Trends* (ed. I. Nachamkin, M.J. Blaser and L.S. Tompkins), pp. 20–30, American Society for Microbiology, Washington, D.C.

Tytgat, G. (2000) *Helicobacter pylori*: past, present and future. *J. Gastroenterol. Hepatol.* **15**, 30–33.

Van Doorn, O.J., Bosman, D.K., Van't Hoff, B.W., Taminiau, J.A., Ten Kate, F.J. and Van Der Ende, A. (2001) *Helicobacter pylori* stool antigen test: a reliable non-invasive test for the diagnosis of *Helicobacter pylori* infection in children. *Eur. J. Gastroenterol. Hepatol.* **13**, 1061–1065.

Van den Berg, O., Dediste, A., Vlaes, L., Ebraert, A., Retore, P., Douat, N. *et al.* (2001) Prevalence and clinical features of non *jejuni/coli Campylobacter* species and related organisms in stool specimens. [Abstract P-22]. In (ed. J. Hacker), Abstracts of scientific presentations of the *11th International Workshop on Campylobacter, Helicobacter and Related Organisms, Freiburg, Germany, Sept 1–5, 2001. Int. J. Med. Microbiol.* **291**(Suppl 31), 144.

Velazquez, M. and Feirtag, J.M. (1999) *Helicobacter pylori*: characteristics, pathogenicity, detection methods and mode of transmission implicating foods and water. *Int. J. Food Microbiol.* **53**, 95–104.

Von Reyn, C.F., Maslow, J.N., Barber, T.W., Filkinham, J.O. and Arbeit, R.D. (1994). Persistent colonization of potable water as a source of *Mycobacterium avium* infection in AIDS. *Lancet* **343**, 1137.

Von Wulffen, H., Russman, H., Karch, H., Meyer, T., Bitzan, M., Kohrt, T.C. and Aleksic, S. (1989) Verocytotoxin-producing *Escherichia coli* O2:H5 isolated from patients with ulcerative colitis [Letter]. *Lancet* **i**, 1449–1450.

Waldor, M.K. and Mekalanos, J.J. (1996) Lysogenic conversion by a filamentous bacteriophage encoding cholera toxin. *Science* **272**, 1910–1914.

Warner, M., Kuo, K., Williams, L., Adam, B., Langkop, C., Ruben, R., Francis, B. and Haupt, T. (1996) Lake associated outbreak of *E. coli* O157:H7:Illinois. *Morbid. Mortal. Weekly Rep.* **45**, 437–439.

Washburn, J., Jacobson, J.A., Marston, E. and Thorsen, B. (1976) *Pseudomonas aeruginosa* rash associated with a whirlpool. *J. Am. Med. Ass.* **235**, 2205–2207.

Waters, J.R., Sharp, J.C.M. and Dev, V.J. (1994) Infection caused by *Escherichia coli* O157:H7 in Alberta, Canada, and in Scotland: a five-year review, 1987–1991. *Clin. Infect. Dis.* **19**, 834–843.

Wells, J.G., Davis, B.R., Wachsmuth, I.K., Riley, L.W., Remis, R.S. and Sokolow, R. *et al.* (1983) Laboratory investigation of hemorrhagic colitis outbreaks associated with a rare *Escherichia coli* serotype. *J. Clin. Microbiol.* **18**, 512–520.

Wharton, M., Spiegel, R.A., Moran, J.M., Tauxe, R.V., Wells, J.G., Barg, N., Herndon, J., Meriwether, R.A., MacCormack, J.N. and Levine, R.H. (1990) A large outbreak of antibiotic-resistant shigellosis at a mass gathering. *J. Infect. Dis.* **162**, 1324.

Winn, W.C. (1988) Legionnaires' disease: historical perspective. *Clin. Microbiol. Rev.* **1**, 60–81.

Wittenberg, D.F. (1998) The spread of Shigella dysentteriae type I in Africa. *Jpn. J. Med. Sci. Biol. Suppl.* **1**, S36–S42.

Wittenberg, D.F. (1999) Emerging and re-emerging diseases–epidemic enterohaemorrhagic infections 100 years after Shiga. *S. Afr. Med. J.* **89**, 750–752.

Wong, C.S., Jelacic, S., Habeeb, R.L., Watkins, S.L. and Tarr, P.I. (2000) The risk of the hemolytic-uremic syndrome after antibiotic treatment of *Escherichia coli* O157:H7 infections. *N. Engl. J. Med.* **342**, 1930–1936.

3

Emerging infectious waterborne diseases: viral agents

L.H. Nel and J. Weyer

3.1 INTRODUCTION

In many ways viruses epitomize the nature of infectious disease. Not only are these agents the smallest of all microbes, but also they are usually quite stable and easily transmitted – often they are airborne – and viral diseases such as smallpox, polio, measles and rabies have all been global scourges at one point or another. Much progress has been made in the fight against viral diseases in the last few decades, mainly through advances in improvement in the quality of water, sanitation, immunization, education and early diagnosis. However, these control mechanisms are not infallible and, in concert with new patterns of human movement and increased pressures on natural resources, some of the old viral diseases are re-emerging while new viral diseases are recognized with increasing frequency. In addition, viruses can serve both as parasites and as genetic elements in their hosts. Viruses can have a devastating effect on given populations and a natural progression to mutualism is of no comfort to the human race – reaching an equilibrium will inevitably take too long to prevent major suffering (Morse and Schluederberg, 1990). Many of the emerging and re-emerging diseases described in this and other chapters of this book, are directly linked to infections with the human immunodeficiency virus

© IWA Publishing 2004. *Microbial Waterborne Pathogens* edited by T.E. Cloete, J. Rose, L.H. Nel and T. Ford. ISBN: 1 84339 055 8. Published by IWA Publishing, London, UK.

which is itself a new human virus. The reader is referred to the previous chapter for a discussion on the definitions of new, emerging, re-emerging and resurgent infectious diseases and a report by the Institute of Medicine (http://fas.org/promed) provide excellent background information.

A wide variety of viral pathogens are excreted in faeces and are capable of causing waterborne infections. Although viruses cannot multiply in water resources (outside of viable host cells), many have a low infectious dose (typically 1 to 10 infectious units) and can survive for long periods in water. Some of the many waterborne diseases caused by viruses are Adenoviruses (types 40 and 41), Astroviruses, Caliciviruses (including Noroviruses), Enteroviruses (picornaviruses) and Reoviruses (including rotaviruses). Similar to the non-viral waterborne pathogens, the most common medical condition associated with waterborne viral infection is diarrhoea. The medical and economic importance of viral diarrhoea is without doubt very significant and millions of deaths (particularly of children) are caused by these infections. In general, emergence of waterborne viral diseases may be ascribed to the rapid genetic evolution (particularly for the RNA viruses), the development of anti-viral drug resistance, increased zoonoses and/or altering sociological patterns. At present, only a handful of these viral diseases are considered emerging and will be further discussed here, while the diseases, patterns of emergence and incidence, clinical features, control and diagnosis are summarized in Table 3.1.

3.2 SPECIFIC EMERGING WATERBORNE INFECTIOUS DISEASES CAUSED BY VIRAL INFECTION

3.2.1 Aseptic meningitis and various acute Enterovirus syndromes

Enteroviruses (EV) are small (27–30 nm in diameter) positive-sense RNA viruses, that as their name suggests, replicate in the intestinal tract. EV infection is one of the most common types of viral infection in man, but usually does not result in serious disease. Some serotypes, however, may cause serious clinical syndromes (Melnick, 1990). Such syndromes may include acute paralysis, encephalitis, myocarditis and pericarditis, and chronic infection (particularly in immunocompromised individuals). The most common manifestation of EV infection is a non-specific febrile illness, which may be accompanied by upper respiratory symptoms and is referred to as the 'summer flu' (Dagan, 1996). The most dangerous EV syndrome is aseptic meningitis, an infection of the central nervous system which is diagnosed in 75,000–200,000 American patients annually (Rotbart *et al.*, 1998). EV infections often manifest symptoms that are characteristic of other infections, leading to misdiagnosis and treatment of patients. The situation is further complicated since specific EV serotypes cannot be uniquely associated with any syndrome or disease and *vice versa* (Melnick, 1990).

Table 3.1 New and emerging waterborne diseases of viral origin.

Pathogen and disease (reviewed in:)	Geographical incidence	Factors involved in emergence	Clinical features	Detection and control/prevention/treatment
Astroviruses and Astrovirus enteritis (Walter and Mitchell, 2000)	Worldwide (seroprevalence of up to 100% in some locations)	First discovered in 1975, and at present second only to Rotavirus and Norwalk-like viruses as cause of childhood diarrhoea; Social changes may also play a role since outbreaks often occur in day-care centre settings	Incubation period of 1–4 days; Self-limiting watery diarrhoea, lasts for 2–3 days (milder than Rotavirus diarrhoea); Nausea/vomiting, abdominal pain and low-grade fever	Detection by immune electron microscopy (IEM) of stool samples and immunofluorescence (IF) of infected tissue cultures. No vaccines available; proper hygiene can prevent transmission; treatment supportive with oral rehydration fluid (ORF) or intravenous rehydration (IR)
Caliciviruses and Calicivirus enteritis (Lopman *et al.*, 2002)	Worldwide (most people infected by the age of five)	Norwalk-like viruses were identified in 1972; Increased incidence; Zoonotic caliciviruses are likely to continue emerging from the worlds' oceans	Abrupt onset Diarrhoea with nausea and characteristic projectile vomiting; Upper respiratory symptoms and fever less common	Detected in stools by electron microscopy (EM), and IEM No vaccines available; proper sanitation may help to prevent infection; treatment supportive with ORF or IR.

Enteroviruses and aseptic meningitis and various acute enterovirus syndromes (Rotbart *et al.*, 1998)	Worldwide (Probably the most common of viral infections in man)	Rapid genetic evolvement of some new strains of enterovirus; Development of anti-viral drug resistance	Most infections are inapparent; Most commonly manifests as a non-specific febrile illness ('summer flu'); Enterovirus 71 infections lead to blisters in the oral cavity, palms and soles, ulcers, and fever (hand, foot and mouth disease); Complications lead to aseptic meningitis, encephalitis, myocarditis, pericarditis, pneumonia and paralysis	Diagnosis in throat or faecal samples or biopsies. Serology also used but of limited value due to large number of strains; no vaccines available for the non-polio enteroviruses; prevention by proper sanitation; treatment focuses on management of complications
Hepatitis E-like viruses and Hepatitis (Worm *et al.*, 2002)	China, Asia, Africa, Mexico and Central America (may be more prevalent than currently believed)	New waterborne infectious viral agent, first characterized in the 1990s; zoonotic aspects may play a role	Incubation period on an average of 40 days; symptoms are indistinguishable from Hepatitis A infection; include malaise, anorexia, arthralgia, fever. Hepatomegaly, jaundice, vomiting. Diarrhoea less common; Pregnant women are a high risk group with significantly higher mortality rate	Diagnosis based on the epidemiological characteristics of the outbreak and by serological exclusion of Hepatitis A and B viruses. Confirmation by IEM in faeces or liver biopsy; no vaccines available; prevention by improved hygiene and boiling contaminated water; supportive treatment to relief symptoms

EV belongs to the *Enterovirus* genus of the *Picornaviridae* family and includes the polioviruses, group A coxsackieviruses, group B coxsackieviruses, echoviruses and newer enteroviruses (types 68–71) (Van Regenmortel *et al.*, 2000). There are 3 different polioviruses, 61 non-polio enteroviruses, 23 coxsackie A viruses, 6 coxsackie B viruses, 28 echoviruses, and 4 other enteroviruses. Enteroviruses evolve rapidly and emerge as new strains, but others are of emerging importance due to drug resistance. Some enterovirus serotypes have been responsible for serious, large epidemics throughout the world in recent years. Echovirus infections have been responsible for several large epidemics of aseptic meningitis in Japan, Europe, the Middle East and Cuba in recent years (Sarmiento *et al.*, 2001; Kirschke *et al.*, 2002). These epidemics involved different, highly infectious, serotypes of the virus, namely, types 4, 9 and 30, but also new genetic variants of these viral serotypes. Enterovirus 71 (EV71) has caused major outbreaks of hand-, foot-and-mouth disease in North America, Europe, Malaysia, Japan and Australia since 1995 (Ishimaru *et al.*, 1980; Chan *et al.*, 2000; Wang *et al.*, 2002; Ooi *et al.*, 2002; Cardoza *et al.*, 2003). In 1998, this virus infected at least a million individuals in Taiwan with a 20% case fatality among children under 5 years of age (Ho *et al.*, 1999).

Transmission of EVs is mainly through the faecal–oral route, but may in some cases also involve the respiratory route. EVs are found in water and food and have been isolated from shellfish (Abbaszadegan *et al.*, 1993; Richards, 1985; Tougianidou and Botzenhart, 1993). EVs have also been shown to be particularly resistant to chlorine treatment of water (Payment *et al.*, 1985). Measures of improved hygiene are suggested for prevention of infection with EVs. The eradication of poliomyelitis caused by different polioviruses is envisioned for the near future (Jacob, 2000). One of the major contributing factors to the success, of the Global Polio Eradication Initiative is recognized as the effective oral vaccine. Such a vaccine is, however, not available for the non-polio enteroviruses. Some progress towards the development of antiviral therapy for EV infections has been made (reviewed in Rotbart *et al.*, 1998).

3.2.2 Astrovirus enteritis

The aetiological agents of this disease are small, round viruses with a star-like appearance under the electron microscope. These viruses were only identified in 1975 (Madeley and Cosgrove, 1975). Human astroviruses (HAstV) are only second to Rotaviruses and Norwalk-like viruses as the major causative agents of infectious gastrointestinal illness (Putzker *et al.*, 2000). Seroprevalence studies have shown up to 100% exposure of young children in certain locations (reviewed in Walter and Mitchell, 2000).

Simultaneous infections of HAstV, Rotaviruses and Caliciviruses are very common, complicating accurate diagnosis of the infection. Astrovirus-associated gastroenteritis, however, is clinically milder than Rotavirus-induced diarrhoea (reviewed in Walter and Mitchell, 2000). Following a 1–4 day incubation period, the clinical symptoms present as watery diarrhoea that lasts for 3 days. Mainly young children and the elderly are affected. HAstVs have also been detected in the immunocompromised (Giordano *et al.*, 1999).

Transmission of the HAstVs is via the faecal–oral route with contaminated water or food. HAstVs have been also detected from oysters (Kitahashi *et al.*, 1999). Person-to-person transmission is also reported during outbreaks, being frequently reported from child day-care centres, hospitals and nursing homes (reviewed in Walter and Mitchell, 2000). HAstVs are now recognized as the major pathogen involved in nosocomial diarrhoea of young children (Gaggero *et al.*, 1998; reviewed in Walter and Mitchell, 2000). These viruses may also be present in municipal drinking water, since the residual chlorine concentration of drinking water, as suggested by the United States Environmental Protection Agency (EPA) guidelines, are insufficient to inactivate astroviruses (Gofti-Laroche *et al.*, 2003). HAstVs have also been detected in natural water sources (Taylor *et al.*, 2001).

3.2.3 Calicivirus enteritis

This disease is very common and is caused by a group of small (27–40 nm in diameter), non-enveloped, single-stranded RNA viruses – serological evidence indicates that most people are infected by the age of five (Dolin *et al.*, 1972; Kapikian *et al.*, 1972; Pang *et al.*, 2000). Among the known Calicivirus (HuCVs) pathogens of humans are the Noroviruses (previously known as Norwalk-like viruses) and the Sapporo-like viruses (SLVs) (Berke *et al.*, 1997; Chiba *et al.*, 2000; Monroe *et al.*, 2000; Jiang and Pickering, 2002, reviewed in Lopman *et al.*, 2002). Especially, the Norwalk-like virus are a growing public health concern with up to 50% of gastroenteritis outbreaks in the United States, Japan and some European countries attributed to Norwalk-like viruses (Lew *et al.*, 1994; Dedman *et al.*, 1998; Mead *et al.*, 1999; Hedlund *et al.*, 2000; Inouye *et al.*, 2000; Schreier *et al.*, 2000). The symptoms of infection include acute gastroenteritis, nausea, projectile vomiting and abdominal cramps (Adler and Zickl, 1969). Fever and respiratory symptoms occur in a small number of cases. The disease is marked by an abrupt onset, 24–28 h after exposure, but is typically mild and self-limiting, only lasting 12–60 h (Adler and Zickl, 1969; Kaplan *et al.*, 1982). More cases are reported during the colder months, hence the general naming of the disease as 'winter vomiting disease' (Mounts *et al.*, 2000). While Noroviruses cause sporadic and epidemic gastroenteritis in all age groups of humans, loss of fluids and electrolytes are particularly serious in the very young and elderly (Rockx *et al.*, 2002). Studies have shown that a short-term immunity to homologous virus develops after infection, and protects the patients from subsequent infection (reviewed in Matsui and Greenberg, 2000). The immunity however wanes and long-term protection against homologous virus infection is not afforded (Parrino *et al.*, 1977). There also seems to be very little cross protection among members of the caliciviruses (Wyatt *et al.*, 1974).

Although there is little evidence for non-human mammalian reservoirs of caliciviruses, it is believed that these viruses emerged from ocean reservoirs, with subsequent zoonotic and interspecies movement (Smith *et al.*, 1998). It is believed that pathogenic caliciviruses are likely to continue emerging from the world's oceans in various forms. Transmission is primarily through the faecal–oral route but infection after swallowing aerosolized virus is also implicated (Dolin *et al.*, 1971; Dedman *et al.*, 1998; Marks *et al.*, 2000). The infectious dose is very low, and an inoculum of 10-100 particles is enough to cause disease (Green, 1997). Foodborne transmission is also recorded and bivalve shellfish-related outbreaks frequently occur where appropriate standards for proper disposal of human sewage are not applied (Potasman *et al.*, 2002). Contaminated water and ice and raw produce such as eggs, salad ingredients, and some ready-to-eat foods are other sources of infection (Schaub and Oshiro, 2000). The contamination of fresh produce is often linked to washing or spraying of the food with contaminated water. Food handlers are also often implicated in the spread of these viruses (Hedberg and Osterholm, 1993). Contact with contaminated water for recreational purposes (such as public swimming pools) has also been implicated in some cases (Lee *et al.*, 2002). It has been shown that caliciviruses can pass through water purification filters and remain infectious even at standard levels of chlorine in drinking water (Keswick *et al.*, 1985).

Progress towards the development of a vaccine against HuCVs, specifically against NLV infection, has been made, but a lot of research in this field is still required (reviewed in Estes *et al.*, 2000).

3.2.4 Hepatitis

The Hepatitis E virus (HEV) is a small (27–30 nm), spherical, non-enveloped RNA virus, with a single serotype, and was described for the first time in 1983 (Balayan *et al.*, 1983). This virus was then characterized in subsequent years and has recently been placed in its own taxonomic group, 'Hepatitis E-like viruses', within the class IV (+)-sense RNA viruses (Pringle, 1999). The disease is endemic in the developing countries of Asia, the Middle East, North Africa and Mexico (reviewed in Worm *et al.*, 2002). Clemente-Casares *et al.* (2003) and Pina *et al.* (2000), however suggest that HEV may be more prevalent in developed countries than is currently believed. The virus is primarily transmitted through the faecal–oral route in contaminated drinking water (Balayan *et al.*, 1983). Person-to-person transmission of HEV may occur during large epidemics, but this route of transmission appears to be uncommon (Aggarwal and Naik, 1994; Naik *et al.*, 1992). Recent genetic and serological evidence suggest frequent transmission of HEV between animals and humans and indeed, zoonoses may explain the mechanism of HEV maintenance in populations during the periods between epidemic outbreaks (reviewed in Meng, 2000). Various strains of HEV isolated from human patients have greater sequence homology relative to swine HEV than to other HEV strains (reviewed in Meng, 2000).

The incubation period following exposure to HEV is on average a lengthy 40 days, but may range from 15 to 60 days. The symptoms of HEV and the Hepatitis A virus are clinically indistinguishable. The disease manifests as acute, icteric hepatitis with malaise, anorexia, abdominal pain, arthralgia, fever, anorexia, hepatomegaly, jaundice, nausea and vomiting. Diarrhoea is a less common symptom. The disease is most often seen in young to middle-aged adults (15–40 years old) with a relatively low mortality rate of at 0.1–1% (Khuroo, 1980). HEV incidence in uncommonly low in young children who are usually most affected by infectious enteric disease but nevertheless show high susceptibility to the virus (Arora *et al.*, 1999). However, pregnant women appear to be exceptionally susceptible to severe disease, and excessive mortality has been reported in this group (up to 30%) (Khuroo, 1980; Hussaini *et al.*, 1997).

HEV can be prevented by improved hygienic conditions, and boiling contaminated water also prevents transmission. Progress towards the development of recombinant subunit vaccines against HEV has been made (reviewed in Emerson and Purcell, 2001).

REFERENCES

Abbaszadegan, M., Huber, M.S., Gerba, C.P. and Pepper, I.L. (1993) Detection of enteroviruses in groundwater with the polymerase chain reaction. *Appl. Environ. Microbiol.* **59**, 1318–1324.

Adler, J.L. and Zickl, R. (1969) Winter vomiting disease. *J. Infect. Dis.* **119**, 668–673.

Aggarwal, R. and Naik, S.R. (1994) Hepatitis E: intrafamiliar transmission versus waterborne spread. *J. Hepatol.* **21**, 718–723.

Arora, N.K., Panda, S.K., Nanda, S.K., Ansari, I.H., Joshi, S., Dixit, R. and Bathla, R. (1999) Hepatitis E infection in children: study of an outbreak. *J. Gastroenterol. Hepatol.* **14**, 572–577.

Balayan, M.S., Andjapardize, A.G., Savinskaya, S.S., Ketiladze, E.S., Braginsky, D.M., Savinov, A.P. and Poleschuk, V.F. (1983) Evidence for a virus in non-A, non-B hepatitis transmitted via the faecal–oral route. *Intervirology* **20**, 23–31.

Berke, T., Golding, B., Jiang, X., Cubitt, D.W., Wolfaardt, M., Smith, A.W. and Matson, D.O. (1997) Phylogenetic analysis of the caliciviruses. *J. Med. Virol.* **52**, 419–424.

Cardoza, M.J., Perera, D., Brown, B.A., Cheon, D., Chan, H.C., Chan, K.W., Cho, H. and McMinn, P. (2003) Molecular epidemiology of human enterovirus 71 strains and recent outbreaks in the Asia-Pacific region: Comparative analysis of the VP1 and VP4 genes. *Emerg. Infect. Dis.* **9**(4), 461–468.

Chan, L.G., Parashar, U.D., Lye, M.S., Ong F.G., Zaki S.R., Alexander, J.P., Ho, K.K., Han, L.L., Pallansch, M.A., Suleiman, A.B., Jegatheson, M. and Anderson, L.J. (2000) Deaths of children during an outbreak of hand, foot and mouth disease Sarawak, Malaysia: clinical and pathological characteristics of the disease. For the outbreak study group. *Clin. Infect. Dis.* **31**, 678–683.

Chiba, S,. Nakata, S., Numata-Kinoshita, K. and Honma, S. (2000) Sapporo virus: history and recent findings. *J. Infect. Dis.* **181**(Suppl 2), S303–S308.

Clemente-Casares, P., Pina, S., Buti, M., Jardi, R., Martin, M., Bofill-Mas, S. and Girones, R. (2003) Hepatitis E virus epidemiology in industrialized countries. *Emerg. Infect. Dis.* **9**(4), 448–454.

Dagan, R. (1996) Nonpolio enterovirus and the febrile young infant: epidemiology, clinical and diagnostic aspects. *Pediatr. Infect. Dis J.* **15**, 67–71.

Dedman, D., Laurichesse, H., Caul, E.O. and Wall, P.G. (1998) Surveillance of small round structured virus (SRSV) infection in England and Wales, 1990–1995. *Epidemiol. Infect.* **121**, 139–149.

Dolin, R., Blacklow, N.R., DuPont, H., Buscho, R.F., Wyatt, R.G., Kasel, J.A., Hornick, R. and Chanock, R.M. (1972) Biological properties of Norwalk agent of acute infectious nonbacterial gastroenteritis. *Proc. Soc. Exp. Biol. Med.* **140**, 578–583.

Dolin, R., Blacklow, N.R., DuPont, H., Formal, S., Buscho, R.F., Kasel, J.A., Chames, R.P., Hornick, R. and Chanock, R.M. (1971) Transmission of acute infectious nonbacterial gastroenteritis to volunteers by oral administration of stool filtrates. *J. Infect. Dis.* **123**, 307–312.

Emerson, S.U. and Purcell, R.H. (2001) Recombinant vaccines for hepatitis E. *Trends Mol. Med.*.**7**(10), 462–466.

Estes, M.K., Ball, J.M., Guerrero, R.A., Opekun, A.R., Gilger, M.A., Pacheco, S.S. and Graham, D.Y. (2000) Norwalk virus vaccines: challenges and progress. *J. Infect. Dis.* **181**(Suppl 2), S367–S373.

Gaggero, A., O'Ryan, M., Noel, J.S., Glass, R.I., Monroe, S.S., Mamani, N., Prado, V., Avendaño, L.F. (1998) Prevalence of astrovirus infection among Chilean children with a cute gastroenteritis. *J. Clin. Microbiol.* **36**, 3691–3693.

Giordano, M.O., Martinez, L.C., Rinaldi, D., Espul, C., Martinez, N., Isa, M.B., Depetris, A.R., Medeot, S.I. and Nates, S.V. (1999) Diarrhea and enteric emerging viruses in HIV-infected patients. *AIDS Res. Human Retroviruses* **16**, 1427–1432.

Gofti-Laroche, L., Gratacap-Cavallier, B., Demanse, D., Genoulaz, O., Seignerin, M. and Zmirou, D. (2003) Are waterborne astroviruses implicated in acute digestive morbidity (E.MI.R.A study)? *J. Clin. Virol.* **27**(1), 74–82.

Green, K.Y. (1997) The role of human calicivirus in epidemic gastroenteritis. *Arch. Virol.* **52**(Suppl), 419–424.

Hedberg, C.W. and Osterholm, M.T. (1993) Outbreaks of foodborne and waterborne viral gastroenteritis. *Clin. Microbiol. Rev.* **6**, 199–210.

Hedlund, K.O., Rubliar-Abreu, E. and Svensson, L. (2000) Epidemiology of calicivirus infections in Sweden, 1994–1998. *J. Infect. Dis.* **191**(2), S275–S280.

Hussaini, S.H., Skidmore, S.J. and Richardson, P. (1997) Severe hepatitis E infection during pregnancy. *J. Viral. Hepat.* **4**, 51–54.

Ho, M., Chen, E.R., Hsu, K.H., Twu S.J., Chen K.T., Tsai S.F., Wang, J-R. and Shin, S.R. (1999) An epidemic of enterovirus 71 infection in Taiwan. Taiwan enterovirus epidemic working group. *N. Engl. J. Med.* **341**, 929–935.

Inouye, S., Yamashita, K., Yamadera, S., Yoshikawa, M., Kato, N. and Okabe, N. (2000) Surveillance of viral gastroenteritis in Japan: pediatric cases and outbreak incidents. *J. Infect. Dis.* **181**(2), S270–S274.

Ishimaru, Y., Nakano, S., Yamaoka, K. and Takami, S. (1980) Outbreaks of hand, foot and mouth disease by enterovirus 71: high incidence of complication disorders of central nervous system. *Arch. Dis. Child.* **55**, 583–588.

Jacob, J.T. (2000) The final stages of the global eradication of polio. *N. Engl. J. Med.* **343**(11), 806–807.

Jiang, X. and Pickering, L.K. (2002) Update in caliciviruses and human acute gastroenteritis. *Ped. Infect. Dis.* **21**(11), 1069–1070.

Kapikian, A.Z., Wyatt, R.G., Dolin, R., Thornhill, T.S., Kalica, A.R. and Chanock, R.M. (1972) Visualization by immune electron microscopy of a 27-nm particle associated with acute infectious nonbacterial gastroenteritis. *J. Virol.* **10**, 1075–1081.

Kaplan, J.E., Feldman, R., Campbell, D.S., Lookabaugh, C. and Gary, G.W. (1982) The frequency of a Norwalk-like pattern of illness in outbreaks of acute gastroenteritis. *Am. J. Public Health* **72**, 1329–1332.

Keswick, B.H., Satterwhite, T.K. and Johnson, P.C. (1985) Inactivation of Norwalk virus in drinking water by chlorine. *Appl. Environ. Microbiol.* **50**, 261–264.

Khuroo, M.S. (1980) Study of an epidemic of non-A, non-B hepatitis. Possibility of another human hepatitis virus distinct from post-transfusion non-A, non-B type. *Am. J. Med.* **68**, 818–824.

Kirschke, D.L., Jones, T.F., Buckingham, S.C., Craig, A.S. and Schaffner, W. (2002) Outbreak of aseptic meningitis associated with echovirus 13. *Ped. Infect. Dis. J.* **11**, 1034–1038.

Kitahashi, C.K., Tanaka, T. and Utagawa, E. (1999) Detection of HAV, SRSV and astrovirus genomes from native oysters in Chiba City, Japan. *Kansenshogaku Zasshi* **73**, 559–564.

Lee, S.H., Levy, D.A., Craun, G.F., Beach, M.J. and Calderon R.L. (2002) Surveillance for waterborne disease outbreaks – United States, 1999–2000. *Mortal. Morb. Wkly Rep.* **51**(SS08), 1–28.

Lew, J.F., Valdesuso, J., Vesikari, T., Kapikian, A.Z., Jiang, X., Estes, M.K and Green, K.Y. (1994) Detection of Norwalk virus or Norwalk-like virus infections in Finnish infants and young children. *J. Infect. Dis.* **169**, 1364–1367.

Lopman, B.A., Brown, D.W. and Koopmans, M. (2002) Human caliciviruses in Europe. *J. Clin. Virol.* **24**, 137–160.

Madeley, C.R. and Cosgrove, B.P. (1975). 28 nm particles in faeces in infantile gastroenteritis. *Lancet* **2**, 451–452.

Marks, P.J., Vipond, I.B., Carlisle, D., Feakin, D., Fey, R.E. and Caul, E.O. (2000) Evidence for airborne transmission of Norwalk-like virus (NLV) in a hotel restaurant. *Epidemiol. Infect.* **124**, 481–487.

Matsui, S.M. and Greenberg, H.B. (2000) Immunity to calicivirus infection. *J. Infect. Dis.* **181**(Suppl 2), S331–S335.

Mead, P.S., Slutsker, L., Dietz, V., McCaig, L.F., Bresee, J.S., Shapiro, C. *et al.* (1999) Food-related illness and death in the United States. *Emerg. Infect. Dis.* **5**, 607–625.

Melnick, J.L. (1990) Enteroviruses: polioviruses, coxsackieviruses, echoviruses and newer enteroviruses. In *Virology* (eds. B.N. Field and D.M. Knipe), pp. 549–605, New York, Raven Press.

Meng, X.J. (2000) Novel strains of hepatitis E virus identified from humans and other animal species: is hepatitis E a zoonosis? *J. Hepatol.* **33**, 842–845.

Monroe, S.S., Ando, T. and Glass, R.I. (2000) Introduction: Human enteric caliciviruses-an emerging pathogen whose time has come. *J. Infect. Dis.* **181**(Suppl 2), S249–S251.

Morse, S.S. and Schluederberg, A. (1990) Emerging viruses: the evolution of viruses and viral diseases. *J. Infect. Dis.* **162**, 1–7.

Mounts, A.W., Ando, T., Koopmans, M., Bresee, J.S., Noel, J. and Glass, R.I. (2000) Cold weather seasonality of gastroenteritis associated with Norwalk-like viruses. *J. Infect. Dis.* **181**(2), S284–S287.

Naik, S.R., Aggarwal, R., Salunke, P.N. and Mehrotra, N.N. (1992) A large waterborne viral hepatits E epidemic in Kanpur, India. *Bull. WHO* **70**, 572–577.

Ooi, E.E., Phoon, M.C., Ishak, B. and Chan, S.H. (2002) Seroepidemiology of human enterovirus 71, Singapore. *Emerg. Infect. Dis.* **8**(9), 995–997.

Pang, X.-L., Honma, S., Nakata, S. and Vesikari, T. (2000) Human Calicivirus in acute gastoenteritits of young children in the community. *J. Infect. Dis.* **181**(Suppl 2), S288–S294.

Parrino, T.A., Schreiber, D.S., Trier, J.S., Kapikian, A.Z. and Blacklow, N.R. (1977) Clinical immunity in acute gastroenteritis caused by Norwalk agent. *N. Engl. J. Med.* **297**, 86–89.

Payment, P.M., Tremblay, M. and Trudel, M. (1985) Relative resistance to chlorine of polioviruses and coxsackievirus isolates from environmental sources and drinking water. *Appl. Environ. Microbiol.* **49**, 981–983.

Pina, S., Buti, M., Cortina, M., Piella, J. and Girones, R. (2000) HEV identified in serum from humans with acute hepatitis and in sewage of animal origin in Spain. *J. Hepatol.* **33**, 826–833.

Pringle, C.R. (1999) Virus Taxonomy-1999. The universal system of virus taxonomy, updated to include the new proposals ratified by the International Committee on Taxonomy of viruses during 1998. *Arch. Virol.* **144**, 421–429.

Potasman, I., Paz, A. and Odeh, M. (2002) Infectious outbreaks associated with bivalve shellfish consumption: a worldwide perspective. *Clin. Infect. Dis.* **35**, 921–928.

Putzker, M., Sauer, H., Kirchner, G., Keksel, O. and Malic, A. (2000) Community acquired diarrhea: the incidence of astrovirus infections in Germany. *Clin. Lab.* **46**(5–6), 269–273.

Richards, G.P. (1985) Outbreaks of shellfish-associated enteric virus illness in the United States: requisite for development of viral guidelines. *J. Food Prot.* **48**, 815–823.

Rockx, B., de Wit, M., Vennema, H., Vinje, J., de Bruin, W., van Duynhoven, Y. and Koopmans, M. (2002) Natural history of human calicivirus infection: a prospective cohort study. *Clin. Infect. Dis.* **35**, 246–253.

Rotbart, H.A., O'Connell, J.F. and McKinlay, M.A (1998) Treatment of human enterovirus infections. *Antiviral Res.* **38**, 1–14.

Sarmiento, L., Mas, P., Goyenechea, A., Palomera, R., Morier, L., Capó, V., Quintana, I. and Santin M. (2001) First epidemic of echovirus 16 meningitis in Cuba. *Emerg. Infect. Dis.* **7**(5), 887–889.

Schaub, S.A. and Oshiro, R.K. (2000) Public health concerns about caliciviruses as waterborne contaminants. *J. Infect. Dis.* **181**(Suppl 2), S374–S380.

Schreier, E., Doring, F. and Kunkel, U. (2000) Molecular epidemiology of outbreaks of gastroenteritis associated with small round structured viruses in Germany in 1997/98. *Arch. Virol.* **145**, 443–453.

Smith, A.W., Skilling, D.E., Cherry, N., Mead, J.H. and Matson, D.O. (1998) Calicivirus emergence from ocean reservoirs: zoonotic and interspecies movements. *Emerg. Infect. Dis.* **4**(1), 13–20.

Taylor, M.B., Cox, N., Very, M.A. and Grabouw, W.O.K. (2001) The occurrence of Hepatitis A and astroviruses in selected river and dam waters in South Africa. *Water Res.* **35**(11), 2653–2660.

Tougianidou, D. and Botzenhart, K. (1993) Detection of enteroviral RNA sequences in different water samples. *Wat. Sci. Tech.* **27**, 219–222.

Van Regenmortel, M.H.V., Fauquet, C.M., Bishop, D.H.L., Carstens, E.B., Estes, M.K., Lemon, S.M., Maniloff, J., Mayo, M.A., McGeoch, D.J., Pringle, C.R. and Wickner, R.B. (eds) (2000) Virus Taxonomy: Seventh Report of the International Committee on Taxonomy of Viruses, New York, Academic Press.

Walter, J.E. and Mitchell, D.G. (2000) Role of astroviruses in childhood diarrhea. *Curr. Opin. Pediatr.* **12**, 275–279.

Wang, J.R., Tuan, Y.C., Tsai, H.P., Yan, J.J., Lui, C.C. and Su, I.J. (2002) Change of major genotype of enterovirus 71 in outbreaks of hand-foot-and-mouth disease in Taiwan between 1998 and 2000. *J. Clin. Microbiol.* **40**, 10–15.

Worm, H.C., van der Poel, W.H.M. and Brandstätter, G. (2002) Hepatitis E: and overview. *Microb. Infect.* **4**, 657–666.

Wyatt, R.G., Dolin, R. and Blacklow, N.R. (1974) Comparison of three agents of acute infectious nonbacterial gastroenteritis by cross-challenge in volunteers. *J. Infect. Dis.* **130**, 709–714.

4

Emerging infectious waterborne diseases: protozoan agents

L.H. Nel, W. Markotter and J. Weyer

4.1 INTRODUCTION

Protozoal parasites, typically present in potable water supplies as faecal contaminants, are able to survive for long periods in the aquatic environment. These enteric protozoa are highly resistant to most chemicals commonly used in water treatment and large outbreaks of protozoan enteritis are usually associated with surface water supplies that are inadequately flocculated and filtered, given their resistance to chlorine and other chemicals used in water purification. In fact, protozoal agents have become the leading cause of waterborne disease outbreaks for which the aetiologic agent could be identified and they have become equally important for their role in causing new and emerging waterborne diseases. *Cryptosporidium* and *Giardia* cause the most common of these infections, whereas *Cyclospora* has also recently emerged as an enteric protozoan. Resistance to water-purifying chemicals is one factor in the emergence of protozoal waterborne disease, but other important factors generally also include: their highly infectious nature, changes in social patterns (see *Giardia*), emergence in concert with the AIDS pandemic and the development of improved diagnostic tests to identify these pathogens. A summary of the pathogens, disease characteristics and factors involved in emergence are presented in Table 4.1.

 Microbial Waterborne Pathogens edited by T.E. Cloete, J. Rose, L.H. Nel and T. Ford. ISBN: 1 84339 055 8. Published by IWA Publishing, London, UK.

Table 4.1 Waterborne new and emerging diseases of protozoan origin.

Pathogen and disease	Geographical incidence	Factors involved in disease emergence	Clinical features	Treatment and detection
Cryptosporium parvum Cryptosporidiosis (Reviewed in Quintero-Betancourt *et al.*, 2002; Rose *et al.*, 2002; Tzipori and Ward, 2002)	Occurs worldwide.	AIDS epidemic. Chlorine resistance. Improved diagnostics	Diarrhoea with abdominal cramps, anorexia, fever and vomiting.	Acid-fast staining, immuno-fluorescence microscopy, ELISA and molecular methods No specific treatment, supportive treatment by oral rehydration fluid (ORF) or intravenous rehydration.
Cyclospora cayetanensis Cyclosporiasis (Reviewed in Herwaldt, 2000; Ortega *et al.*, 1998)	Occurs worldwide but most common in tropical and subtropical areas.	AIDS epidemic Improved diagnostics	Prolonged watery diarrhoea, abdominal cramping, weight loss, anorexia, myalgia, and occasionally vomiting and/or fever.	Light microscopy, sporulation assay, molecular methods Treatment by rehydration and with trimethoprim and sulfamethoxazole.
Microsporidia Microsporidiosis	Occurs worldwide	AIDS epidemic	Chronic diarrhoea, dehydration and severe weight loss	Light microscopy, transmission microscopy, immuno-fluorescence assays, molecular methods albendazole and fumagillin
Isospora belli Isosporasis (Reviewed in Marshall *et al.*, 1997 and Goodgame, 1996)	Worldwide (particularly in tropical and subtropical regions)	Emergence coupled to the AIDS pandemic	Self-limiting watery diarrhoea, malabsorption, anorexia; Symptomless cases occur frequently	Microscopy and epi-flourescence Drug treatment available

Giardia duodenalis, (syn. *G. lamblia* or *G. intestinales)* Giardiosis (Reviewed in Thompson, 2000)	Worldwide (most common intestinal parasite of humans and found in many domestic and wild animals)	Social changes, with increasing number of children in day-care facilities are implicated; Zoonotic aspects which involves altered host–parasite relationship (for example, AIDS)	Majority of infections are asymptomatic; Incubation time 1–2 weeks; Symptoms include diarrhoea, loose or watery stool, abdominal cramps, weight loss and dehydration	Detection by direct light microscopy (wet mount), staining (trichrome or iron haematoxylin), or direct fluorescent antibody detection. Enzyme linked immunosorbent assay (ELISA)-based kits are available; Proper hygiene habits for prevention and veterinary vaccines available; Drug treatment available
Toxoplasma gondii Toxoplasmosis (Reviewed in Tenter *et al*. 2000)	Worldwide (up to 100% prevalence in some locations)	*Toxoplasma* is regarded as an important emerging pathogen in the AIDS patient population	Inapparent infections in the immunocompetent host; mild symptoms may include flu-like symptoms, myalgia and neuralgia; More serious toxoplasmic encephalitis in the immunocompromized	Effective diagnosis with serological tests. Biopsies also often performed; Veterinary vaccine available; Drug treatment available but often have toxic effects
Acanthamoeba spp. Amoebic encephalitis and keratitis (Reviewed in Marshall *et al.*, 1997)	Probably worldwide but exact prevalence unknown	A newly described pathogen; Increasing number of cases in contact lens wearers; Opportunistic pathogen of the immunocompromized host	Severe encephalitis in immunocompromizsed individuals (headaches, hallucinations and seizures, etc.); Chronic ulceration of the ocular tissue in contact lens wearers	Microscopic study of corneal scrapings and biopsies. Immunofluorescence assays used; prevention of keratitis through the use of sterile, commercial solvents for contact lens care; Drug treatment available for keratitis

4.2 SPECIFIC EMERGING WATERBORNE INFECTIOUS DISEASES CAUSED BY PROTOZOAN AGENTS

4.2.1 Cryptosporidiosis

The aetiological agent of this persistent diarrhoeal disease is *Cryptosporidium parvum*, a protozoal parasite that is now encountered worldwide (Kosek *et al.*, 2001). Together with the AIDS epidemic, cryptosporidiosis emerged in the early 1980's as a zoonotic disease (Current *et al.*, 1983; Casemore *et al.*, 1985), after having been known as a veterinareal disease for over a century. Outbreaks are seasonal with an increase in infection rates during the warm humid months. Apart from diarrhoea and persistent diarrhoea, symptoms may include abdominal pain, nausea, low-grade fever, vomiting and significant weight loss. Infectious oocysts may still be excreted long after diarrhoeal disease has ended (up to five weeks). The life cycle of the parasite begins when spores or oocysts are ingested. The infected host excretes sporulated oocysts, containing four sporozoites and these are ingested and sporulation occurs in the host, producing more oocysts that are again infective upon excretion (Tzipori and Ward, 2002). Improved diagnostics for this parasite has contributed to its recognition as an emerging pathogen. Acid-fast staining can identify *Cryptosporidium.* Commercially available kits using EIA and DFA techniques are also available (Quintero-Betancourt *et al.*, 2002).

Although not only associated with immunocompromised indivivuals (for example a waterborne outbreak involving 400,000 people occurred in Milwaukee in 1994 (McKenzie *et al.*, 1994), this highly infectious disease has become life threatening in the growing number of HIV-infected individuals worldwide. Compounding the threat of this disease is the fact that the pathogen is difficult to filter from water resources (Goldstein *et al.*, 1996), is resistant to chlorine (Carpenter *et al.*, 1999), is ubiquitous in many animals, can survive in infected water for long periods of time and is highly infectious (Bagley *et al.*, 1998). It is therefore not surprising that *C. parvum* has now become the major protozoan pathogen of humans. Although no effective treatment exists, the disease is usually self-limiting in immunocompetent individuals. The wide spectrum of water supplies implicated in the spread of cryptosporidiosis include contaminated and inefficiently treated/flocculated/filtered drinking water such as from water plants and boreholes or from swimming pools, wade pools, hot tubs, jacuzzis, fountains, lakes, rivers, springs, ponds or streams (Fayer *et al.*, 2000; Smith and Rose, 1998; LeChevallier *et al.*, 1991). A high rate of secondary person-to-person transmission in households and institutions has also been documented.

4.2.2 Cyclosporiasis

This food and waterborne protracted, relapsing gastroenteritic disease is similar to cryptosporidiosis (Ortega *et al.*, 1998) and is also caused by a coccidian parasite, in this case *Cyclospora cayetanensis* (Ortega *et al.*, 1994). The protracted disease, persistent for an average of 7 weeks can be relapsing with symptoms that mimic those of cryptosporidiosis. Complications can result in reactive arthritis syndrome (Connor *et al.*, 2001), Guillain–Barre Syndrome (Richardson *et al.*, 1998) and acalculous cholecystitis (Zar *et al.*, 2001). *Cyclospora* emerged in the 1990s as a serious and widespread gastrointestinal parasite although the first isolation of *Cyclospora* from humans was in 1979 in Papua New Guinea (Ashford, 1979) and *C. cayetanensis* was first isolated from the intestines of moles as long ago as 1870 (Curry and Smith, 1998). There have been reports of isolation from rodents and snakes (Ortega *et al.*, 1994) but there is no definite evidence that these organisms occur in other hosts. Cyclosporiasis is most common in tropical and subtropical regions and endemic to North, Central and South America, Southeast Asia, the Caribbean islands, India and parts of Eastern Europe (Sterling and Ortega, 1999), but large outbreaks have recently been reported in the United States (Mead *et al.*, 1999) and Canada, with reports also from the United Kingdom and various African countries including South Africa (Sterling and Ortega, 1999). Apart from faecally contaminated water, various types of fresh produce such as raspberries, basil, field greens and salad mixes have been indicated as sources of the pathogen (Herwaldt and Ackers, 1997). Although *Cyclospora* oocysts, like *Cryptosporidium* oocysts, are resistant to chlorine, they are twice the size of *Cryptosporidium* oocysts and may be more easily removed by flocculation and filtration treatment methods used in water plants (Herwaldt, 2000). Further, due to the non-infectious nature of newly formed oocysts (in contrast to *Cryptosporidium*, excreted oocytes are non-infectious and unsporulated), secondary person-to-person spread of *Cyclospora cayetanensis* is a much less likely route of transmission than is the case for cryptosporidiosis (Smith *et al.*, 1997). How these agents become infectious, the influence of the environment on the rate of sporulation and the mode of pathogenesis remains unclear at present (Smith *et al.*, 1997).

Like the case with cryptosporidiosis, the emergence of this parasite was influenced by the AIDS epidemic and by somewhat improved diagnostics. Diagnostic laboratories have not routinely tested for *cyclospora* and they are present in very low numbers, so that spool samples have to be treated to concentrate the parasites. Wet mounts can be used to identify the parasite by microscopy, and acid-fast staining but the development and implementation of more sensitive and specific diagnostic techniques would be useful in improving control over this disease (Ortega *et al.*, 1998; Herwaldt, 2000). *Cyclospora* responds well to treatment with trimethoprim-sulfamethoxazole (Guerrant *et al.*, 2001).

4.2.3 Microsporidia

Microsporidia is a term used to describe organisms in the order Microsporidia of the phylum Microspora. Five genera are associated with diseases: *Encephalitozoon*, *Enterocytozoon*, *Septata*, *Pleistophora* and *Vittaforma* (Curry and Smith, 1998). The most common symptoms are diarrhoea, weight loss, abdominal pain, vomiting and fever. They are obligate intracellular protozoa that form highly specialized, environmentally resistant spores (chlorine treatment, for example, has little effect on Microsporidia). The mode(s) of transmission is still unclear but there has been mounting evidence of the involvement of water (Sparfel *et al.*, 1997). Due to the worldwide increase in immunosupressed individuals, these organisms have been isolated as the cause of diarrhoea in several cases. While infection is indeed primarily associated with HIV-positive individuals (Sobottka *et al.*, 1998; Leder *et al.*, 1998), they can also cause disease in immunocompetent patients. Chronic diarrhoea, dehydration and severe weight loss are the predominant symptoms in most hosts (Weber *et al.*, 1994). No effective treatment is available, but albendazole and fumagillin can be used with some success. Because of the small size of microsporidia they can only be detected in tissue or in stool samples by using different staining techniques together with light microscopy or transmission microscopy (Wolk *et al.*, 2000). Immunofluorescence assays and molecular methods are being developed and may already be under evaluation at the time of this publication.

4.2.4 Giardiasis

The aetiological agents of Giardiosis are the flagellated protozoans known as *Giardia duodenalis* (syn. *Giardia intestinalis*; *Giardia lamblia*). Giardiasis is a major public health concern particularly in developing countries, and waterborne outbreaks often occur (Furness *et al.*, 2000). *G. duodenalis* has a worldwide distribution and is indeed the most common intestinal parasite of humans worldwide (Marshall *et al.*, 1997).

Giardia spp. non-invasively inhabits the lumenal surface of the small intestine of possibly all invertebrate species, including reptiles, avians and amphibians. Cysts, which are resistant to the elements, are defaecated by the infected host. Transmission of these cysts is through the faecal–oral route. The factors that contribute to the re-emergence of Giardiasis, are recognized as social changes in the world population and some zoonotic aspects which alter the host–parasite relationship (Thompson, 2000). The gathering of young children in day-care centres provides the ideal setting for transmission of infectious diseases. Bad hygiene is a major cause for this phenomenon and today *G. duodenalis*, together with Rota- and Astroviruses and *Cryptosporidium*, are considered the major causative agents of diarrhoea in child-care centres (Ferson, 1997; Thompson, 1994). The zoonotic aspects include the high prevalence of *Giardia*, which has been noted in dairy cows, and may provide a reservoir for human infections (Thompson, 2000). Giardia in domestic cats and dogs also are a frequent source of infection to humans. In Australia, *G. duodenalis* is the most common enteric parasite of domestic dogs (Bugg *et al.*, 1999).

The majority of *G. duodenalis* infections are asymptomatic, but diarrhoea, abdominal pain and rapid weight loss may be observed (Wolfe, 1992). These symptoms are especially severe in AIDS patients (Faubert, 1996). Veterinary vaccines for *Giardia* are commercially available (Olson *et al.*, 2000).

4.2.5 Toxoplasmosis

Toxoplasmosis is caused by the tissue cyst-forming coccidium, *Toxoplasma gondii*. The parasite has a complex life cycle, involving different stages in definitive and intermediate hosts (Tenter *et al.*, 2000). Members of the Felidae family, particularly domestic cats, are implicated as definitive hosts of the parasite and intermediate hosts include possibly all mammals. Even though other stages of the *T. gondii* life cycle are also infectious, only the oocyst form of the parasite persists in the environment. The unsporulated oocysts are defaecated by the definitive host, sporogony occurs and the infectious oocysts are then transmitted to a host via the oral route. Toxoplasmosis outbreaks have been linked to contaminated municipal drinking water (Bowie *et al.*, 1997).

T. gondii are found all around the world and seroprevalence studies have shown up to 100% presence in some populations (Jackson and Hutchinson, 1989; Tenter *et al.*, 2000). Toxoplasmosis poses not only a medical problem but also has veterinary importance. Toxoplasmosis can cause congenital disease and abortion in both humans and animals. *T. gondii* infections are usually inapparent in immunocompetent individuals (reviewed in Josephson, 1987). Mild symptoms may include lymphadenopathy, and in extreme cases encephalitis, sepsis shock and myocarditis. An increasing number of toxoplasmosis cases have been reported in recent years, since *T. gondii* are responsible for opportunistic infections in immunocompromised individuals (Luft and Remington, 1992). *T. gondii* infections of the AIDS patient group manifest as lymphadenopathy, central nervous system disorders, myocarditis, pneumonitis and cranial lesions (Josephson, 1987). Toxoplasmosis-induced abscesses are often found in the brain tissue of AIDS patients. *T. gondii*-induced encepahalitis are found in up to 40% of AIDS patients and are reported as the cause of death in up to 30% of these patients (Luft *et al.*, 1984; Luft and Remington, 1992).

The treatment of the disease is troublesome due to the toxic effect of available drugs and treatment does not prevent re-infection (Josephson, 1987). A veterinary vaccine, in the form of a live attenuated oocyst, has been developed, but a vaccine for human application is not yet available (Buxton and Innes, 1995). Progress towards this goal has however been made (Bhopale, 2003).

4.2.6 Amoebic encephalitis and keratitis

Acanthamoeba are microscopic free-living amoebae. *Acanthamoeba* spp. have been isolated from various environments, including soil, dust, fresh and salt water. Isolations from air-conditioning units, humidifiers, dialysis units and dentistry equipment have also been reported. The organism is ubiquitous and is probably found worldwide (Marshall *et al.*, 1997).

Infection with A*canthamoeba* manifests as granulomatous amoebic encephalitis in immunocompromised hosts, which includes headaches, stiff neck, nausea, vomiting, malaise, confusion and hallucinations, loss of balance and control of movement and seizures. *Acanthamoeba* infection often has a fatal outcome. *Acanthamoeba* is also a leading cause of destructive keratitis, a chronic ulceration, in contact lens wearers (Moshari *et al.*, 2001). The use of non-sterile saline solutions and tap water for rinsing purposes are implicated as sources of ocular *Acanthamoeba* infections (Kilvington, 1990). *Acanthamoeba* keratitis has also been reported in contact lens wearers after exposure to lake and pond water (Marshall *et al.*, 1997).

Human-to-human transmission has not yet been reported and various routes of transmittance are implicated, but remain to be elucidated. *Acanthamoeba* is resistant to various disinfecting agents and antimicrobials, and has been found in municipal drinking water and even bottled mineral water (Kilvington, 1990; Borazjani *et al.*, 2000; Miltner and Bermudez, 2000).

Acanthamoeba is of emerging importance for two reasons. *Acanthamoeba* is considered an emerging pathogen, since it is a relatively newly described pathogen. A rise in *acanthamoeba* keratitis in contact lens wearers has also been noted in recent years and *Acanthamoeba* spp. are opportunistic pathogens of AIDS patients. On the other hand, *Acanthamoeba* may serve as a host for bacterial agents such as *Legionella pneumophila.* The latter is considered as an emerging waterborne infection especially recognized in health care settings. *Acanthamoeba* protects the bacterial agents against chlorination and other disinfection measures (Emmerson, 2001).

4.2.7 Isosporiasis

Isosporiasis is caused by the coccidian parasite, *Isospora belli*. Many species of *Isospora* are now known, but only *Isopora belli* infects humans (Goodgame, 1996). Cases of *Isospora natalensis* infections in humans have also been reported but are extremely rare (Eldson-Dew, 1953). *Isospora* spp. are found worldwide, but are the most common in tropical and subtropical regions of the Middle East, Africa, South Asia and South America (Lindsay *et al.*, 1997). Precise prevalence rates and details of the epidemiology are vague.

Isospora spp. are transmitted as oocysts, that are excreted in the faeces by an infected host. *I. belli,* are transmitted by person-to-person contact and in contaminated water and food. Isosporiasis in immunocompetent individuals are characterized by self-limiting, watery diarrhoea, malabsorption and anorexia (Mirdha *et al.*, 2002). Inapparent infections are also found and are common with intestinal spore-forming protozoa (Goodgame, 1996). In patients with AIDS, diarrhoea may be more severe and chronic and acalculous cholecystitis are also observed (Velásques *et al.*, 2001, Mirdha *et al.*, 2002). *I. belli* is an opportunistic pathogen and has been detected in 8–20% of AIDS patients in studies in Haiti and Africa (Mannheimer and Soave, 1994). Higher prevalences are noted in developing countries where AIDS is more common (Goodgame, 1996).

4.3 CONCLUSIONS

In considering the reasons for emergence of the specific waterborne infectious diseases of bacterial, viral and protozoal origin, some common elements may be identified. These include:

(1) Genetic adaptation of pathogens, including the acquisition of resistance plasmids, toxin genes and other virulence factors. Rapid genetic evolution is particularly common in the groups of RNA viruses, as discussed in Chapter 3.

(2) Overcrowding, combined with poor sanitation and deteriorating or inadequate public health infrastructure. Collapse of economies, war and natural disasters can be contributing factors here.
(3) Increased exposure to animal reservoirs (as a result of geographical expansion, overpopulation and competition for dwindling water resources) and continual zoonoses. Increased utilization of virgin soil for agriculture and new agricultural practices may also contribute.
(4) Rising incidence of diseases or other factors that affect immune competence in groups or populations, including HIV/AIDS, malnutrition, etc.
(5) Improved diagnostic techniques and tools, including more sensitive molecular tools designed with the use of PCR.

It is therefore conceivable that infectious diseases will continue to emerge and re-emerge. To successfully intervene, improved diagnostics, surveillance, vaccines and drugs are needed. However, improved education and hygiene, proper water sanitation and the responsible and controlled application of antibiotics and other antimicrobials are just as crucial in this continuing battle.

REFERENCES

Ashford, R.W. (1979) Occurrence of an undescribed coccodian in man in Papua New Guinea. A. *Trop. Med. Parasitol.* **73**, 497–500.

Bagley, S.T., Auer, M.T., Stern, D.A. and Babiera, M.J. (1998) Sources and fate of *Giardia* cysts and *Cryptosporidium* oocysts in surface waters. *J. Lake Res. Manage.* **14**, 379–392.

Bhopale, G.M. (2003) Development of a vaccine for toxoplasmosis: current status. *Microb. Infect.*, **5**(5), 457–462.

Borazjani, R.N., May, L.L., Noble, J.A., Avery, S.V. and Ahearn, D.G. (2000) Flow cytometry for determination of the efficacy of contact lens disinfecting solutions against *Acanthamoeba* sp. *Appl. Environ. Microbiol.* **66**, 1057–1061.

Bowie, W.R., King, A.S., Werker, D.H., Isaac-Renton, J.L., Bell, A., Eng, S.B. and Marion, S.A. (1997) Outbreak of toxoplasmosis associated with municipal drinking water. *Lancet* **350**, 173–177.

Bugg, R.J., Robertson, I.D., Elliot, A.D. and Thompson, R.C.A. (1999) Gastrointestinal parasites of urban dogs in Perth, Western Australia. *Vet. J.* **157**, 295–301.

Buxton, D. and Innes, E.A. (1995) A commercial vaccine for ovine toxoplasmosis. *Parasitology* **110**, S11–S16.

Carpenter, C., Fayer, R., Trout, J. and Beach, M.J. (1999) Chlorine disinfection of recreational water for *Cryptosporidium parvum*. *Emerg. Infect. Dis.* **5**, 579–584.

Casemore, D.P., Sands, R.L. and Curry, A. (1985) *Cryptosporidium* species: a new human pathogen. *J. Clin. Pathol.* **38**, 1321–1336.

Connor, B.A., Johnson, E. and Soave, R. (2001). Reiter syndrome following protracted symptoms of *Cyclospora* infection. *Emerg. Infect. Dis.* 7, 453–454.

Current, W.L. and Garcia, L.S. (1991) Cryptosporidiosis. *Clin. Microbiol. Rev.* **4**, 325–358.

Curry, A. and Smith, H.V. (1998) Emerging pathogens: *Isospora, Cyclospora* and *Microsporidia*. *Parasitology* **117**, S143–S159.

Elsdon-Dew, R. (1953) *Isospora natalensis* (sp. nov) in man. *J. Trop. Med. Hyg.* **56**, 149–150.

Emmerson, A.M. (2001) Emerging waterborne infections in health-care settings. *Emerg. Infect.* **7**(2), 272–276.

Faubert, G.M. (1996) The immune response to *Giardia*. *Parisitol. Today* **12**, 140–145.

Fayer, R., Morgan, U. and Upton, S.J. (2000) Epidemiology of *Cryptosporidium:* transmission, detection and identification. *Int. J. Parasitol.* **30**, 1305–1322.

Ferson, M.J. (1997) Infection control in childcare settings. *Commun. Dis. Intell.* **21**, 333–337.

Furness, B.W., Beach, M.J. and Roberts, J.M. (2000) Giardiasis surveillance – United States, 1992–1997. *MMWR Morb. Mortal. Wkly Rep.* **49**(SS07), 1–13.

Goldstein, S.T., Juranek, D.D., Ravenholt, O., Hightower, A.W., Martin, D.G., Mesnik, J.L., Griffiths, S.D., Bryant, A.J., Reich, R.R. and Herwaldt, B.L. (1996) Cryptosporidiosis: an outbreak associated with drinking water despite state-of-the-art water treatment. *Ann. Intern. Med.***124**, 459.

Goodgame, R.W. (1996) Understanding intestinal spore-forming protozoa: *Cryptosporidia, Microsporidia, Isospora* and *Cyclospora. Ann. Int. Med.* **124**(4), 429–441.

Guerrant, R.L., Van Gilder, T., Steiner, T.S., Thielman, N.M., Slutsker, L., Hennessy, T., Griffin, P.M., DuPont, H., Sack, R.B., Tarr, P., Neill, M., Nachamkin, I., Reller, L.B., Osterholm, M.T., Bennish, M.L. and Pickering, L.K. (2001) Practice guidelines for the management of infectious diarrhea. *Clin. Infect. Dis.* **32**, 331–350.

Herwaldt, B.L. (2000) *Cyclospora cayetanensis*: a review, focusing on the outbreaks of cyclosporiasis in the 1990s. *Clin. Infect. Dis.* **31**, 1040–1057.

Herwaldt, B.L. and Ackers, M.L., *Cyclospora* Working group (1997) An outbreak in 1996 of cyclosporiasis associated with imported raspberries. *N. Engl. J. Med.* **336**, 1548–1556.

Jackson, M.H. and Hutchinson, W.M. (1989) The prevalence and source of *Toxoplasma* infection in the environment. *Adv. Parasitol.* **28**, 55–105.

Josephson, S.L. (1987) Toxoplasmosis. *Clin. Microb. News.* **9**(21), 165–168.

Kilvington, S. (1990) Activity of water biocide chemicals and contact lens disinfectants on pathogenic free-living amoebae. *Int. Biodet.* **26**, 127–138.

Kosek, M., Alcantara, C., Lima, A.A. and Guerrant, R.L. (2001) Cryptosporisiosis: an update. *Lancet* **i**, 262–269.

LeChevallier, M.W., Norton and Lee, R.G. (1991) Occurrence of *Giardia* and *Cryptosporidium* spp. in surface water supplies. *Appl. Environ. Microbiol.* **57**, 2610–2616.

Leder, K., Ryan, N., Spelman, D. and Growe, S.M. (1998) Microsporidial disease in HIV-infected patients: a report of 42 patients and review of the literature. *Scand. J. Infect. Dis.* **30**, 331–338.

Lindsay, D.S, Dubey, J.P. and Blagburn, L. (1997) Biology of *Isospora* spp. from humans, nonhuman primates and domestic animals. *Clin. Micr. Rev.* **10**, 19–34.

Luft, B.J. and Remington, J.S. (1992) Toxoplasmosis encephalitis in AIDS. *Clin. Infect. Dis.* **15**, 211–222.

Luft, B.J., Brooks, R.G., Conley, F.K., McCabe, R.E. and Remington, J.S. (1984) Toxoplasmic encephalitis in patients with acquired immune deficiency syndrome. *J. Am. Med. Assoc.* **252**, 913–917.

Mannheimer, S.B. and Soave, R. (1994) Protozoal infection in patients with AIDS. *Infect. Dis. Clin. N. Am.* **8**, 483–497.

Marshall, M.M., Naumovitz, D., Ortega, Y. and Sterling, C.R. (1997) Waterborne protozoan pathogens. *Clin. Micr. Rev.* **10**(1), 67–85.

McKenzie, W.R., Hoxie, N.J., Proctor, M.E. *et al.* (1994) A massive outbreak in Milwaukee of *Cryptosporidium* infection transmitted through the public water supply. *N. Engl. J. Med.* **331**, 161.

Mead, P.S., Slutsker, L., Dietz, V., McCaig, L.F., Bresee, J.S., Shapiro, C., Griffin, P. and Tauxe, R.V. (1999) Food-related illness and death in the United States. *Emerg. Infect. Dis.* **5**, 607–625.

Miltner, E.C. and Bermudez, L.E. (2000) *Mycobacterium avium* grown in *Acanthamoeba castellanii* is protected from the effects of antimicrobials. *Antimicro. Agents Chemother.* **44**, 1990–1994.

Mirdha, B.R, Kabra, S.K. and Samantray, J.C. (2002) Isosporiasis in children. *Ind. Pediatr.* **39**, 941–944.

Moshari, A., McLean, I.W., Dodds, M.T., Damiano, R.E. and McEvoy, P.L. (2001) Chorioretinitis after keratitis caused by *Acanthamoeba*: case report and review of the literature. *Ophthalmology* **108**, 2232–2236.

Olson, M.E., Ceri, H. and Morck, D.W. (2000) *Giardia* vaccination. *Parasitol. Today* **16**(5), 213–217.

Ortega, Y.R., Gilman, R.H. and Sterling, C.R. (1994) A new coccidian parasite (Apicomplexa: Eimeridae) from humans. *J. Parasitol.* **80**, 625–629.

Ortega, Y.R., Sterling, C.R. and Gilman, R.H. (1998) *Cyclospora cayetanensis*. *Adv. Parasitol.* **40**, 399–418.

Quintero-Betancourt, W., Peele, E.R. and Rose, J.B. (2002) *Cryptosporidium parvum* and *Cyclospora cayetanensis*: a review of laboratory methods for detection of these waterborne parasites. *J. Microbiol. Methods* **49**, 209–224.

Richardson, R.F., Remler, B.F., Katirji, B. and Murad, M.H. (1998) Guillain-Barre syndrome after *cyclospora* infection. *Musc. Nerve*, 669–671.

Smith, H.V., Paton, C.A., Mtambo, M.M.A. and Girdwood, R.W.A. (1997) Sporulation of *Cyclospora* sp. Oocysts. *Appl. Environ. Microbiol.* **63**, 1631–1632.

Smith, H.V. and Rose, J.B. (1998) Waterborne cryptosporidiosis: current status. *Parasitol. Today* **14**, 14–22.

Sobottka, I., Schwartz, D.A., Schottelius, J., Visvesvara, G.S., Pieniazed, N.J., Schmetz, C., Kock, N.P., Laufs, R. and Albrechts, H. (1998) Prevalence and clinical significance of intestinal microsporidiosis in human immunodeficiency virus-infected patients with and without diarrhea in Germany: a prospective coprodiagnostic study. *Clin. Infect. Dis.* **26**, 475–480.

Sparfel, J.M., Sarfati, C., Liguory, O., Caroff, B., Dumoutier, N., Gueglio, B., Billaud, E., Raffi, F., Molina, J.M. and Miegeville, M. (1997) Detection of *Microsporidia* and identification of *Enterocytozoon bienensi* in surface water by filtration followed by specific PCR. *J. Eukary. Microbiol.* **44**, 78S.

Sterling, C.R. and Ortega, Y.R. (1999) *Cyclospora*: an enigma worth unraveling. *Emerg. Infect. Dis.* **5**, 48–53.

Tenter, A.M., Heckeroth, A.R. and Weiss, L.M. (2000) *Toxoplasma gondii*: from animals to humans. *Int. J. Parisitol.* **30**, 1217–1258.

Thompson, R.C.A. (2000) Giardiasis as a re-emerging infectious disease and its zoonotic protential. *Int. J. Parasitol.* **30**, 1259–1267.

Thompson, S.C. (1994) *Giardia lamblia* in children and the childcare setting: a review of the literature. *J. Ped. Child Health* **30**, 202–209.

Tzipori, S. and Ward, H. (2002) Cryptosporidiosis: biology, pathogenesis and disease. *Microb. Infect.* **4**, 1047–1058.

Velásques, J.N., Carnevale, S., Mariano, M., Kuo, L.H., Caballero, A., Chertcoff, A., Ibáñez, C. and Bozzini, J.P. (2001) Isosporosis and unizoite tissue cysts in patients with acquired immunodeficiency syndrome. *Human Pathol.* **32**, 500–505.

Weber, R., Bryan, R.T., Schwartz, D.A. and Owen, R.L. (1994) Human Microsporidial infections. *Clin. Micr. Rev.* **7**, 426–461.

Wolfe, M.S. (1992) Giardiasis. *Clin. Micr. Rev.* **5**, 93–100.

Wolk, D.M., Johnson, C.H., Rice, E.W., Marshall, M.M., Grahn, K.F., Plummer, C.B. and Sterlin, C.R. (2000) A spore counting method and cell culture model for chlorine disinfection studies od *Encephalitozoon* syn. *Septata intestinalis*. *Appl. Environ. Microbiol.* **66**, 1266–1273.

Zar, F.A., El-Bayoumi, E. and Yungbluth, M.M. (2001) Histological proof of acalculous cholecystitis due to *Cyclospora cayetanensis*. *Clin. Infect. Dis.* **33**, E140–E141.

5

Legionella

W.F. McCoy

Legionellae are uniquely the twentieth century bacteria. Were it not for affordable air-conditioning and easy access to indoor hot water plumbing, these bacteria might remain unknown to us, still. And if the most cost-effective way to reject waste heat was not by evaporative transfer to the atmosphere from droplets of recirculating water in cooling towers, they might never have bothered us.

In the twentieth century, *Legionella* burst forth from its ancient undiscovered freshwater niche to emerge as a significant new human pathogen. Before that century it most likely caused little, if any, human suffering.

In the twenty-first century, legionellosis is a serious preventable disease. It has been detrimental in hundreds of thousands of lives.

5.1 THE BIOLOGY

5.1.1 Cell biology of *Legionella*

Legionellae are γ-proteobacteria. They are Gram-negative, heterotrophic, aerobic, transitionally motile, catalase-positive rods. Typically about 1 μm long and half as wide, they can also grow filamentously to about 20-μm length (WHO 2002). They require pre-formed amino acids for carbon and energy sources, perhaps an

 Microbial Waterborne Pathogens edited by T.E. Cloete, J. Rose, L.H. Nel and T. Ford. ISBN: 1 84339 055 8. Published by IWA Publishing, London, UK.

astonishing amount of iron, they do not oxidize or ferment carbohydrates and they do not form spores (Fields, 2002).

Their natural habitat is now indisputably proved to be parasitic to aquatic protist hosts. They incidentally, and disastrously, infect and grow within mammalian phagocytic cells including most importantly, human alveolar macrophage and alveolar epithelium. Although they survive outside this intracellular environment, apparently it is merely transient in the search to find new hosts to infect, for growth otherwise appears to be ecologically insignificant (Fields, 1989, 1993; Abu Kwaik, 1998; Newsome, 1998; Atlas, 1999).

Legionellae are sensitive to many environmental and antimicrobial factors. They do not survive above 60°C nor can they grow below 20°C. They are unable to proliferate in aquatic environments without their hosts or perhaps complex diverse biofilms (Surman, 2002). Outside their hosts and apart from protection in microbial biofilms (Costerton, 1995), they succumb to typical antimicrobials at typical use concentrations (Wireman, 1993; Walker, 2003). But they do seem somewhat more tolerant of certain non-oxidizing antimicrobials and completely tolerant, at typical use-doses, to certain other antimicrobials compared to typical Gram-negative aquatic bacteria (Kim, 2002).

Legionellae possess genetic elements that impart an ability to infect and grow within phagocytic protist host cells, like amoeba, and also within mammalian phagocytes in the immune system, like human macrophage (Segal, 1999). The first of these genes discovered was the *m*acrophage *i*nfectivity *p*otentiator (*mip*). The *mip* gene and the Mip protein that it encodes are virulence-associated factors in *L. pneumophila*. Infections of guinea pig, mammalian phagocytic cells, and protozoan cells depend on a functional *mip* gene and Mip protein. This protein is profusely shed from the surface of *Legionella* inside phagosomes of the protist *Acanthamoeba* (Helbig, 2002). Although Mip is an active peptidyl-prolyl-*cis/trans* isomerase, this enzymatic function does not seem to be necessary in pathogenesis; the mode of action for the Mip virulence factor is unknown.

Legionella exploit the type IV protein secretion system for virulence (Nagal, 2003). Type IV systems transport proteins or nucleic acid across the bacterial cell wall directly into the host (Lee, 2001). In *Legionella*, there are 24 known genes located in two separate loci on the bacterial chromosome that encode proteins secreted through the type IV machinery. These genes are referred to as the *dot/icm* (*d*efective for *o*rganelle *t*rafficking/*i*ntra*c*ellular *m*ultiplication) gene complex. So far, every mutant of the *dot/icm* family of genes in *Legionella* is defective for evasion of the endocytic pathway and therefore, these mutants are ineffective in early survival after phagocytosis (Swanson, 2000).

After entrance into the host phagosome, several *dot/icm* dependent factors immediately isolate the phagosome from its normal course; this allows *Legionella* to survive the first few minutes of infection because fusion of phagosome with lysosome is delayed. During the next several hours, the phagosome-enclosed parasite remains in lag phase, isolated from the lytic endosomal trafficking process and the pathogen becomes acid resistant (this acid resistance attribute is the basis for the selectivity step with HCl in the ISO analytical method, see section 5.3.3.1). Factors encoded by *dot/icm* and secreted through type IV machinery in *L. pneumophia* cause the phagosome to associate with the endoplasmic reticulum (Fields, 2002; Swanson and Bachman, 2002).

The next phase in the *Legionella* life cycle is characterised by rapid growth of the pathogen in corrupted phagosomes. It involves exploitation of autophagy in the host. Autophagy is a response to nutrient deprivation and it is ubiquitous among eukaryotic cells. A distinguishing feature of autophagy is the dynamic rearrangement of cellular membrane to sequester cytosol and organelles for delivery to the lysosome or vacuole wherein the contents are degraded into simple nutrients. Thus, the host phagocyte consumes portions of itself for survival during times of nutrient deprivation. *Legionellae* exploit this process to support rapid intracellular growth in the host phagocyte. However, nutrients become limiting to the pathogen within hours as the host cell is consumed. Nutrient depletion in the corrupted phagosome leads to the next phase in the pathogenic life cycle of *Legionella*.

The next phase is characterised by the response of exponentially growing *Legionella* to amino acid starvation as the nutrient content of the host is spent. The depletion of amino acids available to *Legionella* causes accumulation of uncharged tRNA inside the rapidly growing bacterial cells. This condition activates the ribosome-associated enzyme RelA. The RelA regulator converts guanosine triphosphate (GTP) to guanosine 3′,5′-bis-pyrophosphate (ppGpp) which positively activates the stationary phase sigma factor, σ, RpoS (Hammer 1999). This cascade of events triggers virulence traits in *Legionella* including protist cytotoxicity factors and synthesis of bacterial flagella. A protein, IcmT, is necessary for pore-forming activity in the host cell membrane that leads to death of the host (Molmeret, 2002; Alli, 2003). Other cytotoxins are also apparently necessary including legiolysin (encoded by *lly*) and another cytotoxic protein called Msp, which is a zinc metalloprotease. As *Legionella* becomes vigorously motile within massive intracellular vesicles, the spent host becomes completely overwhelmed by the infection and eventually it bursts to release motile and virulent progeny that seek opportunities to repeat this life cycle by infecting new hosts (Swanson, 2000; Swanson and Bachman, 2002).

Legionellae in search of new hosts to infect present hair-like structures on their surfaces called pili that base in the inner membrane, extend through the periplasmic space and onward through the outer membrane into the extracellular environment (Stone 1998). Pili are used for attachment to surfaces and/or transmission of genetic elements such as plasmids. *Legionella* process pilin protein subunits attached to adhesive type IV pili with prepilin peptidases, as do other Gram-negative pathogenic bacteria such as *Pseudomonas aeruginosa* and *Neiserria gonorrhoeae*. Type IV pili are characterized by an adhesive hydrophobic tip and they are retractable. The *L. pneumophila* prepilin peptidase, protein PilD, is necessary for attachment to host cells and biofilms. Mutants deficient in the *pilD* gene are not as infective. The gene also appears to be necessary for efficient intracellular growth in macrophage and protozoa (Helbig, 2002).

The inner membrane-bound PilD protein is also related to the type II protein secretion system common in many Gram-negative bacteria. The proteins these genes encode are necessary for intracellular growth of *Legionella* in their natural protist hosts and also in alveolar macrophages. They include lipases/esterases, phospholipase A, acid phosphatases, and proteases (Rossier, 2001; Rossier, 2002; Aragon, 2002). Some of these proteins may be responsible for pathogenicity traits in Legionnaires' disease (see section 5.2.2).

Attachment of *Legionella* to macrophages can occur at the major outer membrane protein (MOMP), which is encoded by *ompS.* The MOMP binds complement component C3 and thereby mediates uptake via the CR1 and CR3 macrophage receptors (Heuner, 2002). Another *Legionella* surface protein implicated in attachment is the heat shock protein (Hsp60) encoded by *htpB*. Hsp60 effects attachment and entry to HeLa epithelial cells (Faulkner, 2002). Since MOMP binding is complement-dependent and is therefore mediated by the immune response, there appears to be no ecological role for this attachment mechanism in protist hosts. Likewise, Hsp60 binding also appears to be due to its antigenic attributes and an ecological role in protist infection is unknown.

Legionella require a lot of iron. To get them to grow in culture media, astonishing concentrations are necessary ($>10\ \mu M$) as ferric pyrophosphate, ferric citrate, or ferric chloride. Stationary phase liquid cultures present a dark brown colour due to a pigment called homogentisic acid (HGA)-melanin that is iron-dependent, haemolytic and may act as an extracellular sink for excess iron. Mutational analysis of iron assimilation deficiency proves that iron acquisition is necessary for intracellular parasitic growth of *Legionella.* To assimilate iron, they may produce as many as three siderophores (iron-carrying molecules) with biochemical evidence for a unique legiobactin and genetic evidence for hydroxamate-like and pyoverdin-like iron chelators. *Legionellae* can bind and use hemin and therefore, they may use haeme-containing proteins as iron sources. *Legionella* may have the capacity to transport ferrous iron across its cell membrane suggesting that life is possible for them perhaps in a reduced environment. Fully oxidized ferric iron is certainly transported across the cell membrane and assimilated intracellularly. This impressive diversity in iron assimilation may account for *Legionella* survival in certain harsh aquatic habitats (Cianciotto, 2002).

The cell biology of *Legionella* is summarized schematically in Figure 5.1.

There are at least 48 species of *Legionella* (but 5 have been designated *Legionella*-like amoebic pathogens) and all of them should be thought of as potentially pathogenic. Although *L. pneumophila* serogroup 1 is thought to be the most virulent, it is first among many other potentially virulent serogroups. There are 15 serogroups of *L. pneumophila* and as many as 70 in total among all the species in the genus. Two other genera have been proposed: *Fluoribacter* for blue-white fluorescing species such as *L. bozemanii* and *Tatlockia* for the species *L. micadadei*. No further taxonomic differences are made in this review because all species and serogroups should be considered as potentially pathogenic in humans.

Legionellae are phylogenetically close relatives of *Coxiella burnetti*, the bacterial pathogen that causes Q fever. These parasites have been shown to possess

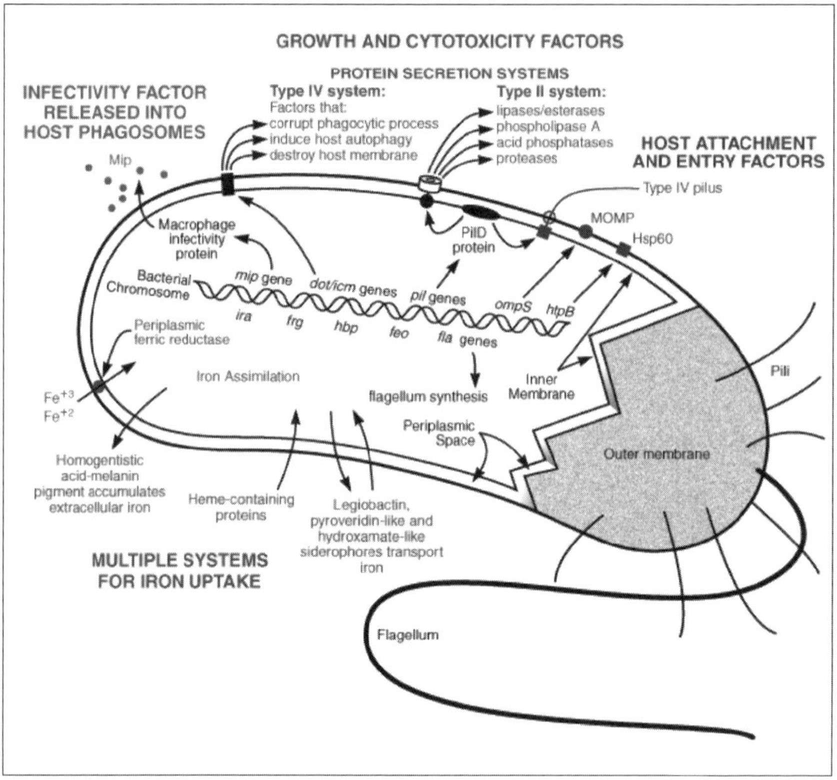

Figure 5.1 A schematic summary of the cell biology of *Legionella.* The architecture of the cell wall is consistent with that of other Gram-negative bacteria. Virulence factors are encoded by the *dot/icm* genes, the *mip* gene, the *pil* genes, iron-uptake genes (*ira, frg, hbp, feo*), and for flagellum synthesis, the *fla* genes. Attachment to certain phagocytic hosts and to surfaces is effected by the type IV pilus, the major outer membrane protein (MOMP) and the heat shock protein (Hsp60). An important infectivity factor is the macrophage infectivity protein (Mip). There are two protein secretion systems necessary for survival inside the host. Proteins secreted by the type IV system corrupt the phagocytic process and provide the pathogen an environment inside the host in which to grow and then later, to escape by forming pores in the host cell membrane. Proteins secreted by the type II system are necessary for growth within the corrupted phagosome. A uniquely diverse iron assimilation capability is necessary for intracellular growth.

many interesting similarities in modes of survival within phagocytic cells (Maruin, 1999; Swanson and Fernandez-Moreia, 2002).

5.1.2 Parasitology of *Legionella*

Legionellae present a parasitic life cycle with definable stages: Host infectivity, intracellular growth, motility, destruction of the host, and release of progeny into the environment. Ultrastructurally distinct cell differentiation occurs during the

Legionella life cycle. After intracellular replication, the pathogen differentiates into the infectious *m*ature *i*ntracellular *f*orm (MIF) that is motile and can survive prolonged starvation and harsh conditions when released back into the environment. Upon gaining entrance by infection to the intracellular parasitic environment of a new host, several ultrastructurally distinct intermediate forms differentiate finally into the delicate non-motile yet prolific replicative form of *Legionella*. Thus, it is clear that the *Legionella* life cycle is bimodal and highly pleomorphic (Faulkner, 2002).

The amoebic and ciliated phyla (groups) of protozoa appear to be the most susceptible natural hosts to *Legionella* but very little is known about the range of this parasitic association; there may be many other undiscovered protist hosts.

Protists exhibit an enormous diversity in form and are mostly microscopic ranging in size from 10–200 μm. At least 60,000 species have been described but almost surely this is but a fraction of the total number of protist life forms. There are plant-like protists (the algae), fungus-like protists (the slime molds) and animal-like protists (the protozoa). There are four phyla of animal-like protists: amoebic (Sarcodina), ciliated (Ciliophora), nonciliated (Sporozoa), and flagellated (Mastigophora). At least 14 genera of protozoa are certainly susceptible to infection by *Legionella.* The best-known protozoan hosts are *Acanthamoeba, Hartmannella, Naegleria* and *Tetrahymena.* It is also clear that *Legionella* infect at least one fungus-like protist of the slime mould variety, *Dictyostelium discoideum* (Steinert 2002). Infection occurs in the haploid amoebic form of this micro-organism, which grows as separate, independent cells but then they interact to form multicellular structures when challenged by adverse conditions such as starvation. Individual cells signal each other by releasing the chemoattractant cAMP and aggregate together by chemotaxis to form a mound that is surrounded by an extracellular matrix. Formation of a multicellular organism in this way is unique in nature.

Many protozoa make their living by consuming bacteria, especially in biofilms (Fields, 1983; Baker, 1992; Fields, 1993 Watnick, 2000; Donlan, 2002). They can be microscopically observed doing so in essentially all natural and engineered water systems. These grazers can quickly decimate bacterial biofilms and when they run out of biofilm, they form cysts to wait for biofilm redevelopment (Murga, 2001). *Legionellae* present a special problem because instead of being consumed along with other bacteria in the biofilm, these parasites can grow within and then overwhelm their protozoan hosts.

Legionellae infect protists by corrupting the phagocytic process. In the early 1980s it was observed that they can enter the host by coiling phagocytosis and this has been confirmed (Bozue, 1996). *Legionella* end up in phagosomes after entering the host but the parasite inhibits early fusion with the lysosome mostly because of factors derived from the *dot/icm* gene complex. Bacterial lysis is thus avoided. Instead, *Legionella* causes the phagosome to associate with endoplasmic reticulum, exploit autophagy, and combine with lysosomes from the endocytic pathway. The pathogen, now in its ultrastructurally distinct replicative form, then prolifically multiplies within these acidified vesicles. After only a few hours, thousands of

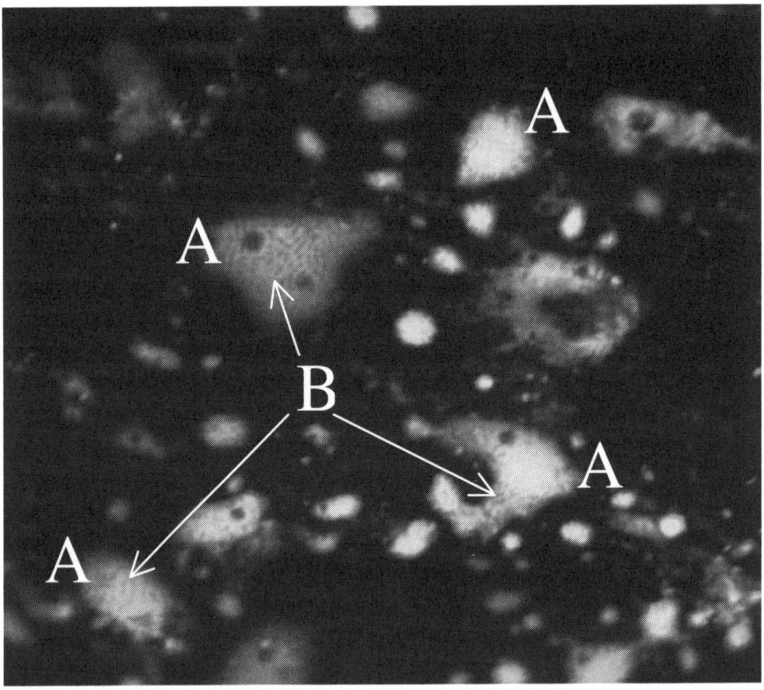

Figure 5.2 Amoebae (A) on biofilms developed during experiments to simulate real-world conditions in domestic water service systems (Loret, 2003). Viable bacteria (B) can be seen inside infected amoebae. The microscope used was a Zeiss LSM 510 (laser scanning microscope) with water immersion lens. Specimens were stained with the BacLight bacterial viability kit (Molecular Probes, USA). The size of the micrograph is $80 \times 80\ \mu$m. Micrograph by V. Thomas and Y. Lévi, Université Paris Sud, T. Bouchez, CEMAGREF Antony, from experiments run by J.F. Loret and S. Robert, of CIRSEE, SUEZ Environment, Paris, France.

lively motile bacteria can be observed, with microscopy, to have arisen as a result of intracellular growth within the host protozoan. They swarm around voraciously until within as little as 48h, the host cell becomes completely overwhelmed by the infection (Figure 5.2).

During these catastrophic biological events that ultimately lead to cell death, protozoa appear to attempt a clearing of the infection by expelling pathogen-laden vesicles from the cytoplasm. These small vesicles (about 5 μm diameter), are surrounded by membrane and are thought to be potentially infective in mammals since they can contain hundreds of virulent *Legionella* cells. They certainly appear to be relevant because transmission of the pathogen by this vehicle in the environment can occur (Berk, 1998).

An important aspect of this host-parasite association is that *Legionella* are known to survive inside protozoan cysts. This undoubtedly results in resistance of the pathogen to heat, drying, and antimicrobial chemicals (Steinert, 1998; Fields, 1996; Kilvington, 1990).

The ecological result of protist cell death is to release into the aquatic environment many hundreds or thousands of motile virulent bacteria in search of new hosts to infect. In low nutrient aquatic environments, *Legionella* loses the ability to replicate but not the ability to infect new hosts. After 125 days in sterilized tap water, *L. pneumophila* JR32 cells were still *v*iable *b*ut were *n*on*c*ulturable (VBNC) on standard growth medium; these survivors were resuscitated when *Acanthamoeba castellanii* was added to the microcosm and were still infective (Steinert 1997). In contrast to this, infection of mammalian host cells appears to be incidental (or accidental); the virulent progeny of these infections ultimately can benefit only in an ecologically insignificant way, as person-to-person transmission has never been observed.

Legionella was the first bacterial pathogen shown to grow inside common free-living amoebae (*Acanthamoeba* and *Naeglaria*) from aquatic environments (Rowbotham, 1980). Today, many potentially pathogenic intracellular bacterial parasites of amoeba are associated with human diseases (Table 5.1). This list suggests that abatement of bacterial pathogen hosts (the protozoa) from engineered water systems is fundamental to reducing public health risk.

A further motive to abate pathogen hosts in water systems comes from evidence that at least one pathogenic yeast-like fungus (*Cryptococcus neoformans*) can cause meningitis in immunocompromised patients, with increased virulence as a result of interactions with aquatic amoeba (Steenbergen, 2001). Still another potentially important finding is that *Vibrio cholerae* (the causative agent of cholera) survived its association with fresh water amoeba and proliferated in laboratory microcosms much more effectively in the presence of amoeba and could be re-isolated from encysting *Naegleria grubei* (Thom, 1992). In fact, coliforms and other bacterial pathogens survived within protozoa during chlorination; chlorine efficacy against bacteria significantly decreased for protozoan-ingested *Escherichia, Citrobacter, Enterobacter, Klebsiella, Sallmonella, Shigella, Legionella, and Campylobacter*; they all survived ingestion by protozoa; free chlorine was at least 50 × less effective against bacterial pathogens ingested by the ciliated protozoan *Tetrahymena*; bacterial resistance to digestion by predatory protozoans was proposed as a precursor of pathogenicity and it may be a mechanism for survival in aquatic environments (King, 1988).

Due to the similarities between phagocytic protists and human phagocytes, another motive to control exposure of susceptible humans to freshwater amoeba from engineered water systems is to protect against the emergence of new communicable diseases. Even after many hundreds of thousands of legionellosis cases worldwide since emergence of the disease in the twentieth century, there has never been a single documented case of person-to-person transmission of legionellosis (no secondary cases). This indicates that the pathogen has not yet adapted to sustainable life within the human respiratory system and confirms that the natural host for the parasite could not be alveolar macrophage. Eventually however, after enough exposure to susceptible human immune systems, *Legionella* or other parasites of freshwater amoeba (see Table 5.1) could sufficiently adapt to achieve communicable transmission such as occurs person-to-person by *Mycobacterium tuberculosis*, the causative agent of tuberculosis.

Table 5.1 Pathogenic bacterial parasites of *Acanthamoeba* and other aquatic protozoa

Bacterial parasites	Human diseases	References
Afipia spp.	Unknown. Possible link to cat scratch disease (*A. felis*), septic arthritis, bone marrow infection (*A. broomeae*), bone infection (*A. clevelandensis*)	La Scola, 2000
Several novel spp. in the *Rickettsiales*	Endosymbionts of amoeba related to *Rickettsia* that cause infections transmitted by ticks and other insects	Fritsche, 1999
Burkholderia psuedomallei and other spp.	Melioidosis (infections that can lead to pulmonary, bloodstream, and/or organ disease)	Inglis, 2000
Candidatus Odyssella thessalonicensis	Unknown. A newly described genus and species of obligate intracellular *Acanthamoeba* parasite	Birtles, 2000
Chlamydia pneumoniae	Community-acquired pneumonia and/or bronchitis; possible association with atherosclerotic vascular disease and with asthma. Causative agent in psittacosis.	Essig, 1997
Coxiella burnetti	Q fever. Community-acquired pneumonia	Maurin, 1999
Ehrlichia-like organisms	Unknown. Possible link to ehrlichosis (bacterial infections transmitted by ticks)	Michel, 1995
Francisella tularensis	Tularaemia produces symptoms that can be confused with typhoid fever or other *Salmonella* infections, brucellosis, malaria, tuberculosis, Q-fever, psittacosis, or legionellosis.	Abd, 2003
Helicobacter pylori	Peptic ulcer disease (90% of all ulcers)	Winieck, 2002
Legionella spp.	Legionellosis (Legionnaires' disease)	Rowbotham, 1980 Barker, 1992
Legionella-like amoebic parasites (LLAP)	Unknown. Potentially causes community-acquired pneumonia	Adeleke, 1996
Listeria monocytogenes	Listeriosis (food poisoning)	Ly, 1990
Mycobacterium spp.	Bacteremia, multi-organ bacterial disease, pulmonary infections, especially in HIV/AIDS patients	Steinert, 2001
Neochlamydia harmannellae	Unknown. Potential *Chlamydia*-like intracellular pathogen	Horn, 2000
Parachlamydia acanthamoeba	Possible community-acquired pneumonia	Greub, 2002
Simkania negevensis (in the bacterial order *Chlamydiales*)	Associated with bronchiolitis in infants and pneumonia in adults. *Chlamydia*-like intracellular pathogen	Kahane, 2001

Free-living amoebae appear to proliferate readily in domestic water such as hot water systems in hospitals (Rohr, 1998). Amoebae were observed in 52% (29 out of 56) of samples from six hospital hot water systems and 47% (23 out of 49) of samples from moist areas in the same hospitals. *Hartmannella*, *Echinamoeba*, *Saccamoebae*, *Vahlkampfia*, (in hot water) and *Acanthamoeba*, *Naegleria, Vanella* (in moist areas) were identified in this study. Further to their role as pathogen hosts, *Acanthamoeba* themselves are causative agents in potentially serious diseases such as granulomatous amebic encephalitis, amebic keratitis (infections of the eye), and sinusitis (Marciano-Cabral, 2003).

5.2 LEGIONELLOSIS

Legionellosis is the condition of being infected by *Legionellae*. It can result in the pneumonia referred to as Legionnaires' disease and/or the flu-like symptoms known as Pontiac Fever.

5.2.1 Etiology and epidemiology

Legionellae are common in freshwater aquatic environments and prevalent in engineered water systems. Survey data of such systems indicates a close association of the opportunistic pathogen with humans in habitation (Table 5.2).

Legionellae have been shown to survive all phases of wastewater treatment and they apparently survive after discharge into the ocean (Palmer, 1993, 1995). One species (*L. longbeachae*) is associated with potting soil mixes, aggressively infects *Tetrahymena* (a ciliated protozoan), and is known to have caused legionellosis in Australia and in Japan (Steele, 1996; Broadbent, 2002).

It is now clear that virulence of pathogenic *Legionellae* is enhanced by proliferation inside their protozoan hosts (Abu Kwaik, 1998). Thus the natural host–parasite relationship appears to be directly causative in the disease. Person-to-person transmission of the pathogen is not known to occur and there is only sketchy evidence of causes involving exposure other than from deeply aspirated aerosols contaminated with virulent *Legionella*.

Data are now available to describe the extent of legionellosis disease. Surveillance data from Europe (Joseph, 2002) and the USA (CDC, 2003) indicate that on average the incidence rate of reported cases is at least six cases per million population. Legionellosis is a notifiable disease in Australia. In New South Wales, Australia there were seven outbreaks of legionellosis between 1991 and 2000 but most cases (>90%) were sporadic (not associated with an outbreak); the reported case rate was 9.3 per million population and the mortality rate was 10% (McAnulty, 2001).

However, severe underreporting of legionellosis is acknowledged. In countries with the best surveillance and mandated reporting requirements (for example, Denmark, Netherlands, Belgium), incidence rates were about 15 cases per million population in 1999 (Joseph, 2002). There is very poor surveillance and/or no

Table 5.2 Occurrence of *Legionella* in engineered water systems

Engineered water system	% +	Location/Notes	Reference
Residential water supply			
Drinking water	3–33	USA	Rusin, 1997
Groundwater	83	24 samples from 12 wells	Riffard, 2001
Potable water taps (domestic and institutional)	61	96% were <1000 CFU/ml	Atlas, 1999
Single-family residences	6	Canada	Marrie, 1994
Multi-family residences	25	Canada	Marrie, 1994
Water heaters	21–79	Europe	Tiefenbrunner, 1993
Showers, spas, and fountains			
Spa pools	2	Singapore	Heng, 1997
Decorative Fountains	15–19	Singapore	Heng, 1997
Showers	8–92	Europe	Tiefenbrunner, 1993
Warm or hot water systems			
Cooling towers	51	USA	Miller, 1993
Cooling towers	36	Singapore	Heng, 1997
Cooling towers	47	Finland	Kusnetsov, 1997
Cooling towers	90	Detection by PCR	Koide, 1993
Hospitals	47	France (Paris)	Nahapetian, 1991
Hospitals	68	SE Germany	Lück, 1993
Hospitals	70	Saxony	Habicht, 1988
Outpatient clinics	50	SE Germany	Lück, 1993
Dental offices	58	SE Germany	Lück, 1993
Public buildings	85	SE Germany	Lück, 1993
Hotels	18	Saxony	Habicht, 1988
Private households	65	Mostly <100 CFU/ml	Lück 1993
Apartment buildings	30	Mostly in showers	Zacheus, 1994

reporting requirement in many countries and in certain states of the United States. There may be more than 8000 cases (>30 cases per million population) each year in the United States but extrapolations calculated from incidence rates in one state indicate that there may be up to 18,000 cases per year (Marston, 1994; Marston, 1997). Thus, it is not overstated to say that worldwide, hundreds of thousands of people are affected by the disease each year.

The demographics of legionellosis are now well characterised. For example, during 2001 in the United States, 87% of all reported cases were age >40 years; men were nearly twice as likely as women to be afflicted; children under 15 years rarely (less than 1% of reported cases) contract legionellosis (CDC, 2003). These demographics are very similar in Australia (McAnulty, 2001) and in Europe. Many factors contribute to the susceptibility of disease including age, gender, immunosuppression, smoking, alcoholism, and autoimmune disease. Most legionellosis is associated with domestic hot water service and industrial water systems such as evaporative cooling towers. A significant travel-associated case rate is apparent in the data. Most reported legionellosis is sporadic community-acquired pneumonia with less than about 5% reported from outbreaks of the disease (Marston, 1994).

The incidence rate of community-aquired pneumonia (CAP) requiring hospitalization in two Ohio counties in 1991 was 2668 per million population with an overall case-fatality rate of 8.8%. These results were used in a model to extrapolate that there are about 485,000 cases of CAP requiring hospitalization annually in the United States (Marston, 1997). About 15% of all CAP cases in the United States are caused by the 'atypical pathogens' (*Legionella*, *Chlamydia, Coxiella* and *Mycoplasma*) whereas 85% of CAP is caused by the 'typical bacterial pathogens' (*Steptococcus*, *Haemophilus* and *Moraxella*); in the United States, the causative agent of CAP is not determined in about 30% of cases (Cunha, 2003). Legionellosis may cause 4.1%–20.1% of community-acquired pneumonia (CAP) cases requiring hospitalization and most legionellosis is associated with building water systems (reported in Rusin, 1997).

Pneumonia is the leading cause of death due to infectious diseases in the United States (Marston, 1997). There are more than 4 million episodes of community-acquired pneumonia every year in the United States but only about 10% require hospitalization; the estimated total cost of treating community-acquired pneumonia in the United States is at least $10 billion dollars (Lave, 1999). But the economic burden is far greater than this. For every $1 spent on pneumonia healthcare costs, employers spent another $12 on costs related to the workers' pneumonia; in fact, the annual total cost to employers for workers who contract pneumonia is about 5 times higher than for other workers (Birnbaum, 2001).

Although most legionellosis is sporadic and is associated with building water in close proximity to susceptible populations, explosive rare outbreaks of legionellosis result in far more public notice and governmental regulatory activity. The most spectacular outbreak of legionellosis occurred in July 2001 at Murcia, Spain (García-Fulgueiras, 2003). The source of the outbreak was proven to be from improperly controlled contaminated hospital cooling towers in the north of the city; there were 800 suspected cases, 449 confirmed. Hospitalization was necessary for 64% of the cases. There were six deaths directly related to this outbreak (less than 2% of confirmed cases). The remarkably low fatality rate was due to quick detection of the outbreak, early diagnosis of the disease, and appropriate treatment of patients. The explosive qualities of the outbreak prompted clinicians to diagnosis accurately and initiate treatments immediately. This outbreak precipitated the first regulatory document with specific legal status in Spain (Spain 2001; Table 5.4) and it appeared just 20 days after onset of the outbreak.

Previously, in Victoria, Australia an outbreak also resulted in quick publication of government-approved code of practice supported by enacted legislated penalties for non-compliance after 93 confirmed cases and 3 deaths occurred in April 2000. In this outbreak, hundreds of people were exposed to *Legionella*-laden mist from an improperly treated cooling tower at the newly opened Melbourne Aquarium.

Prior to the Murcia, Spain outbreak the largest and most significant was that which occurred in The Netherlands at the Westfriese floral show in February 1999. One hundred eighty-eight people who visited the fair were afflicted with legionellosis pneumonia (133 confirmed cases, 55 probable), 163 (87%) were hospitalized, 24 (21%) required mechanical ventilation and 21 (11%) people died. The source of the outbreak was a contaminated whirlpool spa on exhibition at the flower show; another spa and a sprinkler system were also shown to be contaminated but in a careful epidemiological study, the probable source of the outbreak was the new spa (Den Boer, 2002).

Although relatively rare, outbreaks of legionellosis have been an important factor in public awareness from the beginning. Indeed, the Stafford, UK outbreak of 1985 lead to the first government-approved code of practice in 1987. And legionellosis pneumonia bears the name Legionnaires' disease as a result of the first known outbreak that occurred in 1976 at the American Legion convention held in the Bellevue-Stratford hotel of Philadelphia; in this famous outbreak, 221 persons were afflicted and 34 (15%) people died.

The World Health Organization classifies legionellosis as a serious and preventable disease (WHO, *Legionella* and the Prevention of Legionellosis, *in preparation* to publish in 2004, www.who.int/water_sanitation_health). It is preventable because proliferation and transmission of the pathogen is the result of correctable conditions in engineered water systems that are in close proximity to susceptible populations. Legionellosis disease is serious because it can result in severe acute pneumonia and, even in patients who recover, it can cause long-term debilitating health effects.

5.2.2 Pathogenicity

Legionellosis is a respiratory disease. Aspiration of infective particles into the alveoli can lead to mild flu-like symptoms or severe pulmonary disease. Aerosol droplet size is key in pathogenicity because particles larger than about 8 μm diameter are excluded from alveoli by the terminal bronchiole. In severe disease, damage to the lungs occurs as the result of bacterial growth inside alveolar macrophage (white blood cells) and type 1 alveolar epithelium. Figure 5.3 provides an anatomical sketch of the primary infection site for *Legionella*.

The alveolar epithelium consists of two cell types, the alveolar type I and alveolar type II cells. Although both cells are found in approximately equal numbers, type II cells are typically found in the corners of the alveoli, whereas type I cells are flattened and cover 90% of the alveolar surface area. Type II epithelium has been associated with much of the physiology of lung epithelium (e.g., surfactant secretion, salt movement, and progenitors for new epithelium). Type I cells facilitate gas exchange between air from the atmosphere and blood cells in capillaries. They also provide a protective barrier to blood capillaries and are endocytic. However, the primary protective cells in the alveolar sac, the alveolar macrophage (white blood cells), are phagocytic. These cells and type I alveolar epithelium are susceptible to infection and intracellular growth by the *Legionellae* (Abu Kwaik, 1998).

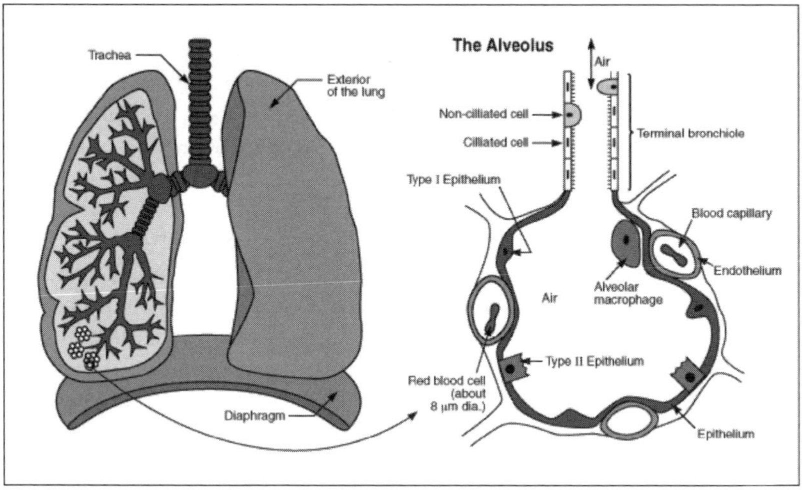

Figure 5.3 Legionellosis pathogenicity depends on deep aspiration of infective particles through the trachea, the bronchial passages, the terminal bronchiole and finally into alveoli, the site of gas exchange between air and blood during respiration. The delicate structure of the alveolus is protected from particles larger than about 8 μm in diameter by ciliated cells in the terminal bronchiole. Particles less than this diameter may enter into the alveolar sac. In healthy individuals, phagocytic cells in the alveolus eliminate micro-organisms and viruses that make it through all of the other defences in the lung. Alveolar macrophage and type I epithelium are susceptible to infections by *Legionellae*. (Adapted from Mochitate and Furnyama, Env. Health Sci. Div., Natl. Inst. for Environ. Studies, Japan).

Infectivity factors in *Legionellae* that lead to disease appear to be associated mostly with *dot/icm* genes that are involved in assemblage of the type IV protein secretion system; these 24 genes are central to pathogenesis. Biphasic killing of mammalian cells begins with apoptosis (programmed cell death). Induction of necrosis and lysis of the host then ensues. A protein, IcmT, specified by a gene in the *dot/icm* gene complex is essential for pore-forming activity and release of bacteria from the infected host cell. Pore-forming activity is important in pulmonary cytotoxicity and inflammation (Molmeret, 2002; Alli, 2003).

An interesting hypothesis is that phospholipase activity from the type II protein secretion in *Legionella pneumophila* (see Cell Biology of *Legionella*) might cause damage to alveolar epithelial cells during infections of the lung by *Legionella*. Phospholipase from *L. pneumophila* degrades phosphatidylglycerol and phospatidylcholine, the major constituents of lung surfactant, to the haemolytic agent lysophosphatidylcholine, which can induce acute respiratory distress syndrome and elicit inflammation (Flieger, 2002).

Free-living amoebae appear to be very important in legionellosis disease. Using laboratory animals (A/J and BALB/c mice) to model legionellosis disease, inhaled *Legionella*-infected *Hartmannella vermiformis* were significantly infective particles (Brieland, 1996, 1997a). Inhaled amoebae enhanced intrapulmonary growth of *Legionella* by providing a niche for bacterial replication (Brieland, 1997b). Two *Acanthamoeba* species copiously expelled 2–6 μm diameter membrane-bound protozoan vesicles laden with virulent bacteria; the amoebae remained viable, they produced a continuous source of *Legionellae*-filled vesicles by shedding dozens of potentially infective particles per amoeba in a 24-h period depending on conditions (Berk, 1998).

The infective dose of *Legionella* is not known and of course, it varies depending on the susceptibility of the host. For this reason, it is not possible to set a scientifically defensible "safe" concentration of *Legionella*. Nor is it therefore, scientifically rational to set action (or 'trigger') levels based upon detected concentrations of *Legionella* in environmental samples (further discussion in section 5.3.3.3).

In the early days of legionellosis diagnosis and treatment, mortality rates were very high approaching 20% and sometimes greater than 50% in immunosuppressed patients. Treatments with macrolide or quinolone antimicrobial agents that concentrate within macrophage cellular vesicles such as lysosomes have resulted in much lower mortality rates (Edelstein, 2002). But it has now become clear that the severity of the disease can cause irreparable damage to the lung. Figure 5.4 shows chest X-ray analyses of a legionellosis patient during clinical presentation of symptoms and then after recovery. These radiographic images reveal significant and permanent damage to the lungs (Bartlett, 1986).

5.2.3 Clinical diagnosis and therapeutics

Legionellosis may present in a variety of ways indicative of atypical community-acquired pneumonia (CAP) and produce clinical symptoms including anorexia, vomiting, non-productive cough, diarrhoea, myalgia (muscle pain), headache, rising fever and chills, bradycardia (slowing of heart rate) and/or confusion/delerium. Radiographs are non-specific; most infiltrates are alveolar not interstitial. Blood work abnormalities include elevated liver enzymes, hyponatraemia (low sodium) and hypophosphotaemia (low phosphate). In clinic, extrapulmonary involvement (such as elevated liver enzymes) is characteristic for atypical CAP making it clinically distinct from typical CAP.

Bradycardia (slowing of the heart rate) is a cardinal feature of Legionnaires' disease; relative bradycardia clinically distinguishes legionellosis from atypical CAP caused by *Mycoplasma pneumoniae* (Cunha, 2003). However, bradycardia does occur in atypical CAP caused by *Coxiella burnetti* and *Chlamydia pneumoniae*. It is interesting to note that *Coxiella* is phylogenetically closely related to *Legionella* and both *Coxiella burnetti* and *Chlamydia pneumoniae* are now known to be natural parasites of freshwater amoeba such as *Acanthamoeba* (see Table 5.1, Maruin, 1999 and Essig, 1997).

There are several methods for legionellosis diagnosis. Isolation of bacterial cultures from lung sputum samples is the preferred method because speciation and typing of the isolate can give important information (section 5.3.3.1). Urinary antigen testing has become widely used in the clinic even though it is not very

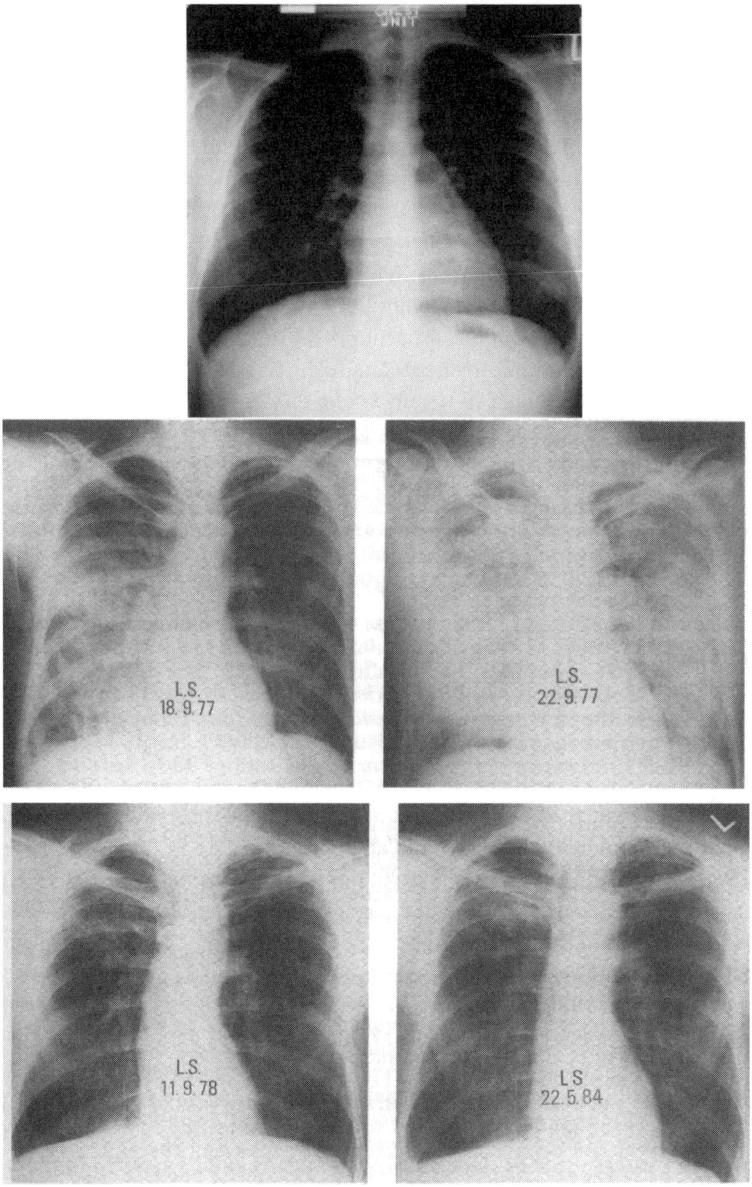

Figure 5.4 Chest X-rays of a normal healthy individual (top) and a legionellosis patient on admission (middle left), day 4 (middle right), one year post-infection (bottom left) and 7 years post-infection (bottom right). Long-term effects can be seen in the post-infection radiographic images (radiographic images were originally published in Bartlett (1986)).

sensitive to *Legionellae* other than *L. pneumophila* serogroup 1 (Benson, 2000, Hayden, 2001). Polymerase chain reaction (PCR) detection and other molecular methods show promise for clinical diagnosis of legionellosis (section 5.3.3.2).

There is no international case definition for legionellosis. The recommended presumptive case is presence of specific *Legionella* antigen in respiratory secretion or urine, direct fluorescent antibody staining of the organism in respiratory secretions or lung tissue, or at least a 4× rise in specific serum antibody titre to *Legionella* species. The recommended confirmed case is (one or more of the following) isolation of *Legionella* from respiratory secretions, lung tissue, pleural fluid, blood or other normally sterile fluids, at least 4× rise in specific serum antibody to *Legionella pneumophila* serogroup 1 (WHO 1997). In New South Wales, Australia, the case definition for Legionnaires' disease is a person with the signs and symptoms of pneumonia in whom *Legionella* has been isolated; or who has a positive *Legionella* antigen in their urine or respiratory secretions; or who has a four-fold or greater rise in antibody titre between acute and convalescent sera (McAnulty, 2001).

Several antibiotics are effective *in vivo* against *Legionella* especially those that concentrate intracellularly within lysosomes. The macrolide/azalide and quinolone drugs have the greatest efficacy. Erythromycin, the first macrolide drug ever commercialized (1950s), is very effective in treatment of legionellosis. Two newer macrolide antimicrobials, azithromycin and clarithromycin, are now prefered. Quinolone drugs (especially ciprofloxacin and levofloxacin) are just as effective (Dedicoat 1999). Empiric therapy for CAP (when the causative agent is not known) should be directed at typical and atypical pathogens: A respiratory quinolone or a cephalosporin (such as ceftriaxone) plus either a macrolide or a ketolide provides optimal coverage; doxycycline is especially useful because it covers the pathogens of both typical and atypical CAP (Cuhna, 2003).

Several treatment options for legionellosis are given in Table 5.3 (the table is adapted from Edelstein, 2002; Dedicoat, 1999; Stout, 1997, and Srinivasan at www.hopkins-heic.org/infectious_diseases/legionella.htm; the dosages given are for patients with normal renal and hepatic function).

The newer macrolides and quinolones are considered drugs of choice. Tetracycline and doxycyline antibiotics have a broad spectrum of activity, are relatively safe, and can be recommended; they may even have antiprotozoal activity. Rifampicin is very active against extracellular and intracellular *Legionella* and is synergistic (*in vitro*) with erythromycin (Dedicoat, 1999). Typical duration of recommended treatments is 14 days in normal patients and 21 days in immunosuppressed patients.

In contrast, the β-lactam drugs (e.g., penicillin and ampicillin) do not penetrate the intracelluar compartments (phagosomes and lysosomes) in which *Legionella* grow and this may be why they are ineffective in treatment of legionellosis disease (Edelstein, 2002). Further to resistance, many *Legionella* species (e.g., *L. pneumophila*) but not all species (e.g., *L. micdadei*) produce β-lactamase which inactivates penicillin (Marre, 1982). Although aminoglycoside antimicrobials

Table 5.3 Preferred treatments for legionellosis (adapted from Edelstein, 2002; Dedicoat, 1999; Stout, 1997; Srinivasan at www.hopkins-heic.org/infectious_diseases/legionella.htm)

Antibiotic	Dosage	Notes
Azithromycin	1000 mg once followed by 500 mg 2×/day	5–10 days of treatment appears sufficient, IV or orally
Ciprofloxacin	400 mg IV or 750 mg orally 2×/day	7–10 days
Levofloxacin	500 mg 1×/day	Newer quinolones (gatifloxacin, moxifloxacin, etc.) also appear effective *in vitro*. IV or orally
Clarithromycin	500 mg 2×/day	IV formulation not available in the United States.
Erythromycin	1000 mg IV 4×/day or 500 mg orally 4×/day	Less often prescribed due to significant GI and ototoxicity (ear)
Doxycycline	100 mg 2×/day	200 mg load is recommended
Tetracycline	500 mg 4×/day	IV or orally
Rifampin	300–600 mg 2×/day	Should only be used in combination with macrolide or quinolone in severely ill patients.

penetrate lysosomal compartments, accumulation of the drug is very slow which is perhaps why they are ineffective in treatment of legionellosis (Butler, 1997).

In a retrospective review of the first recognised outbreak in Philadelphia in 1976, patients treated with erythromycin or tetracycline had a 50% lower mortality rate compared with patients treated with β-lactam drugs (Fraser, 1977).

5.3 REDUCING THE RISK OF LEGIONELLOSIS

Since legionellosis is a preventable disease, it is by definition therefore possible to substantially control it. Since legionellosis is a serious disease, there is significant motive to do so. Most legionellosis results from exposure of susceptible people to engineered water systems. Twenty-seven years after the first outbreak in 1976 there is now ample evidence that prevention and control of legionellosis is both practical and effective.

For instance, prevention and control measures performed during two long-term studies at hospitals were very successful (WHO, 2002); the frequency of *L. pneumophila* isolation from patients with pneumonia was reduced 99.4% (Grosserode, 1993) and in immunocompromised patients it was reduced 98.9% (Junge-Mathys, 1994). The activities employed in these hospitals included disinfection of the domestic water service system (the plumbing), pathogen monitoring of the water, clinical monitoring for signs of infection, and education of staff regarding the risk of legionellosis. Documented success in a United States healthcare facility was notable when mechanical, operational, and chemical improvements were applied in a multifaceted-engineering approach to reduce the risk of legionellosis (Neumer, 2002).

5.3.1 Global Code of Practice

There is now an extensive body of published guidance on the subject of legionellosis prevention and control. Taken *en masse*, these documents comprise a performance standard, i.e., a code of practice. Codes of practice (also called guidelines, standards, or best practices) are the collaborative work of government, academic, and industrial experts. These guidance documents define a standard of care against which the owner of an engineered water system could be judged in an accusation of negligence. Thus, guidance publications are extremely important even in countries with no specific regulatory penalty or special legal status to enforce an approved code of practice (Table 5.4).

Codes of practice are not laws nor are they regulations. Rather, they give advice on how to comply with laws and regulations. For example, the best-known advice and guidance comes from the UK Health and Safety Commission (HSC 2000) in their *Approved Code of Practice and Guidance* (often referred to in the UK as "L8"). It is advice on how to comply with the *Control of Substances Hazardous to Health Regulations* (1999, regulation 6–9 and 12) and the *UK Health and Safety at Work Act* (1974, sections 2, 3, 4 and 6). 'L8' is the result of guidance revisions (EH48 → HS(G)70 → L8) beginning in 1987 following the Stafford, UK outbreak of 1985.

Analogously in the United States, the *OSHA Technical Manual on Legionnaires' Disease* is widely cited. It gives guidance in regard to the legal responsibility of employers to protect the health and safety of employees, customers, and the surrounding community according to the *Occupational Health and Safety Act* and the *Code of Federal Regulations* (CFR Title 40 §142.52) in the United States (OSHA, 1998, CFR, 1998). Complying with this guidance can mitigate against an accusation of negligence in the case of a legionellosis incident or outbreak (an outbreak is defined as two or more occurrences in the same locale within a brief period of time).

Taking further example from the approved code of practice in the UK, the document is in two parts: The actual code of practice and, in the second, specific guidance on how to comply with the code of practice. It is in this second part, published in November of 2000, that advice on pathogen monitoring and action levels appear.

The code of practice in Victoria, Australia (DHS, 2002) is crafted for emphasis on developing a site-specific risk management plan. This guidance was the rapid direct result of a legionellosis outbreak in which several government officials, including the Australian Finance Minister, were confirmed as having contracted legionellosis during a reception at the Melbourne Aquarium in April 2000. Hundreds of other notable political figures including the Prime Minister were put at risk and

Table 5.4 Codes of practice for reducing the risk of legionellosis

Place	Year	Title	Reference
Victoria, Australia	2002	Code of practice for water treatment service providers.	DHS 2001, 2002
Singapore	2001	Code of practice for the control of legionella bacteria.	Singapore 2001
France	2001	Sanitary and hygiene criteria for preventing and controlling legionellosis	France 2001
Spain	2001	Guide to good practice: *Legionella* in cooling towers	Spain 2001
USA	2001	Environmental infection control in healthcare facilities	CDC 2001
UK	2000	Legionnaires' Disease: The control of legionella bacteria in water systems, approved code of practice and guidance (often referred to as 'L8')	HSC 2000
Australia/ NewZealand	2000	AU/NZ standard: Performance-based maintenance of cooling water systems.	Australia 2000
Netherlands	2000	The temporary mains water (*Legionella* prevention) regulation	Netherlands 2000
USA	2000	Guideline: Minimizing the risk of legionellosis associated with building water systems	ASHRAE 2000
USA	1998	Technical manual, Ch 7: Legionnaires' disease	OSHA1998
Germany	1997	Technical improvements for the reduction of the growth of *Legionella*	DIN, DVGW 1996, 1997
Australia/ NewZealand	1995	AU/NZ standard: operation and maintenance; design, installation and commissioning	Australia 1995a,b
UK	1991 rev. 1995	The control of legionellosis, approved code of practice	HS(G) 70 HSC 1995
UK	1987	Legionnaires' disease	EH48, 1987

the opposition leader underwent tests for the bacteria after experiencing symptoms. A total of 93 people were confirmed to have contracted the disease, seven were hospitalised in critical condition, and three people died.

Today in Victoria, owners of land upon which a cooling tower system is situated are required to: (1) register the system, (2) develop, implement and review a risk management plan, (3) implement an improved maintenance program and (4) keep records of maintenance and test results for audit/inspection purposes. Specific legislation to enforce this code of practice was established by the government.

The maximum penalty for failing to complete a risk management plan in Victoria, Australia is AU$6000 per cooling water system. The maximum penalty for failing to register a cooling tower system is AU$12,000 (DHS, 2001a, DHS, 2001b, DHS, 2002).

Taking regulatory requirements in the Netherlands as another example (Netherlands, 2000), owners of collective water installations, e.g., in hospitals, hotels, campsites and swimming pools, are obliged to conduct a risk analysis and if necessary take measures to prevent health risks posed by *Legionellae*. Actions to correct deficiencies and reduce risk must be specified in a control (risk management) plan.

In the United States, guidance has been published by the Occupational Health & Safety Administration (OSHA, 1998), the American Association of Heating, Refrigerating, and Air-Conditioning Engineers (ASHRAE, 2000), the Cooling Technology Institute (CTI, 2000), the Centers for Disease Control and Prevention (CDC, 2001), the American Society of Testing and Materials (ASTM, 2002), and others. This guidance taken together establishes a performance standard in the United States to which cooling water users should comply in order to protect their employees, customers, and the community from legionellosis. Compliance to this guidance mitigates against an accusation of negligence. In a litigious society, motive of this kind to comply with a code of practice can be extremely compelling.

Most legionellosis is associated with domestic water (potable water) service (Cordes, 1981; Shands, 1985; Schlech, 1990; Lee, 2002; WHO, 2002). However, industrial water systems are also significantly associated with legionellosis and must be properly managed to reduce public health risk (Dondero, 1980; Cordes, 1980; Garbe, 1985; Miller, 1993; Keller, 1996). The goal of industrial water treatment is to ensure that equipment runs effectively and economically. If this goal is expanded to include reducing the risk of legionellosis, further practice (beyond conventional water treatment) is necessary (WHO, 2004, in preparation).

Common elements of good practice become apparent in these standards. These common elements can be thought of as technical consensus opinions on how to reduce the risk of legionellosis (Table 5.5).

Table 5.5 Common elements in codes of practice for reducing the risk of legionellosis

Practice	Purpose
Risk assessment	Assess risk, identify improvement needs, assign communication and management responsibilities for implementing a risk management plan
Pathogen analytical	Verify control measures and identify potential risk factors
Cleaning and disinfection	Remedial in cases of incidence or outbreak; preventative for long-term risk reduction
Monitoring and control	Long-term maintenance of low risk in engineered water systems

Figure 5.5 Categories of practice to reduce the risk of legionellosis are linked. Many published codes of practice recommend starting with the risk assessment since it is the essential first step in developing a risk management plan. The other practices indicated should be performed to implement the risk management plan. Reassessment should be performed periodically to update the risk management plan (e.g., every 2 years or when significant changes to the system occur).

Further to Table 5.5, studying codes of practice *en masse* reveals that these categories of practice should be linked to a site-specific risk management plan as shown in Figure 5.5.

Risk assessment, pathogen analysis, cleaning with disinfection, and monitoring with control are helpful in reducing the risk of legionellosis when performed in a co-ordinated well-planned way. Performing these activities outside the context of a risk management plan is not advised.

5.3.2 Risk Assessment

Risk assessment is the essential first step in establishing a risk management plan (e.g., HSC, 2000; JCAHO, 2001; DHS, 2001a). Although for some hazards the risk assessment process can be quantitative (Percival, 2000), the following is required: Quantification of the hazard through dose–response assessments, determination of the extent and duration of human exposure and characterisation of the possible consequences resulting from exposure. Since knowing these is not currently possible for legionellosis, the risk assessment process is always qualitative.

No permissible exposure levels can be set for *Legionella* because: (1) the susceptibility of exposed persons varies greatly, (2) the mixture of biological contaminants is complex and varied and, (3) interactions among all biological agents present in the environment make exposure assessment very difficult.

Qualitative risk assessment is nevertheless the essential first step in developing a legionellosis risk management plan. The purpose of the risk assessment is to document site-specific conditions and recommendations for reducing the risk, if necessary, and to establish assignments for actions, communications, training, and management responsibilities. Risk assessment thus becomes the basis for formulating a risk management plan.

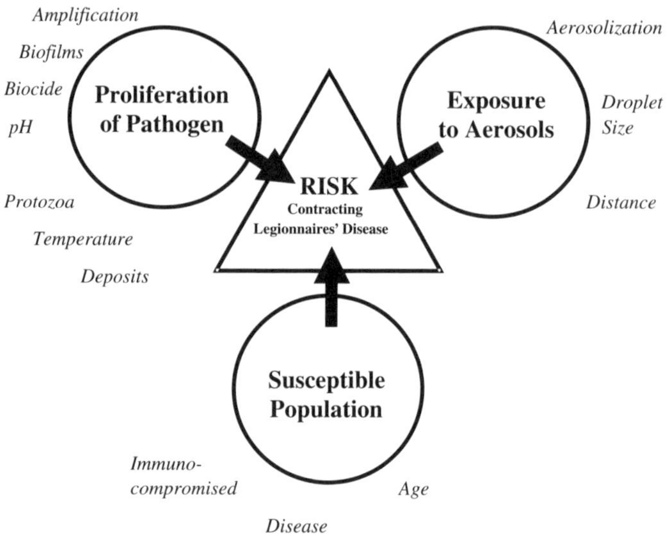

Figure 5.6 Model to qualitatively assess the risk of legionellosis. The risk of contracting legionellosis is the result of three potentials: Proliferation potential, exposure potential and susceptible population potential (McCoy, 2003a; Cooper, 2004).

Legionellosis risk can be described in terms of three factors: pathogen proliferation potential, exposure potential and population susceptibility (Figure 5.6).

Pathogen proliferation potential exists in nearly all water systems. Many factors are involved but most importantly are the presence of microbial biofilms, the degree of microbial diversity, and the availability of nutrients especially in warm water systems. Managing the microbial fouling process to reduce the risk of legionellosis principally consists of controlling biofilms and limiting microbial diversity within the entire system. Exposure to *Legionella* occurs by aspiration (deep inhalation into the lungs) of microscopic ($<8\ \mu$m diameter) aerosols. Certain industrial water systems produce aerosols as an inherent function of their operation. Minimising aerosol dissemination from such systems can directly reduce health-related risk. Assessment of exposure potential focuses on the proximity of aerosol-generating systems to human populations. The most susceptible individuals in any population are elderly males, heavy smokers, alcoholics, patients with chronic pulmonary disease, and immunosuppressed people (Arnow, 1982; Bollin, 1985; Addiss, 1989; Breiman, 1993; Bentham, 2002; McCoy, 2003b).

It can be difficult to document progress in reducing legionellosis risk and in comparing one system to another since the assessment is qualitative. Therefore, schemes have been designed to provide risk ranking and risk scores. Table 5.6 is adapted from a published code of practice and provides an example of one way to rank legionellosis risk (DHS, 2001a; DHS, 2001b; DHS, 2002).

An important reason to classify risk is as an aid in linking the assessment with specific recommendations for reducing risk. For example, if a system is ranked

Table 5.6 Legionellosis risk classification scheme (adapted from DHS, 2001a)

Critical risk	Higher risk →			Lower risk
Risk classification	**A**	**B**	**C**	**D**
Criteria	If the system matches any of the responses below	Matches any of the responses below and none in **A**	Matches any of the responses below and none in **A** or **B**	Matches any of the responses below and none in **A**, **B**, or **C**
Stagnant water	System is idle more than a month. Re-circulation pump not timed. Dead legs prevalent.	Same as **A** but with timed re-circulation pump	Any single factor in **A**	System operates continuously and there are no dead legs
Nutrients and growth	Contaminated water. No corrosion control. Wet surfaces open to sunlight.	Any two factors in **A**	Any one factor in **A**	No factors in **A**
Poor water quality	No automated biocide dosing protocol. No water treatment programme	No automated biocide dosing. Water treatment programme.	Automated biocide dosing. No water treatment programme.	Both automated biocide dosing and water treatment programme.
Poor equipment design or operation	No system design review. No system operation and performance review. No drift eliminators.	Drift eliminators for cooling towers in place.	Same as **B** with at least one review lacking.	Both reviews in place. Good drift eliminators.
Location and access	Located in a healthcare or residential care facility or large numbers of people are exposed.	Same as **A** but moderate numbers of people exposed.	Same as **B** but few people exposed.	Not located near susceptible people.

in the A category (higher risk), then more frequent service in pathogen analysis, cleaning with disinfection, and monitoring with control should be specified compared to a system that is assessed into a lower risk classification. Progress in reducing risk can be documented by reassessing the system after implementation of the risk management plan. When improvements result in a lower risk classification for the system, then less frequent service to maintain control should be required.

Another approach to dealing with the qualitative nature of legionellosis risk assessment is to use a scoring system (e.g., DHS, 2001b). These schemes can be helpful especially for water systems in which population susceptibility to legionellosis is always high (e.g., the domestic water supply in hospital). The key to making risk scores work is to establish pre-determined criteria for each critical risk factor assessed and then, to apply these criteria uniformly across all systems assessed. The best service providers are able to consistently and efficiently practice this risk assessment art globally.

Maintaining focus on the purpose of risk assessment is essential. Risk assessments should link identification of risks to recommendations for reducing risk by establishing management responsibilities and clear lines of communication. The risk assessment should be the basis for a site-specific risk management plan.

5.3.3 Pathogen analysis

A commonly misapplied activity in legionellosis control is analysis to determine pathogen concentrations in samples from engineered water systems. In the worst cases, pathogen analysis results are used wastefully, counter-productively and therefore, dangerously.

The essential problem statement is this: *Legionella* concentrations in water systems cannot been correlated to quantitative determinations of legionellosis risk. Negative *Legionella* test results from water systems are not necessarily equivalent to a low risk of contracting legionellosis. Positive *Legionella* test results cannot be used alone to determine the risk of legionellosis.

Most scientists agree that it is not currently possible to determine the risk of legionellosis from measurements of *Legionella* concentrations because (1) the infective dose of pathogen is not known, (2) the infective dose varies according to the susceptibility of the exposed population, (3) the test method requires 10 days and is therefore only a retrospective analysis of the pathogen concentration when sample analysis began, (4) effects due to sample transit *en route* to the lab are not controlled and (5) the risk of legionellosis also depends upon the risk of population exposure and inhalation of aerosols containing *Legionella*.

However, when results are properly interpreted within the context of a risk management plan, pathogen analytical services can be very helpful. Pathogen analytical services should be used to confirm that control measures have been effectively applied. Also, results can be used diagnostically to identify contaminated or infested areas in a water system. Confirmation that control measures are working is a necessary part of every risk management plan.

5.3.3.1 The standard culture method

The first isolations of *Legionellae* outside guinea pigs and from embryonated chicken eggs (McDade, 1977) were on Mueller-Hinton agar supplemented with haemoglobin and L-cysteine (Feeley, 1978). The method was refined and developed into the Feeley–Gorman medium. By this innovation, quantitative analysis of viable *Legionella* in all sorts of environments was made possible. Starch was replaced with charcoal to detoxify the medium and the amino acid source was changed to yeast extract. This led to use of the medium known as buffered charcoal yeast extract (BCYE) and is today the basis for all analytical culture methods.

Several refinements and improvements resulted in the currently preferred BCYE medium that is enriched with α-ketoglutarate (Edelstein BCYE-α medium) with and without selective antimicrobial agents and indicator dyes. In order to isolate certain species, this medium can be supplemented with bovine serum albumin (Fields, 2002).

The standard method ISO 11731 has been developed (ISO 1998) which specifies use of the BCYE-α medium supplemented with ammonia-free *g*lycine, *v*ancomycin, *p*olymyxin B, and *c*ycloheximide (GVPC). In addition to these supplements, GVPC contains ferric pyrophosphate, L-cysteine, α-ketoglutarate. This method is generally consistent with the original method developed by the Centers for Disease Control and Prevention (B. Fields, personal communication) and with standard methods used in Australia and Singapore (AU/NZ, 3896). A method that is substantially similar to these is used in France (AFNOR T90-431).

Selectivity steps are required to inhibit competition from faster growing bacteria that may overwhelm *Legionella* in the sample. Samples are split in three; one is plated with no treatment, one sample is heated (30 min at 50°C) and the third is acidified with HCl (pH 2.2 for 5 min). Concentration of the sample by centrifugation (6000*g* for 10 min) or membrane filtration is optional depending on detection limit requirements.

Standard culture methods specify pathogen confirmation steps that require comparative growth on BCYE-α agar and blood agar (or nutrient agar) not supplemented with iron and L-cysteine. Colonies that grow on GVPC but fail to grow on BCYE without L-cysteine and blood (or nutrient) agar are confirmed *Legionella spp*.

Serogroup typing is then performed on confirmed *Legionella* colonies by latex bead agglutination to differentiate serogroup 1 from serogroup 2–14 and from other serogroups. With the suitable range of antisera available, typing and subtyping to full species and sub species is possible.

Several factors limit the culture method. Analyst experience with the method directly correlates with pathogen recovery and higher concentrations. The method requires 10 days to yield confirmed results. Preparation of the medium is tricky and requires expensive quality control. The pathogen is sensitive to factors that are difficult to control during sample transit. Concentration steps used to achieve lower detection limits are notoriously inefficient and irreproducible (<50% of viable *Legionella* are recovered during sample concentration processing and variation among labs is high).

5.3.3.2 Molecular methods

Molecular methods are faster, less expensive, less subjective, more sensitive, and can be performed in the field. They all suffer a critical limitation: None of the commercially available molecular methods can differentiate viable *Legionella* among the background of nonviable bacteria in environmental samples (however, real-time polymerase chain reaction methods show promise). Thus, in practice only the culture method can detect the effect of disinfection and this is an essential requirement for a pathogen analysis method to confirm effective control in engineered water systems.

Molecular methods include nucleic acid detection using the polymerase chain reaction (PCR) or fluorescence *in situ* hybridization (FISH), and serologic methods by antigen/antibody reactions detected with enzyme linked immuno-specific assays (ELISA) or differential fluorescent antibody direct cell counting (Cloud, 2000; Fields, 2002; Campos, 2003). Most scientists agree that these detection systems are useful in the clinical laboratory for diagnosis and sero-grouping. In fact, dependence on the culture method in the clinical laboratory has been substantially reduced because the molecular methods are superior in so many ways. Advances in these methods have improved clinical diagnosis of legionellosis. Quantitative differentiation of viable *Legionella* is not a requirement in most clinical applications.

For environmental or industrial samples, nucleic acid or serological methods should be used only as a rapid screen to identify those samples free of any *Legionella*, whatsoever. This service can be helpful in some risk reduction situations. However unfortunately, this service does not eliminate the need for culturing samples to quantify viable *Legionella* in samples from water systems.

5.3.3.3 Using pathogen analytical results effectively

Given the ability to measure pathogen concentrations in water systems, a natural tendency then is for people to demand action (or 'trigger') levels based on these analyses. The overwhelming consensus of scientific opinion is that such action levels are arbitrary. They are not established on any scientifically relevant or experimentally verifiable basis.

Action level 'triggers' based on pathogen analytical results appear in codes of practice because industry experts (the practitioners of hygiene services) and their clients (the people with a legal responsibility to provide safe conditions for their employees, customers, and the surrounding community) have insisted on having them. Thusly, the old axiom is proved that "further regulation is always driven by the regulated".

The justification for these action levels is purely practical. Action levels provide a numeric framework for practice in the field and they are a basis for expedient communications. This is helpful. Table 5.7 gives a compilation of action levels for cooling tower systems. A key feature is that actions specified are generally the same even though trigger levels vary as much as 3 orders of magnitude in codes of practice from around the world.

Table 5.7 Action levels ('triggers') specified in guidance for reducing the risk of legionellosis in cooling water systems

Legionella concentration (CFU/l or CFU/ml)	Actions required
≤ 100 CFU/l[a] ≤ 100 CFU/ml[b] ≤ 1000 CFU/l[c]	Acceptable control. No remedial action required.
100 to 1000 CFU/l[a] 100 to 1000 CFU/ml[b] 10^3 to 10^5 CFU/l[c]	Immediately review treatment program and risk assessment to develop remedial risk management plan. Confirm results immediately by re-sampling
≥ 1000 CFU/l[a] ≥ 1000 CFU/ml[b] ≥ 10^5 CFU/l[c]	Immediately implement corrective action by shot-dose antimicrobial. Immediately resample. Review risk assessment and control programme. Reassess the system.

[a]UK and much of Europe; [b]US, AU and Singapore; [c]France.

The differences in these trigger concentrations are at first disconcerting and then confusing. But since they are arbitrary anyway, the practical consequence of differences is irrelevant so long as trigger concentrations and the actions required are applied consistently in a given region.

An interesting aspect about these limits is that they are based in part upon what was the perceived detection limits for the assay at the time the limits were established. Thus, the UK action levels (which are 3 orders of magnitude lower than those specified in Australia, Singapore, United States) are based in part on the notion that a concentration of the sample by centrifugation could achieve such a low detection limit. However, the concentration step is controversial because recovery of *Legionella* from centrifuged environmental samples is low (less than 50% and sometimes not even 30% recovered). Similarly, another limit is based on sample concentration by membrane filtration that also fails to reliably recover the pathogen from environmental samples. In regions where the concentration step is not required, the detection limit of the assay is much higher and so are the action levels.

Similarly, action levels for domestic water service systems, such as potable water for buildings, are also specified in guidance documents and they are also arbitrary (data not given here). But they are similarly linked to consistent action recommendations, regardless of the absolute value of the trigger level, and therefore they serve a useful practical purpose.

One rational suggestion to improve this situation is that there should be a global consensus on arbitrary action levels based on a scientifically defensible detection limit for *Legionella* in the standard culture method. This would reduce confusion especially for international organisations that seek to apply consistent criteria in their facilities, worldwide. Due to the uncertainties that arise when environmental samples are concentrated by centrifugation or membrane filtration, it would be best to not perform these operations, in the author's opinion. If that technical point were conceded, then the higher action levels would stand as the most defensible. Since there would be no need to concentrate samples, the ISO 11731 method (without a concentration requirement) could be recommended, worldwide.

5.3.4 Cleaning and disinfection

Most guidance documents recommend regular cleaning and disinfection to reduce the risk of legionellosis in cooling water systems. Emphasis in this guidance is to make sure systems are clean (free of debris, scale, corrosion product) so that they can be properly disinfected. Note that disinfection does not mean sterilization; it is neither practical nor advisable to attempt sterilization in most water systems. Exceptions are systems in contact with extremely susceptible populations such as occurs in certain hospital situations or for dental care in which relatively small volumes of water are used in a way that could allow considerable aspiration of aerosol.

Many codes of practice recommend cleaning and disinfection of cooling water systems twice per year (e.g., OSHA, 1998; HSC, 2000). Cleaning and disinfection should be done before putting an out-of-service system back online. These are not regulations or laws, but rather, they are guidance recommendations on how best to comply with approved code of practice (or standards, or best practices, etc.), which are, in turn, advice documents on how to comply with health and safety laws. For example, a code of practice will specify that the risk of exposure to the pathogen should be prevented or controlled and that precautions should include maintaining the cleanliness of the system and the water in it. This advice implies that regular cleaning and disinfection procedures should be performed. Therefore, appendices are provided in the codes of practice that give specific guidance for cleaning and disinfection of cooling water systems and they generally look something like that given in Table 5.8.

The UK guidance (HSC, 2000) on cleaning and disinfection recommends removal of cooling tower packing ('fill'), where practicable, at least once a year preferably twice a year for manual cleaning. This recommendation is controversial because in practice, compliance is expensive, labour-intensive, and in many cases not necessary or counter-productive (Dave Geary, personal communication, D.F. Geary Consultants, dfgeary2@aol.com). Most other guidance documents do not recommend removal of the cooling tower fill.

Guidance is given for domestic water service cleaning and disinfection procedures (not included here). Thermal disinfection of domestic water service is often recommended but supplemental disinfection with chlorine, chloramines, or chlorine dioxide is preferably recommended in most guidance documents.

5.3.5 Monitoring and control

Monitoring and control is an essential component of every legionellosis risk management plan.

5.3.5.1 Monitoring

The purpose of the monitoring program should be to confirm that control measures are being applied consistently. Typically, the parameters monitored are microbiological quality, water temperature, and oxidant concentration.

Table 5.8 General description of cleaning and disinfection recommendations for cooling water systems. This is not detailed guidance nor is it intended as a recommendation. For that, the reader is advised to refer to actual codes of practice (see Table 5.3)

Step	Procedure
Pre-cleaning disinfection	Take the system off-line. Wearing personal protective equipment (PPE), add disinfectant (chlorine, bromine, chlorine dioxide are preferred) to achieve 5 ppm free residual oxidant. Circulate for several hours with the fan off.
Cleaning	Wearing PPE, manually clean surfaces as much as possible, remove debris, use side-stream filtration where practical, but avoid methods that generate excessive spray. If excess scale is apparent, treatment to de-scale the facility may be necessary. This procedure should be done in a way that limits exposure to people. Suggestions include performing it after working hours or excluding people from the area as much as possible. After cleaning, where possible, the water should be drained from the system and rinsed until clear but in large systems this may be neither practical nor advisable.
Post-cleaning disinfection	Wearing PPE, apply disinfectant again to achieve 5ppm free residual oxidant for several hours. Drain and refill where practical. For large systems, reduce the oxidant with sulfite or thiosulfate before discharge. Refill the system; passivate the metal; begin water treatment programme to control scale and corrosion; restart microbial fouling control program and then, bring the system back online
Documentation	Document details of the work. Provide a certification of completion. Check to make sure the procedure was done in accord with the risk management plan. If there is no risk management plan, advise in writing that such a plan should be developed immediately. The first step in developing a risk management plan is to properly perform and document a site-specific risk assessment.

Total aerobic heterotrophic bacteria (TAHB) counts from plates or dip slides are often used to help confirm microbial fouling control in industrial water systems. This microbiological monitor does not directly measure the real target of control (the biofilms) but nevertheless, it correlates generally with microbial fouling in many systems. These analyses are simple and inexpensive to perform in the field. Various codes of practice specify action (trigger) levels based on TAHB counts and are useful in establishing a general basis for field practice. Table 5.9 gives examples.

These action levels are arbitrary and do not correlate directly to microbial fouling nor to risk of legionellosis. Similarly, arbitrary action levels are given in guidance for domestic water service systems (examples not given).

It is important to further emphasise that these action levels, as well as the action levels given for *Legionella* concentrations in Table 5.6 are not established on any scientifically defensible basis that correlates to actual risk. However, such action level guidance is useful in the field to motivate actions and to expedite communications.

Table 5.9 Examples of microbiological quality monitoring and action level specifications for cooling water systems

Total aerobic heterotrophic count (CFU/ml)	Actions required
≤ 10,000[a] or ≤ 100,000[b]	Acceptable control. No remedial action required.
10,000 to 100,000[b] Greater than 1,000,000	Immediately review treatment program and risk assessment to develop remedial risk management plan. Confirm pathogen analytical immediately by resampling

[a]UK guidance; [b]US and AU guidance.

5.3.5.2 Control

Control methods for *Legionella* in water systems have been extensively reviewed (Kim, 2002; WHO, 2002; Loret, 2003). For industrial water systems, chlorine, bromine, stabilized bromine, and chlorine dioxide are the preferred oxidants used for microbial control. Also used effectively for industrial water systems are the 'nonoxidizing' antimicrobials. The preferred active ingredients are dibromonitrilopropionamide, glutaraldehyde, and isothiazolones, but there are many others to chose from (Kim, 2002). These products are very effective when used properly.

As has been previously discussed in section 5.1.2, control of the pathogen host (the protozoa) is critical in reducing the risk of legionellosis. Currently, the best practical means is to control the biofilms upon which protozoa graze (Donlan *et al.*, 2002; Donlan, 2002). This can be very difficult especially in dynamic systems.

Industrial water systems are never sterilized as a result of microbial control programs applied. Even if enough chemical or other remedy could be added to achieve sterilization, the system would rapidly become re-inoculated with microorganisms since cooling systems are open to the environment. The most significant practical consequence of this fact is that increasingly tolerant microbial communities comprised of the survivors of the applied antimicrobial treatment are selected for especially in diverse microbial biofilms (Srikanth, 1994; Sutherland, 1996). It is advisable therefore to proactively vary the antimicrobial stresses in the cooling water microbial control program (McCoy, 2003b). For example, a practical way to vary antimicrobial effects is to alternate between continuous-dosing and slug-dosing an oxidizing antimicrobial chemical. Another effective strategy is to alternate use of an oxidizing antimicrobial with a non-oxidizing antimicrobial to ensure that different modes of antimicrobial effect are applied. Proactive variation of antimicrobial stresses depends on performance-based monitoring to assess the extent of microbial control achieved.

Domestic water systems can become infected with high levels of *Legionella* if the water is allowed to stagnate or achieve optimum growth temperatures of 30–40°C. Risk management plans for these systems should place emphasis on maintaining hygienic conditions, good water circulation and avoidance of optimum growth temperatures for the pathogen.

It is often necessary to apply secondary disinfection in domestic water systems even though the water may have been previously treated to potable water standards (primary disinfection) before entering the distribution system. In the United States, primary drinking water regulations establish the goal of zero (below detectable) *Legionella* (CFR 1998, EPA 1999a) and this standard should be maintained through to the end user. In buildings with complex water distribution systems and hot water systems (which are, of course, nearly all buildings), microbiological quality can significantly degrade if disinfectant residual is depleted and if biofilms develop in the system and/or in deadlegs, for example.

Control in domestic water systems is best achieved by maintaining residuals of chlorine, chlorine dioxide, chloramines or combinations of these preferred treatments throughout the facility. Selection criteria and application guidance for disinfection have been given (EPA 1999b). A practical and helpful comparison summary of disinfection options for hospital systems in the United States is available at *www.hopkins-heic.org/infectious_diseases/legionella.htm.*

Treatments have been compared in well-controlled experiments to simulate real world conditions in domestic water service (Loret 2003); results are given in Figure 5.7.

Control can also be achieved with copper/silver ionisation (Lin 2002), ozone, UV light at 254 nm disinfection (Eckmanns 2002), and thermal disinfection but treatments are variously effective and should always be used in conjunction with one or more of the preferred methods of control in domestic water systems. Other biofilm control methods for water systems in dental offices and other healthcare facilities have been recently evaluated including iodine, enzymes (proteinases and carbohydrases), surfactants, peroxyacids (such as hydrogen peroxide, peracetic acid, butylhydroperoxide, and combinations of these), hydrogen peroxide/silver, and superoxidized water (Walker, 2003; Shetty, 1999; Zinkevich, 2000; Thantasha, 2002).

5.4 CONCLUSIONS

The cell biology and parasitology of the *Legionellae* reveal that they are naturally intracellular parasites of aquatic protist hosts, mostly the amoebic protozoa. They are incidental parasites of mammalian phagocytic cells.

Legionellae are known today as causative in severe preventable pulmonary disease that can cause permanent damage even in patients who recover from the acute illness.

At least about six legionellosis cases per million population are reported annually but the actual number of cases is very likely 10× that number. This indicates that many hundreds of thousands of people are affected by the disease each year.

Legionellosis disease is serious and preventable. The risk is the result of susceptible population exposure to microscopic water aerosols from inadequately controlled warm water plumbing and from comfort or industrial cooling water heat transfer towers.

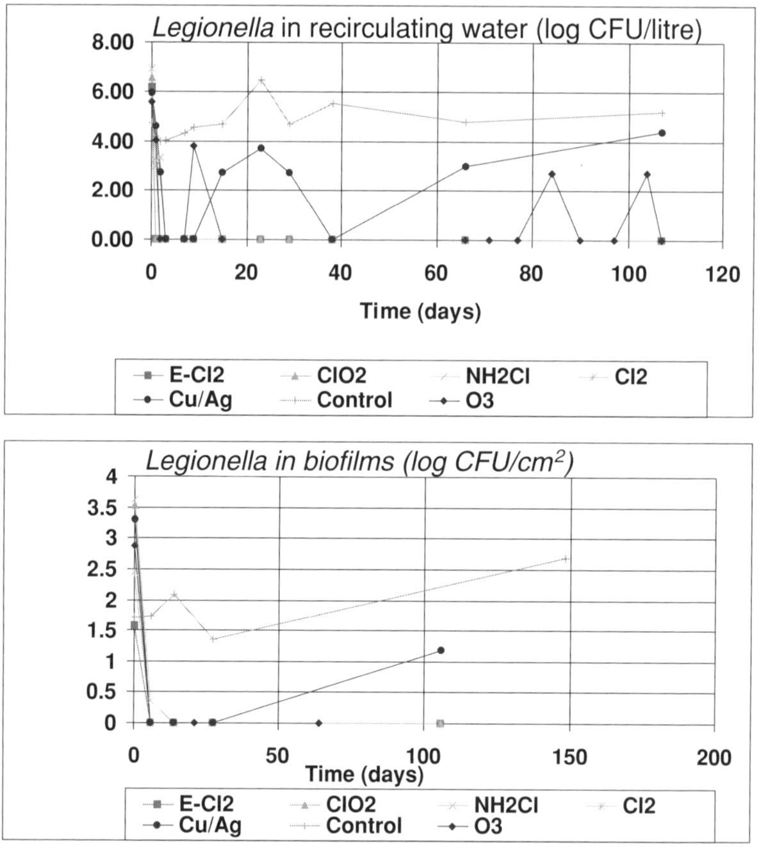

Figure 5.7 Comparison of treatments for *Legionella* control in a scale-model domestic water system. Mixed oxidants electrolytically generated from brine (E-Cl_2), chlorine dioxide (ClO_2), chloramines (NH_2Cl), and chlorine (Cl_2) controlled *Legionella* in biofilms and in the bulk recirculating water. Copper/silver ionisation and ozone were not as effective (Loret 2003; Lorret, 2004).

A global code of practice for reducing the risk of legionellosis can be derived from the extensive published guidance, worldwide. Common elements from these codes of practice are risk assessment, pathogen analysis, cleaning and disinfection, monitoring and control.

Acknowledgements

Andrew Cooper helped in assembling Table 5.2 and the discussion on risk assessment. Jean François Loret explained and interpreted data in Figures 5.2 and 5.6. Tom Makin provided interpretations of radiographic images in Figure 5.3. Howard Barnes helped in assembling Tables 5.7, 5.9 and gave perspective on control measures. Noel Christopher provided clarification of the regulatory history in the UK and perspectives on control. Their contributions are gratefully acknowledged.

REFERENCES

Abd, H., Johansson, T., Golovliov, I., Sandström, G., Forsman, M. (2003) Survival and growth of *Francisella tularensis* in *Acanthamoeba castellanii*. *Appl. Environ. Microbiol.* **69**(1), 600–606.

Abu Kwaik, Y., Gao, L.Y., Stone, B.J., Venkataramna, C. and Harb, O.S. (1998) Invasion of protozoa by *Legionella pneumophila* and its role in bacterial ecology and pathogenesis. *Appl. Environ. Microbiol.* **64**(9), 3127–3133.

Addiss, D.G., Davis, J.P., LaVenture, M., Wand, P.J., Hutchinson, M.A. and McKinney, R.M. (1989) Community-acquired Legionnaires' disease associated with a cooling tower: evidence for longer distance transport of *Legionella pneumophila*. *Am. J. Epidemiol.* **130**, 557–568.

Adeleke, A., Pruckler, J., Benson, R., Rowbotham, T., Halablab, M. and Fields, B.S. (1996) *Legionella*-like amoebal pathogens - phylogenetic status and possible role in respiratory disease. *Emerg. Infect. Dis.* **2**, 225–229.

Alli, O.A.T., Zink, S., von Lackum, N.K. and Abu-Kwaik, Y. (2003) Comparative assessment of virulence traits in *Legionella* spp. *Microbiology* **149**, 631–641.

Aragon,V., Kurtz, S., McClain, M., Engleberg, N.C. and Cianciotto, N.P. (2002) Analysis of acid phosphatase and esterase/lipase mutants of *Legionella pneumophila*, In *Legionella* (eds R. Marre *et al.*), Chap. 3, pp. 18–21, ASM Press, American Society for Microbiology, Washington, D.C.

Armon, R., Starosvetsky, J., Arbel, T. and Green, M. (1997) Survival of *Legionella pneumophil*a and *Salmonella typhimurium* in biofilm systems. *Water Sci. Tech.* **35**(11–12), 293–300.

Arnow, P.M., Chou, T., Weil, D., Shapiro, E.N. and Kretzschmar, C. (1982) Nosocomial Legionnaires' disease caused by aerosolized tap water from respiratory devices. *J. Infect. Dis.* **146**, 460–467.

ASHRAE (Am. Soc. of Heating, Refrigeration and Air-Conditioning Engineers) (2000), Guideline 12-2000: Minimizing the Risk of Legionellosis Associated with Building Water Systems.1791 Tullie Circle, N.E., Atlanta, GA 30329. www.ashrae.org.

ASTM (Am. Soc. of Testing and Matls.) D5952-96 (2002) Standard Guide for Inspecting Water Systems for *Legionellae* and Investigating Possible Outbreaks of Legionellosis (Legionnaires' Disease or Pontiac Fever). P.O. Box C700, Conshohocken, PA 19428-2959 www.astm.org.

Atlas, R.M. (1999) *Legionella*: from environmental habitats to disease pathology, detection and control. *Environ. Microbiol.* **1**(4), 283–293.

Australian/New Zealand Standard. (2000) Air-handling and water systems of buildings-microbial control, part 3: performance-based maintenance of cooling water systems. *AS/NZS* 3666.1.

Australian/New Zealand Standard. (1995) Air-handling and water systems of buildings-microbial control, part 1: design, installation and commissioning. *AS/NZS* 3666.1.

Australian/New Zealand Standard. (1995) Air-handling and water systems of buildings-microbial control, part 2: operation and maintenance. *AS/NZS* 3666.2.

Bartlett, C.L.R., Macrae, A.D. and Macfarlane, J.T. (1986) Clinical aspects and diagnosis of *Legionella*, In *Legionella Infections*, Chap. 3, p. 43, Fig. 3.1 (a to d). Edward Arnold Publishers, London, UK.

Barker, J., Brown, M.R.W., Collier, P.J., Farrell, I. and Gilbert, P. (1992) Relationship between *Legionella pneumophila* and *Acathamoeba polyphaga*: physiological status and susceptibility to chemical inactivation. *Appl. Environ. Microbiol.* **58**, 2420–2425.

Benson, R.F., Tang, P.W. and Fields, B.S. (2000) Evaluation of the Binax and Biotest antigen kits for detection of Legionnaires' Disease due to multiple serogroups and species of *Legionella*. *J. Clin. Microbiol.* **38**(7), 2763–2765.

Bentham, R.H., Pradham, M., Hakendorf, P. and Wilmot, P. (2002) Using geographical information systems for risk assessment and control of Legionnaires' disease associated with cooling towers. In *Legionella*, (eds R. Marre *et al.*), Chap. 63, pp. 318–320, ASM Press, American Society for Microbiology, Washington, D.C.

Berk, S.G., Ting, R.S., Turner, G.W. and Ashburn, J. (1998) Production of respirable vesicles containing live *Legionella pneumophila* cells by two *Acanthamoeba* spp. *Appl. Environ. Microbiol.* **64**(1), 279–286.

Birnbaum, H.G., Morley, M., Greenberg, P.E., Cifaldi, M. and Colice, G.L. (2001) Economic burden of pneumonia in an employed population *Arch. Intern. Med.* **161**, 2725–2731.

Birtles, R.J., Rowbotham, T.J., Michel, R., Pitcher, D.G., La Scola, B. Alexiou-Daniel, S. and Raoult, D. (2000) *Candidatus Odyssella thessalonicensis* gen. nov., sp. nov., and obligate intracellular parasite of *Acanthamoeba* species. *Intl. J. Sys. Evol. Microbiol.* **50**, 63–72.

Bollin, G.E., Plouffe, J.F., Para, M.F. and Hackman, B. (1985) Aerosols containing *Legionella pneumophila* generated by shower heads and hot water faucets. *Appl. Environ. Microbiol.* **50**, 1128–1131.

Breiman, R.F. (1993) Modes of transmission in epidemic and nonepidemic *Legionella* infection: directions for further study. In *Legionella: Current Status and Emerging Perspectives.* (eds J.M. Barbaree, R.F. Breiman and A.P. Dufour), pp. 129–136, American Society for Microbiology, Washington, D.C.

Brieland, J.K., Fantone, J.C., Remick, D.G., LeGendre, M., McClain, M. and Engleberg, N.C. (1997a) The role of *Legionella pneumophila*-infected *Hartmannella vermiformis* as an infectious particle in a murine model of Legionnaires' disease. *Infect. Immun.* **65**(12), 5330–5333.

Brieland, J., McClain, M., LeGendre, M. and Engleberg, N.C. (1997b) Interpulmonary *Hartmannella vermiformis*: a potential niche for *Legionella pneumophila* replication in a murine model of legionellosis. *Infect. Immun.* **65**(1), 4892–4896.

Brieland, J., McClain, M. Heath, L. Chrisp, C., Huffnagle, G., LeGendre, M., Hurley, M., Fantone, J. and Engleberg, C. (1996) Coinoculation with *Hartmannella vermiformis* enhances replicative *Legionella pneumophila* lung infection in a murine model of Legionnaires' disease. *Infect. Immun.* **64**(7), 2449–2456.

Broadbent, C. (1993) *Legionella* in cooling towers: Practical research, design, treatment, and control guidelines. In *Legionella: Current Status and Emerging Perspectives* (eds J.M. Barbaree, R.F. Breiman and A.P. Dufour), pp. 217–222 American Society for Microbiology, Washington, D.C.

Broadbent, C. (2002) Australian risk management approaches to control of legionella in cooling water systems, In *Legionella*, (eds R. Marre *et al.*), Chap. 371, pp. 371–375, ASM Press, American Society for Microbiology, Washington, D.C.

Bozue, J.A. and Johnson, W. (1996) Interaction of *Legionella pneumophila* with *Acanthamoeba castellanii*: Uptake by coiling phagoytosis and inhibition of phagosome-lysosome fusion. *Infect. Immun.* **64**(2), 668–673.

Butler, J.C, Fields, B.S. and Brieman, R.F. (1997) Prevention and control of legionellosis. *Infect. Dis. Clin. Pract.* **6**(7), 458–464.

Campos, C., Loret, J.F., Cooper, A.J. and Kelly, R.F. (2003) Disinfection of domestic water systems for *Legionella pneumophila. J. Water Supply: Res. Technol. Aqua.* **52**(5), 341–354.

CDC (Centers for Disease Control and Prevention), Healthcare Infection Control Practices Advisory Committee (2001) Draft Guideline for Environmental Infection Control in Healthcare Facilities. http://www.cdc.gov/ncidod/hip/enviro/guide.html.

CDC (Centers for Disease Control and Prevention), U.S. Department of Health and Human Services (2003) Summary of notifiable diseases – United States, 2001. *Morbid. Mortal. Wkly Rep.* **50**(53), May 2, 2003.

CFR (US Code of Federal Regulations). (1998) Title 40 §141.52 Maximum contaminant level goals for microbiological contaminants. www.epa.gov.

Cianciotto, N.P., Kurtz, S., Krcmarik, K., Mody, S., Prasad, U., Robey, M., Salerno, J. and Viswanathan, V.K. (2002) Iron requirements of and acquisition of iron by *Legionella pneumophila*, In *Legionella* (eds R. Marre *et al.*), Chap. 6, pp. 31–37, ASM Press, American Society for Microbiology, Washington, D.C.

Cloud, J.L., Carroll, K.C., Pixton, P., Erali, M. and Hillyard, D.R. (2000) Detection of *Legionella* species in respiratory specimens using PCR with sequencing confirmation. *J. Clin. Microbiol.* **38**(5), 1709–1712.

Cooper, A.J., Barnes, H.R. and Myers, E.R. (2004) Assessing risk of *Legionella*. *ASHRAE J.* **46**(4), 22–27.

Cordes, L.G., Faser, D.W., Skailiy, P., Perlio, C.A., Elsea, W.R., Mallison, G.F. and Hayes, P.S. (1980) Legionnaires' disease outbreak at an Atlanta, Georgia country club: evidence for spread from an evaporative condenser. *Am. J. Epidemiol.* **111**, 425–431.

Cordes, L.G., Weisenthal, A.M., Gorman, G.W., Phair, J.P., Brown, A., Yu, V.L., Magnussen, M.H., Myer, R.D., Wold, J.S., Shands, K.N. and Fraser, D.W. (1981) Isolation of *Legionella pneumophila* from hospital shower heads. *Ann. Intern. Med.* **94**, 195–197.

Costerton, J.W., Lewandowski, Z., Caldwell, D.E., Korber, D.R. and Lappin-Scott, H.M. (1995) Microbial biofilms. *Annu. Rev. Microbiol.* **49**, 711–745.

CTI (Cooling Technology Institute) (2000) Legionellosis. Guideline: Best practices for control of *Legionella*. Cooling Technology Institute. Houston, TX 77273 www.cti.org

Cunha, B.A. (2003) Community-aquired pneumonia. *Infect. Med.* **20**(1), 27–30.

Dedicoat, M. and Venkatesan, P. (1999) The treatment of Legionnaires' disease. *J. Antimicrob. Chemother.* **43**, 747–752.

Den Boer, J.W., Yzerman, E.P.F., Schellekends, J., Lettinga, K.D. *et al.* (2002) A large outbreak of Legionnaires' disease at a flower show, the Netherlands, 1999. *Emer. Inf. Dis.* **8**(1), 37–43.

DHS (Department of Human Services, Public Health Division). (2001a) A guide to developing risk management plans. Government Dept of Human Services. Melbourne, Victoria, Australia www.legionella.vic.gov.au.

DHS (Department of Human Services, Public Health Division) (2001b) Supplementary notes for hospitals. Public Health Group, Victorian Government Department of Human Services, Melbourne, Victoria. www.legionella.vic.gov.au.

DHS (Department of Human Services, Public Health Division) (2002) Code of practice for water treatment service providers (cooling tower systems). Public Health Group, Victorian Government Department of Human Services, Melbourne, Victoria, Australia www.legionella.vic.gov.au.

DIN (Deutsches Institut fur Normung) (1997). DIN 19643-1 and 38402-14.

Dondero, T.J., Rentdorff, R.C., Mallison, G.F., Weeks, R.M., Levy, J.S., Wong, E.W. and Schaffner, W. (1980) An outbreak of Legionnaires' disease associated with a contaminated air-conditioning cooling tower. *N. Engl. J. Med.* **302**, 365–370.

Donlan, R.M and Costerton, J.W. (2002) Biofilms: Survival mecahnaisms of clinically relevant micro-organisms. *Clin. Microbiol. Rev.* **15**(2), 167–193.

Donlan, R.M. (2002) Biofilms: Microbial life on surfaces. *Emer. Inf. Dis.* **8**, 881–890.

DVGW (Deutscher Verein des Gas- und Wasserfaches e.V) (1996). Hot water systems and pipes; technical improvements for the reduction of the growth of *Legionella*: renovation and operation. Manual W 551 and W 552.

Eckmanns, T., Schwab, F., Posselt, H., Gastmeier, P. and Rüden, H. (2002) UV light for elimination of Legionellae. In *Legionella* (eds R. Marre *et al.*), Chap. 81, pp. 402–405, ASM Press, American Society for Microbiology, Washington, D.C.

Edelstein, P.H. (2002) Chemotherapy of Legionnaires' Disease with macrolide or quinolone antimicrobial agents. In *Legionella* (eds R. Marre *et al.*), Chap. 32, pp. 183–188, ASM Press, American Society for Microbiology, Washington, D.C.

EPA (U.S. Environmental Protection Agency). (1999a) Alternative disinfectants and oxidants guidance manual. EPA 815-R-99-014, Office of Water (4607), April 1999 www.epa.gov.

EPA (U.S. Environmental Protection Agency). (1999b) Microbial and disinfection by-product rules – Simultaneous compliance guidance manual. EPA 815-R-99-014, Office of Water (4607), August 1999 www.epa.gov.

Essig, A., Heinemann, M., Simnacher, U. and Marre, R. (1997). Infection of *Acanthamoeba castellanii* by *Chlamydia pneumoniae*. *Appl. Environ. Microbiol.* **63**, 1396–1399.

Faulkner, G. and Garduño, A. (2002) Ultrastructural analysis of differentiation in *Legionella pneumophila*. *J. Bact.* **184**(24), 7025–7041.

Feeley, J.C., Gorman, R.E., Weaver, R.E., Mackel, D.C. and Smith, H.W. (1978). Primary isolation media for Legionnaires' disease bacterium. *J. Clin. Microbiol.* **8**, 320–325.

Fields, B.S., Sanden, G.N., Barbaree, J.M., Morril, W. E., Windowsky, R.M., White, E.H. and Feeley, J.C. (1989) Intracellular multiplication of *Legionella pneumophila* in amoebae isolated from hospital hot water tanks. *Curr. Microbiol.* **18**, 131–137.

Fields, B.S. (1993) *Legionella* and protozoa: interaction of a pathogen and its natural host. In (eds J.M. Barbaree, R.F. Breiman, and A.P. Dufour (ed.), *Legionella:* current status and emerging perspectives. American Society for Microbiology, Washington, D.C.

Fields, B.S. (1996) The molecular ecology of legionellae. *Trends Microbiol.* **4**, 286–290.

Fields, B.S., Benson, R.F. and Besser, R.E. (2002) *Legionella* and Legionnaires' disease: 25 years of investigation. *Clin. Microbiol. Rev.* **15**(3), 506–526.

Flieger, A., Gong, S., Faigle, M. Stevanovic, S., Northoff, H., Cianciotto, N.P. and Neumeister, B. (2002) *Legionella pneumophila* secretes different phosholipases A, In *Legionella* (eds R. Marre *et al.*), Chap. 4, pp. 22–26, ASM Press, American Society for Microbiology, Washington, D.C.

France (2001) Guide des bonnes pratiques: *Legionella* et tours aéroréfrigérantes. *Ministére de L'Emploi et de la Solidartité, Ministére de L'Economie des Finances et de L'Industrie, Ministére de L'Amenagement du Territoire et de L'Environment.*

Fraser, D.W., Tsai, T.R., Orenstein, W., Parkin, W.E., Beecham, H.J., Sharrar, R.G. *et al.* (1977) Legionnaires' disease: description of an epidemic of pneumonia. *New Engl. J. Med.* **297**, 1189–1197.

Fraser, D.W. (1980) Legionellosis: evidence of airborne transmission. *Ann. N.Y. Acad. Sci.* **353**, 61–66.

Fraser, D.W. (1984) Sources of legionellosis. In *Legionella*: Proceedings of the Second International Symposium. American Society for Microbiology, Washington, D.C.

Fritsche, T.R., Horn, M., Seyedirashti, S., Gautom, R.K., Schleifer, K.-H. and Wagner, M. (1999) *In situ* detection of novel bacterial endosymbionts of *Acanthamoeba* spp. phylogenetically related to members of the Order *Rickettsiales*. *Appl. Environ. Microbiol.* **65**(1), 206–212.

Garbe, P.L., Davis, B.J., Weisfield, J.S,. Markowitz, L, Miner, P., Garrity, F., Barbaree J.M. and Reingold, A.L. (1985) Nosocomial Legionnaires' disease: epidemiologic demonstration of cooling towers as a source. *J. Am. Med. Ass.* **254**, 521–524.

García-Fulgueiras, A. Navarro, C., Fenoll, D., García, J. *et al.* (2003) Legionnaires' disease outbreak in Murcia, Spain, *Emerg. Infect. Dis.* **9**(8) August 2003[11]. Available from: URL: http://www.cdc.gov/ncidod/EID/vol0no8/03-0337.htm

Greub, G. and Raoult, D. (2002) Crescent bodies of *Parachlamydia acanthamoeba* and its life cycle within *Acanthamoeba polyphaga*: An electron micrograph study. *Appl. Environ. Microbiol.* **68**(6), 3076–3084.

Grosserode, M. *et al.* (1993) Continuous huperchlorination for control of nosocomial *Legionella pneumophila* pneumonia: a ten-year follow-up of efficacy, environmental effects, and dosts. In *Legionella*: *Current Status and Emerging Perspectives* (eds J.M. Barbarre, R.F. Breiman A.P. Dufour), pp. 226–229, ASM Press, American Society for Microbiology, Washington, D.C.

Habicht, W. and Muller, H.E. (1988) Occurrence and parameters of frequency of *Legionella* in warm water systems of hospitals and hotels in Lower Saxony. *Praventive Medzin* (written in English). **188**(1), 79–88.

Hanrahan, J.P., Morse, D.L. and Scharf, V.B. (1987) A community hospital outbreak of legionellosis: transmission by potable hot water. *Am. J. Epidemiol.* **125**, 639–649.

Hammer, B.K. and Swanson, M.S. (1999) Co-ordination of *Legionella pneumophila* virulence with entry into stationary phase by ppGpp. *Mol. Microbiol.* **33**, 721–731.

Hayden, R.T., Uhl, J.R., Qian, X., Hopkins, M.K., Aubry, M.C., Limper, A.H., Lloyd, R.V. and Cockerill, F.R. (2001) Direct detection of *Legionella* species from bronchoalveloar lavage and open lung biopsy specimens: Comparison of LightCycler PCR, *in situ* hybridization, direct fluorescence antigen detection and culture. *J. Clin. Microbiol.* **39**(7), 2618–2626.

Helbig, J.H., Uldum, S.A., Lück, P.C., Harrison, T.G. (2002) Detection of *Legionella pneumophila* antigen in urine samples: Recognition of serogroups and monoclonal subgroups. In *Legionella* (eds R. Marre *et al.*), Chap. 37, pp. 204–206, ASM Press, American Society for Microbiology, Washington, D.C.

Helbig, J.H., Lück, C., Jacobs, E. and Witt, M. (2002) Localization of *Legionella pneumophila* Mip protein inside phagosomes of *Acanthamoeba castellanii*. In *Legionella* (eds R. Marre *et al.*), Chap. 9, pp. 49–51, ASM Press, American Society for Microbiology, Washington, D.C.

Heng, B.H., Goh, K.T., Ng, D.L.K. and Ling, A.E. (1997) Surveillance of Legionellosis & *Legionella* bacteria in the built environment in Singapore. *Ann. Acad. Med. Singapore* **26**, 225–565.

Heuner, K., Steinert, M., Dietrich, C., Fischer, G., Köhler, R. and Hacker, J. (2002) Function and expression of *Legionella pneumophila* surace factors. In *Legionella* (eds R. Marre *et al.*), Chap. 8, pp. 43–48, ASM Press, American Society for Microbiology, Washington, D.C.

Hlady, W.G., Mullen, R.C., Mintz, C.S., Shelton, B.G., Hopkins, R.S. and Daikos, G.L. (1993) Outbreak of Legionnaires' disease linked to a decorative fountain by molecular epidemiology. *Am. J. Epidemiol.* **138**, 555–562.

Horn, M., Wagner, M., Müller, K.-D., Schmid, E.N., Fritsche, T.R., Schleifer, K.-H. and Michel, R. (2000) *Neochlamydia hartmannellae* gen. nov., sp. nov. (*Parachlamydiaceae*), an endoparasite of the amoeba *Hartmannella vermiformis*. *Microbiology* **146**, 1231–1239.

HSC (Health and Safety Commission and Executive) (2000) Legionnaires' disease. The control of legionella bacteria in water systems. Approved Code of Practice & Guidance, L8. HSE Books. P.O. Box 1999, Sudbury, UK www.hse.gov.uk.

Inglis, T.J.J., Rigby, P., Robertson, T.A., Dutton, N.S., Henderson, M. and Chang, B.J. (2000). Interaction between *Burkholderia pseudomallei* and *Acanthamoeba* species results in coiling phagocytosis, endamebic bacterial survival, and escape. *Infect. Immun.* **68**(3), 1681–1686.

ISO (1998) Water quality-Detection and Enumeration of *Legionella*. ISO 11731. *International Organization for Standardization*, Geneva, Switzerland.

JCAHO (Joint Commission on Accreditation of Healthcare Organizations). (2001) Utility Systems Management. Std EC.1.7 www.jcaho.org/Onlineform/OnLineForm.asp.

Jernigan, D.B., Hofmann, J., Cetron, M.S. *et al.* (1996) Outbreak of Legionnaires' disease among cruise ship passengers exposed to a contaminated whirlpool spa. *Lancet* **347**, 494–499.

Joly, J.R. (1993) Prevention and control of legionellosis. In *Legionella*: *Current Status and Emerging Perspectives*. (eds J.M. Barbaree *et al.*), American Society for Microbiology, Washington, D.C.

Joshi, A. and Swanson, M.S. (1999) Comparative analysis of *Legionella pneumophila* and *Legionella micadadei* virulence traits. *Infect. Immun.* **67**(8), 4134–4142.

Joseph, C. (2002) Surveillance of Legionnaires' Disease in europe, In *Legionella* (eds R. Marre *et al.*), Chap. 62, pp. 311–317, ASM Press, American Society for Microbiology, Washington, D.C.

Junge-Mathys, E. and Mathys, W. (1994) Die Legionellose - ein Beispiel für umweltbedingte Infektionen [Legionellosis - an example of environmentally caused infections] *Intensiv* **2**, 29–33.

Kahane, S., Dvoskin, B., Mathias, M. and Friedman, M. (2001). Infection of *Acanthamoeba polyphaga* with *Simkania negevensis* and *S. negevensis* survival with amoebal cysts. *Appl. Environ. Microbiol.* **67**(10), 4789–4795.

Keller, D.W., Hajjeh, R., DeMaria, A. *et al.* (1996) Community outbreak of Legionnaires' disease: an investigation confirming the potential for cooling towers to transmit *Legionella* species. *Clin. Infect. Dis.* **22**, 257–261.

Kilvington, S. and Price, J. (1990) Survival of *L. pneumophila* within cysts of *Acanthamoeba polyphaga* following chlorine exposure. *J. Appl. Bacteriol.* **68**, 519–525.

Kim, B.R., Anderson, J.E., Mueller, S.A., Gaines, W.A. and Kendall, A.M. (2002) Literature review-efficacy of various disinfectants against *Legionella* in water systems. *Water Res.* **36**, 4433–4444.

King, C.H., Shotts Jr, E.B., Wooley, R.E. and Porter, K.G. 1988. Survival of coliforms and bacterial pathogens within protozoa during chlorination. *Appl. Environ. Microbiol.* **54**, 3023–3033.

Klahre, J. and Flemming, H.-C. (2000) Monitoring of biofouling in papermill process waters. *Water Res.* **34**, 3657–3665.

Koide, M., Saito, A., Kusano, N. and Higa, F. (1993) Detection of *Legionella spp.* in cooling tower water by the polymerase chain reaction method. *Appl. Environ. Microbiol.* **59**, 1943–1946.

Kusnetsov, J.M., Martikainen, P.J., Jousimies-Somer, H.R., Vaisanen, M.L., Tulkki, A.I., Ahonen, H.E. and Nevalainen, A.I. (1997) Physical, chemical and microbiological water characteristics associated with the occurrence of *Legionella* in cooling tower systems. *Water Res.* **27**(1), 85–90.

La Scola, B., Barrassi, L. and Raoult, D. (2000) Isolation of new fastidious α-proteobacteria and *Afipia felis* from hospital water supplies by direct plating and amoebal co-culture procedures. *FEMS Microbiol. Ecol.* **34**, 129–137.

Lave, J.R., Lin, C.J., Fine, M.J. and Hughes-Cromwick, P. (1999) The cost of treating patients with community-aquired pneumonia. *Semin. Respir. Crit. Care Med.* **20**(3), 189–197.

Lee, V.T. and Schneewind, O. (2001) Protein secretion and the pathogenesis of bacterial infections. *Genes Dev.* **15**, 1725–1752.

Lee, J.V., Surman, S.B., Kirby, A., and Seddon, F. (2002) Eleven years of experience with novel strategies for *Legionella* control in a large teaching hospital. In *Legionella* (eds R. Marre *et al.*), Chap. 80, pp. 398–401, ASM Press, American Society for Microbiology, Washington, D.C.

Lin, Y.E., Vidic, R.D., Stout, J.E. and Yu, V.L. (2002) Negative effect of high pH on biocidal efficacy of copper and silver ions in controlling *Legionella pneumophila*. *Appl. Environ. Microbiol.* **68**(6), 2711–2715.

Loret, J-F, *et al.* (2003) Review of *Legionella* control technologies. *J. Water Supply: Res. Technol. Aqua* (in press).

Loret, J.F., Robert, S., Thomas, V., Cooper, A.J., McCoy, W.F. and Levi, Y. (2004) Comparison of disinfectants for biofilm, protozoa and *Legionella* control. *J. Water Health*, submitted.

Loret, J.F., Robert, S., Thomas, V., Levi, Y., Cooper, A.J. and Myers, E.R. (2003) Comparison of treatment efficiency for *Legionella* control in domestic water systems. *AWWA Water Quality Technology Conference Proceedings*. Edition: 2003, 20546.

Luck, P.C., Leupold, I., Hlawitschka, M., Helbig, J.H., Carmienke, I., Jatzwauk, L. and Guderitz, T. (1993) Prevalence of *Legionella* species, serogroups, and monoclonal subgroups in hot water systems in south-eastern Germany. *Zentbl. Hyg. Umweltmed.* **193**, 450–460.

Ly, T.M. and Müller, H.E. (1990) Ingested *Listeria monocytogenes* survive and multiply in protozoa. *J. Med. Microbiol.* **33**, 129–137.

Lye, D., Fout, G.S., Crout S.R., Danielson, R., Thio, C.L. and Paszko-Kolva, C.M. (1997) Survey of ground, surface, and potable waters for the presence of *Legionella* species by enviroampR PCR *Legionella* kit, culture, and immunofluorescent staining. *Water Res.* **31**(2), 287–293.

Mahoney, F.J., Hoge, C.W., Farley, T.A., Barbaree, J.M., Breiman, R.F., Benson, R.F. and McFarland, L.M. (1992) Community-wide outbreak of Legionnaires' disease associated with a grocery store mist machine. *J. Infect. Dis.* **165**, 736–739.

Mangione, E.J., Remis, R.S., Tait, K.A., McGee, W.B., Gorman, G.W., Wentworth, B.B., Baron, P.A., Hightower, A.W., Barbaree, J.M. and Broome, C.V. (1982) An outbreak of Pontiac fever related to whirlpool use, Michigan. *JAMA* **253**, 535–539.

Marre, R., Medeiros, A.A. and Pasculle, A.W. (1982) Characterization of the β-lactamases of six species of *Legionella. J. Bacteriol.* **151**(1), 216–212.

Marrie, T., Green, P., Burbridge, S., Bezanson, G., Neale, S., Hoffman, P.S. and Haldane, D. (1994) *Legionellaceae* in the potable water of Nova Scotia hospitals and Halifax residences. *Epidemiol. Infect.* **112**, 143–150.

Marciano-Cabral, F. and Cabral, G. (2003) *Acanthamoeba* spp. as agents of disease in humans. *Clin. Microbiol. Rev.* **16**(2), 273–307.

Marston, B.J., Lipman, H.B. and Breiman, R.F. (1994) Surveillance for Legionnaires' disease. *Arch. Intern. Med.* **154**, 2417–2422.

Marston, B.J., Plouffe, J.F., File, T.M., Hackman, B.A., Salstrom, S.J., Lipman, H.B., Kolczak, M.S. and Breiman, R.F. (1997) Incidence of community-acquired pneumonia requiring hospitalization. Results of a p[opulation-based active surveillance study in Ophon. The commuity-based pneumonia incidence study group. *Arch. Intern. Med.* **157**(15), 1709–1718.

Mastro, T.D., Fields, B.S., Breiman, R.F., Campbell, J., Plikaytis, B.D. and Spika, J.S. (1991) Nosocomial Legionnaires' disease and use of medication nebulizers. *J. Infect. Dis.* **163**:667–670.

Maurin, M. and Raoult, D. (1999) Q fever. *Clin. Microbiol. Rev.* **12**(4), 518–553.

McAnulty, J., Habib, M. and Delpech, V. (2001). Epireview: Legionairres' Disease, NSW, 1991–2000. *NSW Pub. Health Bull.* **12**(10), 282–285, www.health.nsw.gov.au.

McCoy, W.F. (2003a) Microbial fouling control for cooling water. In *Business Briefing: CPI Technology*, pp. 65–67, World Markets Research Centre, Ltd, WMRC plc, London, UK www.wmrc.com.

McCoy, W.F. (2003b) Microbial-fouling control for industrial systems, In *Biofilms in Medicine, Industry and Environmental Biotechnology: Characteristics, Analysis and Control* (eds P. Lens *et al.*), IWA Publishing, London.

McDade, J.E., Sheppard, C.C., Faser, D.W., Tsai, T.R., Redus, M.A., Dowdle, W.R. (1977) Legionnaires' disease: isolation of a bacterium and demonstration of its role in other respiratory diseases. *N. Engl. J. Med.* **287**, 1197–1203.

Michel, R., Muller, K.D. and Schmid, E.N. (1995) *Ehrlichia*-like organisms (KSL1) observed as obligate intracellular parasites of *Saccamoeba* species. *Endocytobiosis Cell Res.* **11**, 69–80.

Miller, R.D. and Kenepp, K.A. (1993) Risk assessments for Legionnaires' disease based on routine surveillance of cooling towers for Legionellae. In: *Legionella*: *Current Status And Emerging Perspectives*. (eds J.M. Barbaree *et al.*). American Society for Microbiology, Washington, D.C.

Molmeret, M., Alli, O.A.T., Zink, S., Flieger, A., Cianciootto, N.P. and Kwaik, Y.A. (2002) IcmT is essential for pore formation-mediated egress *of Legionella pneumopnila* from mammalian and protozoan cells. *Infect. Immun.* **70**(1), 69–78.

Murga, R., T.S. Forster, E. Brown, Pruckler, J.M, Fields, B.S. and Donlan, R.M. (2001) Role of biofilms in the survival of *Legionella pneumophila* in a model potable-water system. *Microbiology* **147**, 3121–3126.

The Netherlands. (2000) The Temporary Mains Water (*Legionella* Prevention) Regulation. (http://www2.minvrom.nl/pagina.html?id=480)

Nahapetian, K., Challemel, O., Beurtrin, D., Burbrou, S., Gournon, P. and Squinazi, F. (1991) The intracellular multiplication of *Legionella pneumophila* in protozoa from hospital plumbing systems. *Res. Microbiol.* **142**, 677–685.

Nagal, H. and Roy, C.R. (2003) Show me the substrates: modulation of host cell function by type IV secretion systems. *Cell. Microbiol.* **5**(6), 373–383.

Neuner, A.R. (2002) Health system triumphs over *Legionella*. *AFE Fac. Eng. J.l.* Nov/Dec 2002 www.afe.org.

Newsome, A.L., Scott, T.M., Benson, R.F. and Fields, B.S. (1998) Isolation of an amoeba naturally harboring a distinctive *Legionella* species. *Appl. Environ. Microbiol.* **64**(5), 1688–1693.

OSHA (Occupational, Safety & Health Admn.) (1998) Technical Manual, Section II, Chapter 7, Legionnaires' Disease. www.osha-slc.gov/TechMan_data/II_7.html

Palmer, C.J., Bonilla, G.F, Roll, B., Paszko-Kolva, C. Sangermano, L.R. and Fujioka, R.S. (1995) Detection of *Legionella* species in reclaimed water and air with the EnviroAmp Legionella PCR kit and direct fluorescent antibody staining. *Appl. Environ. Microbiol.* **61**, 407–412.

Palmer, C.J., Tsai, Y.L., Paszko-Kolva, C., Mayer, C. and Sangermano, L.R. (1993) Detection of *Legionella* species in sewage and ocean water by polymerase chain reaction, direct fluorescent-antibody, and plate culture methods. *Appl. Environ. Microbiol.* **59**, 3618–3624.

Percival, S.T., Walker, J.T. and Hunter, P.R. (2000) Risk assessment In *Microbiological Aspects of Biofilms and Drinking Water*. Chap 4, pp. 41–48, CRC Press, London

Riffard, S., Douglass, S., Brooks, T., Springthorpe, S., Filion, L.G. and Sattar, S.A. (2001) Occurrence of *Legionella* in groundwater: an ecological study. *Water Sci. Tech.* **43**(12), 99–102.

Rossier, O. and Cianciotto, N.P. (2001) Type II protein secretion is a subset of the PilD-dependent processes that facilitate intracellular infection by *Legionella pneumophila*. *Infect. Immun.* **69**(4), 2092–2098.

Rossier, O., Edelstein, P.H. and Cianciotto, N. (2002) Role of the type II protein secretion pathway in pathogenesis of *Legionella pneumophila*. In *Legionella* (eds R. Marre *et al.*), Chap. 2, pp. 13–17, ASM Press, American Society for Microbiology, Washington, D.C.

Rusin, P.A., Rose, J.B., Haas, C.N. and Gerba, C.P. (1997) Risk assessment of opportunistic bacterial pathogens in drinkng water. *Rev. Environ. Contam. Toxicol.* **152**, 57–83.

Rohr, U., Weber, S., Michel, R., Selenka, F. and Wilhelm, M. (1998) Comparison of free-living amoebae in hot water systems of hospitals with isolates from moist sanitary areas by identifying genera and determining temperature tolerance. *Appl. Environ. Microbol.* **64**(5), 1822–1824.

Rowbotham, T.J. (1980) Preliminary report on the pathogenicity of *Legionella pneumophila* for freshwater and soil amoebae. *J. Clin. Pathol.* **33**, 1179–1183.

Schlech, W.F. (1990) *Legionella* and fountains. *Lancet* **336**, 576 (letter).

Segal, G. and Shuman, H.A. (1999) *Legionella pneumophila* utilizes the same genes to multiply within *Acanthamoeba castellanii* and human macrophages. *Infect. Immun.* **67**, 2117–2124.

Shands, K. *et al.* (1985) Potable water as a source of Legionnaires' disease. *J. Am. Med. Assn.* **253**, 1412.

Shetty, N., Srinivasan, S. Holton, J. and Ridgway, G.L. (1999) Evaluation of microbicidal activity of a new disinfectant: Sterilox 2500 against *Clostridium difficile* spores, *Helicobacter pylori*, vancomuycin resistant *Enterococcus* species, *Candida albicans*, and several *Mycobacterium* species. *J. Hosp. Infect.* **41**, 101–105.

Singapore. Ministry of the Environment, Singapore. (2001) Code of practice for the control of legionella bacteria in cooling towers.

Spain. (2001) 14770 Royal Decree 909/2001 establishing the sanitary and hygiene criteria for preventing and controlling legionellosis. *The Ministry of Health and Consumer Affairs*. Palma de Mallorca, 27 July 2001.

Spitalny, K.C., Vogt, R.L., Orciari, L.A. *et al.* (1984) Pontiac fever associated with a whirlpool spa. *Am. J. Epidemiol.* **120**, 809–817[17].

Srikanth, S. and Berk, S.G. (1994) Adaption of amoebae to cooling tower biocides. *Microbial. Ecol.* **27**, 283–301.

Steele, T.W. and McLennan, A.M. (1996) Infection of *Terahymena pyriformis* by *Legionella longbeachae* and other *Legionella* species found in potting mixes. *Appl. Environ. Microbiol.* **62**, 1081–1083.

Steinert, M., Emody, L., Amann, R. and Hacker, J. (1997) Resuscitation of viable but nonculturable *Legionella pneumophila* Philadelphia JR32 *by Acanthamoeba castellanii. Appl Environ Microbiol.* **63**(5), 2047–2053.

Steinert, M., Birkness, K., White, E., Fields, B. and Quinn, F. (1998*) Mycobacterium avium* bacilli grow saprozoically in cocultrue with *Acanthamoeba polyphaga* and survive within cyst walls. *Appl. Environ. Microbiol.* **64**(6), 2256–2261.

Steinert, M., Haegle, S. Skriwan, C., Grimm, D., Fajard, M., Heuner, K., Schleicher, M., Hentschel, U., Ludwig, W., Marre, R. and Hacker, J. (2002) Interaction of *Legionella pneumophila* with *Dictyostelium discoideum* (eds R. Marre *et al.*), Chap. 28, pp. 161–164, ASM Press, American Society for Microbiology, Washington, D.C.

Stone, B.J. and Kwaik, Y.A. (1998) Expression of multiple pili by *Legionella pneumophila*: Identification and characterization of a type IV pilin gene and its role in adherence to mammalian and protozoan cells. *Infect. Immun.* **66**, 1768–1775

Stout, J.E. *et al.* (1992) Potable water as the cause of sporadic cases of community-acquired Legionnaires' disease. *N. Engl. J. Med.* **326**, 151.

Stout, J.E. and Yu, V.L. (1997) Legionellosis. *N. Engl. J. Med.* **337**, 682–687.

Sutherland, E.E. and Berk, S.G. (1996) Survival of protozoa in cooling tower biocides. *J. Ind. Microbiol.* **16**, 73–78.

Surman, S., Morton, G., Keevil, B. and Fitzgeorge, R. (2002) *Legionella pneumophila* proliferation is not dependent on intracellular replication (eds R. Marre *et al.*), Chap 17, pp. 86–89, ASM Press, American Society for Microbiology, Washington, D.C.

Swanson, M.S. and Hammer, B.K. (2000) *Legionella pneumophila* pathogenesis: A fateful journey from amoebae to macrophages. *Annu. Rev. Microbiol.* **54**, 567–613.

Swanson, M.S. and Bachman, M.A. (2002) The *Legionella pneumophila* life cycle: Connections between growth phase, virulence expression, and replication vacuole biogenesis (eds R. Marre *et al.*), Chap. 15, pp. 74–81, ASM Press, American Society for Microbiology, Washington, D.C.

Swanson, M.S. and Fernandez-Moreia, E. (2002) A microbial strategy to multiply in macrophages: the pregnant pause. *Traffic* **3**, 170–177.

Thantsha, M.S. (2002) Electrochemically activated water as an environmentally safe disinfectant. M.Sc. Thesis. Dept. of Microbiology and Plant Pathology, University of Pretoria, Pretoria, South Africa.

Thom, S., Warhurst, D. and Drasar, B.S. (1992) Association of *Vibrio cholerae* with fresh water amoebae. *J. Med. Microbiol.* **36**(5), 303–306.

Tiefenbrunner, F., Arnold, A., Dierich, M.P. and Emde, K. (1993) Occurrence and distribution of *Legionella pneumophila* in water systems of central European private homes. In *Legionella*: *Current Status and Emerging Perspectives*. J.M. Barbaree, *et al.* (eds.). American Society for Microbiology, Washington, D.C.

Tobin, J.O., Bartlett, C.L.R. and Waitkins, S.A. (1981) Legionnaires' disease: further evidence to implicate water storage systems as sources. *Br. Med. J.* **282**, 573.

Walker, J.T., Bradshaw, D.J., Fulkford, M.R. and Marsh, P.D. (2003) Microbiological evaluation of a range of disinfect products to control mixed-species biofilm contamination in a laboratory model of a dental unit water system. *Appl. Environ. Microbiol.* **69**(6), 3327–3332.

Watnick, P. and Kolter, R. (2000) Biofilm, city of microbes. *J. Bacteriol.* **182**(10), 2675–2679.

Winiecka-Krusnell, J., Wreiber, K., von Euler, A., Engstrand, L. and Linder, E. (2002) Free-living amoebae promote growth and survival of *Helicobacger pylori*. *Scand. J. Infect. Dis.* **34**, 253–256

Wireman, J.E., Schmidt, A., Scavo, C.R. and Hutchins, D.T. (1993) Biofilm formation by *Legionella pneumophila* in a model domestic hot water system. In *Legionella*: *Current Status and Emerging Perspectives*. J.M. Barbaree *et al.* (eds.). American Society for Microbiol., Washington, D.C.

WHO (World Health Organization). (2002) *Legionella*. In: *Guidelines For Drinking-Water Quality*, (contributors: Botzenhart, K., Bartram, J., Butler, J., Havelaar, A.H., Lee, J., Mariee, T., Pastoris, C., Plouffe, J., Tiefenbrunner, F., Yu, V.). World Health Organization, Geneva, Switzerland.

WHO (World Health Organization) (1997) WHO Recommended Surveillance Standards. WHO/EMC/DIS/97.1, pp. 59–60 www.who.int/emc.

Yee, R.B. and Wadowsky, R.M. (1982) Multiplication of *Legionella pneumophila* in unsterilized tap water. *Appl. Environ. Microbiol.* **43**, 1330–1334.

Zacheus, O.M. and Martinkainen, P.J. (1994) Occurrence of legionellae in hot water distribution systems of Finnish apartment buildings. *Can. J. Microbiol.* **40**(12), 993–999.

Zinkevich, V., Beech, I.B., Tapper, R. and Bogdarina, I. (2000) The effect of super-oxidized water on *Eschericia coli*. *J. Hosp. Infect.* **46**, 153–156.

6

The effect of climate change on waterborne disease[1]

T. Scott, E. Lipp and J. Rose

6.1 INTRODUCTION

From the ancient beginnings of the human race and first recordings of infectious diseases a connection has been made between climate or weather and maladies in communities. The Hippocratic treatise on epidemics in fact attributed and associated illness with seasons, waters, winds and regions (Smith, 1979). In the current modern era, global warming has been a subject of a lot of scientific discussion and considerable amount of research, attempting to address climate variability, climate change and the full impact of these on biological systems including humans. From the perspective of water it is easy to understand the relationship between the climate and the quantity of water, and while it is acknowledged that quantity and quality are linked, the role of climatic factors associated with degradation in water quality and waterborne disease has just begun to be elucidated.

Globally, waterborne disease is a significant cause of morbidity and mortality and this threat is now gaining attention worldwide with warnings from international organizations and health communities. We are experiencing a global water crisis.

[1] This chapter was adapted in large part from the American Academy of Microbiology, 2001. *Health, climate and infectious disease: a global perspective*, AAM Press, Washington, D.C.

 Microbial Waterborne Pathogens edited by T.E. Cloete, J. Rose, L.H. Nel and T. Ford. ISBN: 1 84339 055 8. Published by IWA Publishing, London, UK.

According to the World Health Organization, 40% of the world's population (6 billion people) lack sanitation services and 1 billion people lack access to safe water. Up to 2.2 million will die from diarrhoea. By 2025, 48% of the world's population (3.5 billion) will live in water-stressed river basins (Brown, 2002). Therefore, a better understanding of the multitude of factors (including climate) that contribute to the occurrence of waterborne diseases would result in the implementation of better control, prediction and remediation efforts. The ability to predict potential outbreaks, based on susceptible populations and climate variability, would be particularly useful from a public health perspective. However, the goal must be to incorporate climate signals into our current knowledge of infectious disease transmission for predicting waterborne disease. By doing this, a proactive rather than a reactive, approach may be taken to reduce the burden of waterborne disease worldwide via early warning systems designed for use by the water industry.

Weather information is that which the general public and water professionals are accustomed to receive. This is the day-to-day change in atmosphere, including temperature, humidity, precipitation and winds. Predictions can be made in regions on the weather up to a 3–5 day forecast with some degree of accuracy. Climate may be defined as the average meteorological conditions over time frames of months to years to decades, which address variability in: (a) temperature gradients globally for elevation and distance from the equator to the pole, (b) frequency and intensity of extreme events and (c) regional patterns of temperature, precipitation and storm frequency. Seasonal and diurnal normal cycles are well recognized; however, it is climate modes, which cover inter-annual up to decadal cycles and are farther reaching in spatial scales that are of great interest in climate studies. These climate modes, which are complex interactions between recurring pressure and wind patterns, have now been uncovered and through research can now be predicted to one degree or another. The most studied and well-understood mode is the El Niño/Southern Oscillation (ENSO) cycle, referred to often as just El Niño; this is the cycle most often known to non-climatologists (Ropelewski and Halpert, 1987; http: // www.cpc.ncep.noaa.gov/products/analysis_monitoring/impacts/warm.gif). Thus, the global impacts of El Niño and predictions about which part of the world will be dry, wet, warm or cool can be made. Other climate modes include the Pacific North American Oscillation and the North Atlantic Oscillation. Long time scales and large global spatial scales are referred to and this must be reconciled with our usual view of water and waterborne disease.

The complex interactions between changes in climate and waterborne disease have to be identified, studied and understood so that one can develop models that are capable of predicting and preventing the emergence and re-emergence or spread of many diseases. The recognition that certain diseases are related in some way to the 'season' or 'weather' is accepted by the general public (e.g., the flu season); however, the interaction of the environment and climate influencing water quality are not accepted. There is a growing interest in forecasting or predicting climate events, and these may consequently be useful in preventing disease. An understanding of the modes of disease variability vis-á-vis how they relate to climate fluctuations is absolutely necessary to make progress towards control and improved protection of public health in the future. It is obvious, however, that climate is not the only factor contributing to the risk of disease. Demographic, behavioural and socioeconomic factors may override any climatological relationships and can therefore be used for appropriate adaptation to any developed models. In the case of waterborne diseases, however, climate may be a very good proxy or indicator of susceptibility and will allow preventative measures to be put in place.

6.2 CLIMATIC IMPACTS ON WATERBORNE DISEASE

The quality of water, including surface (fresh and marine) and ground waters can be related to climate and weather variability. For example, oceanic and coastal waters are known to harbour and transport micro-organisms that cause disease in humans and other animals. As modulators of climate, oceans also indirectly influence disease patterns and distribution of many pathogens. While certain pathogenic or toxigenic micro-organisms, including toxic phytoplankton and *Vibrio* spp., occur naturally in marine and estuarine waters, anthropogenic contaminants, including enteric bacteria, protozoa and viruses, may be introduced into water as sewage pollution. Waterborne disease agents can impact water supply as a result of heavy rainfall, flooding, changes in temperature and climate modes such as El Niño. The following sections summarize some of the data on waterborne microbial agents of concern and the impact of climate on their transmission and outbreaks of disease. Approaches for a water research agenda to help define knowledge gaps and directions for the future to limit the global burden of waterborne disease are also discussed.

6.2.1 Climate sensitive waterborne diseases

Waterborne disease causing organisms are for the most part in three general categories of micro-organisms, i.e. bacteria, protozoa and viruses. In most cases, these originate from human and/or animal faecal wastes. Bacterial waterborne agents of most concern include *Vibrio* spp. (such as *Vibrio cholerae*) and cyanobacteria (associated with harmful algal blooms; HABs); however, organisms such as *E. coli* O157:H7, which have traditionally been associated with food-borne outbreaks, have now been implicated in a number of drinking water outbreaks (Yarze and Chase, 1999). Viruses, particularly human enteric viruses, are also capable of being transmitted by the water route and are likely responsible for many subclinical and acute gastrointestinal illnesses (AGI) worldwide. Because most cases of viral gastroenteritis do not require medical attention, the occurrence of these outbreaks often goes unreported. The caliciviruses have now been reported as one of the most significant causes of viral food-borne and waterborne diseases (Monroe *et al*., 2000) and, in addition, because of their size, transport to groundwater is of significant concern. The enteric protozoa (*Cryptosporidium* and *Giardia*) can cause a more severe gastrointestinal illness and are the most widely reported cause of parasitic disease.

Many outbreaks have been reported in the periods following heavy rainfall (*E. coli* 017H7; Yarze and Chase, 1999; *Cryptosporidium*, Mackenzie *et al.* 1993; *Giardia;*). One of the main climate factors then is precipitation, mostly as rainfall, associated with increases in rainfall and flooding effecting the transport of pathogens from their source to the well or surface water directly or potentially impacting the distribution system. Temperature fluctuations will have an effect on the survival of these waterborne disease agents, with cooler temperatures leading to greater survival (Rose *et al.*, 2001). Hunter (2003) reported on two major peaks in spring and fall associated with outbreaks in the United Kingdom, likely influenced by rain and cool temperature. Several outbreaks and increased levels of indicator and pathogenic organisms in the environment have been attributed to extreme weather conditions as well as anomalies such as El Niño (Gueri *et al.*, 1986; Lipp *et al*, 2001b).

6.2.2 Impact of extreme events on waterborne disease

6.2.2.1 Water quality and potential for disease

Until recently, the importance of rainfall in association with the transport of contaminants from their sources to drinking water supplies had not been studied on a watershed basis. Rainfall and runoff have been associated with individual outbreaks of waterborne disease and it was clear that pathogens of faecal origin can find their way into water. However, most studies simply examined and reported on indicator bacteria in storm waters.

Cryptosporidium and *Giardia* (oo)cyst occurrence in surface waters have been particularly well studied in relation to heavy precipitation events. High levels of *Cryptosporidium* in Chesapeake Bay were shown to directly correlate with the highest recorded rainfall, the greatest stream flow into the Bay and low water temperatures (Fayer *et al.*, 2002). Overall microbial loading of water supplies was associated with rainfall and snow melt, thus suggesting the potential for a variety of disease-causing organisms to be introduced into source waters (Kistemann *et al.*, 2002). Kistemann *et al.* (2002) used extreme storm events, meterological and hydrological parameters and water quality monitoring to demonstrate that between 4% and 32% of the micro-organism's (faecal bacteria) annual load could be contributed to a waterbody during a single 12-hour event. A single event could contribute about 4% of the annual load of *Cryptosporidium* and *Giardia* (oo)cysts.

In 1998 and 1999, high levels of *Cryptosporidium* and *Giardia* were detected in the water supply and distribution system in Sydney Australia, using a modified IFA technique and flow cytometry (Cox *et al.*, 2003). Described as the "Sydney Water Crisis", an expert report and scientific scrutiny have suggested that the ingress into the water system occurred with contaminated flood waters impacting the reservoir. Episodic detection occurred following flowing stream rises in the catchment (Hawkins *et al.* 2000 and Cox *et al.*, 2003). Studies by PCR revealed genotype II as the only strain, suggesting that animals could have been the source, yet as this genotype can also be found in humans, this was not necessarily the definitive source.

With climate predictions suggesting that storms will be of greater intensity and that the average precipitation event is likely to be heavier, the risk of waterborne disease in the United States due to heavy rainfall events and/or increased flooding is likely to increase.

6.2.2.2 Data on waterborne outbreaks and climate factors

Rainfall and runoff have also been implicated in several waterborne outbreaks in the United States and the United Kingdom (Hunter, 2003). In 1993, *Cryptosporidium* was responsible for the largest documented waterborne outbreak in the United States when over 400,000 people became ill and 54 people died in Milwaukee, WI due to inadequately treated drinking water. This outbreak followed a period of heavy rainfall, which resulted in high turbidity of source waters, and overwhelmed the local drinking water treatment facility (MacKenzie *et al.*, 1994). *Cryptosporidium* also caused outbreaks in the UK in 1992 and 1993 when catchment areas supplying raw water to drinking water treatment facilities became contaminated (at least in one case by cattle faeces) after periods of heavy rainfall (Atherton *et al.*, 1995; Bridgman *et al.*, 1995). An outbreak of cryptosporidiosis in Japan in 1996 was attributed to various natural and artificial factors including a heavy rainfall event, which was preceded by drought conditions and resulted in an increased turbidity of source waters. As a result, the coagulation process in the local drinking water facility was insufficient to remove *Cryptosporidium* oocysts from the finished water (Yamamoto *et al.*, 2000).

Although bacterial waterborne outbreaks are often underreported and many bacterial pathogens are most often associated with foodborne illness, the largest reported outbreak of *E. coli* O157:H7 occurred at a New York City fairground in 1999 when well water was contaminated by cattle manure following a period of heavy rainfall. It is estimated that at least 1000 people became ill and two people died as a result of infection and subsequent haemolytic uraemic syndrome (Yarze and Chase, 2000).

The first study on the direct statistical association between extreme precipitation and waterborne outbreaks showed that 51% of all outbreaks in a 50-year time-frame were preceded by precipitation events above the 90th percentile and that the outbreaks had a spatial relationship associated with watersheds (Curriero *et al.*, 2001; Rose *et al.* 2000). This investigation demonstrated a methodological approach that included: (1) a geographical positioning of the outbreak in time and space (greater resolution was recommended); (2) division of the area into sub-major hydrologic units (watersheds) representing surface drainage; (3) assessment of precipitation data within the watershed for time of the outbreak and the preceding period, from weather stations with extreme precipitation defined by the distribution associated with the 20-year annual monthly precipitation average. Statistical analysis of 548 outbreaks demonstrated that extreme precipitation occurred during the month of the outbreak for surface waters and had a lag time of two months (preceding) for outbreaks associated with ground waters.

6.3 CHOLERA

6.3.1 Cholera: a case for ecological and climate studies

Cholera remains a significant cause of waterborne disease, and while it is understood that this disease is spread through the faecal–oral route via contaminated food and water, the water industry must now address how the environment and climate are playing a role in disease transmission if key efforts are to be made in curbing the disease.

The key new findings include:

- seasonality to the endemic disease, particularly in regard to rainfall,
- ability of the cholera bacteria to grow to high numbers in zooplankton,
- transport of the bacteria in ballast waters.

While *Vibrio* appears to be sensitive to disinfection, without adequate filtration, bacteria protected by turbidity during the rainy season or protected in zooplankton will continue to remain viable in treated water. Thus, a multi-barrier approach must be adhered to and an understanding of the ecology of the cholera organism in key parts of the world is essential.

Eight pandemics of cholera have been documented since 1817 (Faruque *et al.*, 1998). However, it was not until the seventh outbreak of the disease in Asia, Africa and South America in the 1990s where it spread like wildfire that the ecology of this disease was elucidated. Cholera is a classical faecal–oral disease, yet the monitoring of traditional coliform bacteria has provided little understanding of the risks and management options. Cholera cases began in January, 1991 in Peru after a century of its absence, and the epidemic reached every country in South America in one year primarily due to poor sanitation and drinking water treatment (PAHO, 1991). The spread of the disease was associated with ballast water, international food transport and travellers (Tauxe *et al.*, 1994).

It is the understanding of the ecology of the bacterium that has elucidated unrecognized disease transport pathways and the role of climate. The ability of *Vibrio cholerae* to act as a heterotroph in aquatic environments and live and multiply within plankton creates an environmental reservoir of infection. As plankton move upstream from estuarine waters, village water supplies are affected. The non-culturable state of the bacterium requires that specialized methods and tools be used to study this phenomenon including antibody and nucleic acid based approaches (Colwell *et al.*, 1996; Theron and Cloete, 2002). There is now a clear and consistent link between climate and cholera, particularly ENSO for certain regions of the world such as Bangladesh (Rodo *et al.*, 2002). This understanding of seasonal risks and transmission has led to the implementation of a simple filtration approach for reducing the disease in remote areas. (Colwell *et al.*, 2003).

Specific efforts have examined epidemics using molecular epidemiology and thus the strains and their global transmission during such pandemics can be related to specific health risks. New strains with increasing virulence have been characterized, thus suggesting that virulence-associated genetic elements are related to evolutionary/environmental fitness (Farugue *et al.*, 1998). Despite the efforts to investigate and study cholera, the disease remains a major enemy of public health in many parts of the world.

6.3.2 Modelling of cholera dynamics as a function of climate variability

The modelling of cholera in relation to climate is in the first stages of identifying the key environmental and climate parameters associated with disease variability (Lipp *et al.*, 2001c). As for many diseases, in endemic regions such as Bangladesh, the incidence of cholera cases follows seasonal patterns. Further analyses of such patterns reveal a relationship to climate at both seasonal and inter annual time scales. There is now time series evidence for the role of climate in the inter annual variability of cholera that had previously been hypothesized (Pascual *et al.*, 2000). Empirical models have included environmental parameters such as water temperature and rainfall in the prediction of the seasonal distribution of toxigenic *Vibrio cholerae* in the environment and the occurrence of clinical cases of cholera (Huq *et al.*, 1990, Colwell, 1996). Sea surface temperatures and sea surface height have also been directly correlated with cholera epidemics in Bangladesh (Lobitz *et al.*, 2000) and preliminary evidence shows a similar trend in Perú.

While water disinfection and treatment need to be implemented to control cholera in developing nations, the abundance of *Vibrio cholerae* in environmental reservoirs and the feedback from and to the human populations will be part of the disease dynamics in key local and regional areas. Statistical and mathematical models need to include water temperature, rainfall, salinity, and the abundance of plankton populations to define the *Vibrio* population dynamics in aquatic environments. (Lobitz *et al.*, 2000; Franco *et al.*, 1997). In many areas of the world, unless this pathway is considered, explosive entrance of the disease back into populations is likely to reoccur.

6.4 EL NIÑO

Waterborne disease has also been shown to fluctuate as a result of inter annual variability associated with weather patterns such as the El Niño-Southern Oscillation (ENSO). This phenomenon has been proposed to influence cholera outbreaks and levels of diarrhoeal disease in Perú (Colwell, 1996; Checkley *et al.*, 2000) and has also been associated with dynamics of the disease in Bangladesh (Pascual *et al.*, 2000). Levels of anthropogenic pollution (including human viruses) in estuaries of south Florida and have also been shown to be directly affected by ENSO (Figure 6.1) (Lipp *et al.*, 2001a. and b). Studies in Tampa Bay found a statistical decrease in water quality over a 25-year time frame during El Niño winters, which significantly increased rainfall in the Tampa area and improved in the warmer La Niña cycle (Lipp *et al.*, 2001b).

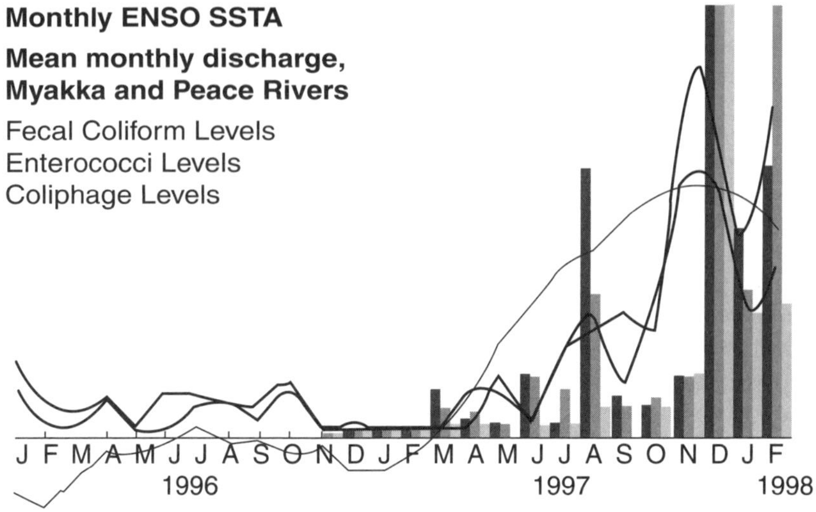

Figure 6.1 El Niño monthly sea surface temperature anomalies, mean monthly discharge from the Myakka and Peace rivers, and water quality indicator levels (normalized to December 1997) from Charlotte Harbor, FL.

In addition to these periodic climatic ‘anomalies’, a signal in long-term sea surface temperature records has recently been identified and is believed to be a manifestation of global warming (Livezey and Smith, 1999). The impact of long-term climate changes will likely include shifts in the distribution of many species, including pathogenic micro-organisms, and may alter host susceptibility to infection. For example, the geographic range of many *Vibrio* spp. would be expected to increase with warming temperatures. Furthermore, changes may affect toxicity in certain organisms such as *Pfiesteria piscicida* in which the toxic zoospore stage is thought to be related secondarily to warm temperatures (the presence of diseased fish is the primary signal for transformation to toxic stage) (Burkholder *et al.*, 1998).

6.5 KNOWLEDGE GAPS AND DIRECTIONS FOR FUTURE RESEARCH

In summary, it is evident that local weather patterns play a role in the dispersal of pathogenic micro-organisms. Events such as extreme rainfall and floods often overburden water treatment facilities and increase storm water run-off. Both of these events may result in the introduction of high levels of enteric pathogens into nearby surface waters and wells if groundwater becomes contaminated. Indeed, heavy rainfall and associated flooding have been statistically related to high incidence of gastrointestinal disease (Gueri *et al.*, 1986).

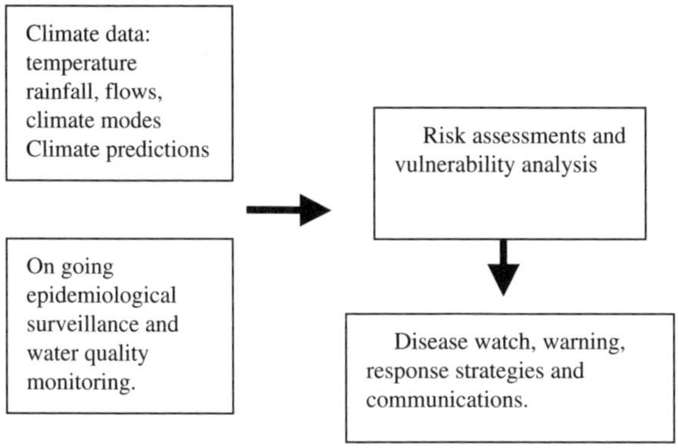

Figure 6.2 Early warning system components (Adapted from NRC, 2001).

The health of the general public remains an important topic throughout the world and waterborne infectious diseases are an increasing threat for both developed and developing countries. Initially, an emphasis must be placed on the identification of at-risk populations and the aetiological agents to which they are likely to be exposed. The medical profession must be educated on what constitutes climate-sensitive diseases and it is crucial that appropriate temporal and spatial scales are evaluated and recorded in reference to diseases and disease statistics.

In the future, the complexity or 'bioclimatology' of any one disease or pathogen should be defined, described and studied within an integrated risk assessment framework to determine susceptibility of particular regions to certain diseases under a given set of climate conditions. This would most likely involve a combination of ecological frameworks and quantitative microbial risk assessment (Haas *et al.*, 1999).

In addition, government agencies responsible for addressing human health and the health of the environment must address the prospect that global warming and climate change is likely to affect regions of the world differently. Warming and cooling trends, changes in precipitation, storm events, humidity, and changes in winds and circulation patterns will affect the ecology of certain pathogenic microorganisms.

For the water industry, the goal would be to define data needs and components for an early warning system that will provide an effective management strategy. Figure 6.2 demonstrates the components and data for such a system (NRC, 2001).

Overall, climatologists, health scientists, ecologists and water professionals including microbiologists must find common ground for successful research and engage in discussions that allow each to know which data are needed and available. In addition, long-term ecological surveillance for pathogens and vectors in conjunction with appropriate collection of disease and climate data is imperative. Uncertainty about the future continues to spark passionate debates regarding risk to the welfare of the human race. In the meantime, tools and techniques, from satellites to molecular probes to sophisticated deterministic models, are now available and can be used to address many of these issues and provide a reassurance that global health and welfare will continue to improve.

REFERENCES

Atherton, F., Newman, C.P. and Casemore, D.P. (1995) An outbreak of waterborne cryptosporidiosis associated with a public water supply in the UK. *Epidemiol. Infect.* **115**, 123–131.

Bridgman, S.A., Robertson, R.M., Syed, Q., Speed, N., Andrews, N. and Hunter, P.R. (1995) Outbreak of cryptosporidiosis associated with a disinfected water supply. *Epidemiol. Infect.* **115**, 555–566.

Burkholder, J.M., Glasgow Jr, H.B. and Lewtus, A.J. (1998) Pysiological ecology of *Pfiesteria piscicida* with general comments of "ambush-predator" dinoflagellates. In *Physiological Ecology of Harmful Algal Blooms* (eds D.M. Anderson *et al.*), NATO ASI Series vol. G. 41, pp. 17–191, Springer-Verlag, Berlin.

Brown, K. (2002) Water security: Forecasting the future with spotty data. *Science* **297**, 926–927).

Checkley, W., Epstein, L.D., Gilman, R.H., Figueroa, D., Cama, R.I., Patz, J.A. and Black, R.E. (2000) Effect of El Niño and ambient temperature on hospital admission and diarrhoeal diseases in Peruvian children. *Lancet* **3**, 442–450.

Colwell, R.R., Huq, A., Islam, M.S., Aziz, K.M., Yunus, M., Khan, N.H., Mahmud, A., Sack, R.B., Nair, G.B., Chakraborty, J., Sack, DA.., Russek-Cohen, E. (2003) Reduction of cholera in Bangladeshi villages by simple filtration. *Proc. Natl. Acad. Sci. U.S.A.* **100**(3), 1051–1055. Epub

Colwell, R.R. (1996) Global climate and infectious disease: the cholera paradigm. *Science* **274**, 2025–2031.

Colwell, R.R., Brayton, P., Herrington, D., Tall, B., Huq, A. and Levine, M.M. (1996) Viable but non-culturable *Vibrio cholerae* 01 revert to a cultivable state in the human intestine. *World J. Microb. Biot.* **12**, 28–31.

Cox, P., Fisher, I., Kastl, G., Jegatheesan, V., Warnecke, M., Angles, M. Bustamante, H., Chiffings, T. and Hawkins, P.R. (2003) Sydney 1998 – lessons from a drinking water crisis. *J. Am. Water Works Assoc.* b95(5). 147161.

Curriero, F.C., Patz, J.A., Rose, J.B. and Lele, S. (2001) The association between extreme precipitation and waterborne disease outbreaks in the United States, 1948–1994. *Am. J. Public Health* 91: 1194–1199.

Faruque, S.M., Alvert, M.J. and Mekalanos, J.J. (1998) Epidemiology, genetics, and ecology of toxigeneic *Vibrio cholerae*. *Microbiol. Mol. Biol. Rev.* **62**, 1301–1314.

Fayer, R., Trout, J.M., Lewis, E.J., Xiao, L., Lal, A., Jenkins, M.C. and Graczyk, T.K. (2002) Temporal variability of *Cryptosporidium* in the Chesapeake Bay. *Parasitol. Res.* **88**, 998–1003.

Franco, A.A., Fix, A.D., Prada, A., Paredes, E., Palomino, J.C., Wright, A.C., Johnson, J.A., McCarter, R., Guerra, H. and Morris, J.G. (1997) Cholera in Lima, Peru correlates with prior isolation of Vibrio cholerae from the environment. *Am. J. Epidemiol.* **146**, 1067–1075.

Gueri, M., Gonzalex, C. and Morin, V. (1986) The effect of the floods caused by "El Nino" on health. *Disasters* **10**, 118–124.

Haas, C.H., Rose, J.B. and Gerba, C.P. (eds). (1999) *Quantitative Microbial Risk Assessment*, John Wiley and Sons, New York, NY.

Hawkins, P., Swanson, P.,Warnecke, M., Shanker, S.R. and Nicholson, C. (2000) Understanding the fate of *Cryptosporidium* and *Giardia* in storage reservoirs: a legacy of Sydney's water contamination incident. *Aquatic Toxicol.* **49**(6), 289–306.

Hunter, P.R. (2003) Climate change and waterborne and vector-borne disease. *J. Appl. Microbiol.* **94**, 37S–46S.

Huq, A., Colwell, R.R., Rahman, R., Chowdhury, A.A., Parveen, S.,l Sack, D.A. and Russek-Cohen, E. (1990) Detection of *Vibrio cholerae* 01 in the aquatic environment by fluorescent monoclonal antibody and culture methods. *Appl. Environ. Microbiol.* **56**, 2370–2373.

Kistemann, T., Classen, T., Koch, C., Dangendorf, F., Fischeder, R., Gebel, J., Vacata, V. and Exner, M. (2002) Microbial load of drinking water reservoir tributaries during extreme rainfall and runoff. *Appl. Environ. Microbiol.* **68**(5), 2188–2197.

Lipp, E.K., Kurz, R., Vincent, R., Rodriguez-Palacios, C., Farrah, S.R. and Rose, J.B. (2001a) The effects of seasonal variability and weather on microbial faecal pollution and enteric pathogens in a subtropical estuary. *Estuaries* **24**, 266–276.

Lipp, E.K., Schmidt, N., Luther, M. and Rose, J.B. (2001b) Determining the effects of El Nino-Southern Oscillation events on coastal water quality. *Estuaries* **24**, 491–497.

Lipp, E.K., Huq, A. and Colwell, R.R. (2001c) Effects of global climate on infectious disease: the cholera model. *Clin. Microbiol. Rev.* **15**, 757–770.

Livzey, R.E. and Smith, T.M. (1999) Covariability of aspects of North American climate with global sea surface temperatures on interannual to interdecadal timescales. *J. Climate* **12**, 289–302.

Lobitz, B.L., Beck, L., Huq, L., Wood, A., Fuchs, B., Faruque, A.S.G. and Colwell, R.R. (2000) Climate and infectious disease: use of remote sensing for detection of *V. cholerae* by indirect measurement. *Proc. Natl. Acad. Sci. U.S.A.* **97**, 1438–1443.

Mac Kenzie, W.R., Hoxie, N.J., Proctor, M.E., Gradus, M.S., Blair, K.A., Peterson, D.E., Kazmierczak, J.J., Addiss, D.G., Fox, K.R., Rose, J.B. and Davis, J.P. (1994) A massive outbreak in Milwaukee of *Cryptosporidium* infection transmitted through the public water supply. *N. Engl. J. Med.* **331**, 161–167.

Monroe, S.S., Ando, T., and Glass, R.I. (2000) International workshop on human caliciviruses. *J. Infect. Dis.* **181**(Suppl 2).

National Research Council (2001) *Under the Weather Climate, Ecosystems, and Infectious Disease*, National Academy Press, Washington, D.C.

Pan American Health Organization (1991) Cholera situation in the Americas. *Epidemiol. Bull.* **12**, 1–4.

Pascual, M., Rodo, X., Ellner, S.P., Colwell, R.R. and Bouma, M.J. (2000) Cholera dynamics and El Nino-Southern Oscillation. *Science* **289**, 1766–1769.

Rodo, X., Pascual, M., Fuchs, G. and Faruque, A.S.G. (2002) ENSO and Cholera: A nonstationary link related to climate change? *Proc. Natl. Acad. Sci. U.S.A.* **99**(20), 12901–12906.

Ropelewski, C.F and Halpert, M.S. (1987) Global and regional scale precipitation patterns associated with the El Niño/Southern Oscillation. *Mon. Wea. Rev.* **115**, 1606–1626.

Rose, J.B., Epstein, P.R., Lipp, E.K., Sherman, B.H., Bernard, S.M. and Patz, J.A. (2001) Climate variability and change in the United States: Potential impacts on water- and foodborne diseases caused by microbiologic agents. *Environ. Health Perspect.* **109**(Suppl 2), 211–221.

Rose, J.B., Daeschner, S., Easterling, D.R., Curriero, F.C., Lele, L. and Patz, J.A. (2000) Climate and waterborne disease outbreaks. *J. Am. Water Works Assoc.* **92**, 77–87.

Smith, W.D. (1979) *The Hippocratic Tradition*, Cornell University Press, Ithaca, N.Y.

Tauxe, R., Seminario, L., Tapia, R., and Libel, M. (1994) The Latin American Epidemic. In *Vibrio cholerae and Cholera: Molecular to Global Perspectives* (eds I.K. Wachsmuth, P.A. Blake and O. Olsvik), American Society for Microbiology Press, Washington, D.C.

Theron, J. and Cloete, T.E. (2002) Emerging waterborne infections: contributing factors, agents and detection tools. *Crit. Rev. Microbiol.* **28**, 1–26.

Yamamoto, N., Urabe, K., Takaoka, M., Nakazawa, K., Gotoh, A., Haga, M., Fuchigami, H., Kimata, I. and Iseki, M. (2000) Outbreak of cryptosporidiosis after contamination of the public water supply in Saitama Prefecture, Japan, in 1996. *Kansenshogaku Zasshi* **74**, 518–526.

Yarze, J.C. and Chase, M.P. (1999) *E. coli* O157:H7 – another waterborne outbreak! *Am. J. Gastroenterol.* **95**(4), 1096.

7

Emerging microbiological detection techniques

J. Theron and T.E. Cloete

7.1 INTRODUCTION

Waterborne microbial diseases, once expected to be eliminated as public health problems, not only remain the leading cause of morbidity from infectious diseases worldwide, but the spectrum of disease is expanding and the incidence of many waterborne microbial diseases once thought conquered is increasing (McGinnis and Foege, 1993; Ford, 1999). These infections, which may be transmitted by contaminated recreational waters, surface water and ground water intended for drinking, place entire communities at risk. Even though major outbreaks of waterborne disease are comparatively rare, there is substantial evidence that human enteric pathogens that are frequently present in domestic water supplies are responsible for low-level incidence of waterborne microbial disease. Although these diseases are rarely debilitating to healthy adults for more than a few hours to a few days, enteric pathogens can cause severe illness, even death, for young children, the elderly or those with compromised immune systems (Ford and MacKenzie, 2000). Due to the numerous infectious agents including enteric bacteria, viruses and protozoa, reservoirs and asymptomatic infected individuals, complete eradication of waterborne diseases may not be possible. However, it should be possible to control these agents

 Microbial Waterborne Pathogens edited by T.E. Cloete, J. Rose, L.H. Nel and T. Ford. ISBN: 1 84339 055 8. Published by IWA Publishing, London, UK.

as long as they can be detected and monitored. Coliforms, the traditional indicators of pathogens, are beginning to fail in some cases by giving misleading information and despite advances being made in microbial testing techniques, some pathogens in water remain undetected. The presence of pathogens in water only becomes evident when a number of people become ill with a waterborne illness and a common source of drinking water is identified. Identification of the specific micro-organism that caused the illness may, however, not always be possible. In fact, no agent is identified as the specific cause in 50% of waterborne disease outbreaks (Moe, 1996). There is thus a great need for alternative testing strategies for waterborne diseases, both for routine monitoring and for investigating disease outbreaks.

7.2 TRADITIONAL VERSUS EMERGING DETECTION TECHNIQUES

Traditionally, prediction of the presence of human enteric pathogens in water has been achieved by monitoring for established microbial 'indicators' of faecal pollution. Not necessarily pathogenic themselves, faecal coliforms, total coliforms,

Table 7.1 Uses and disadvantages of commonly used faecal indicators (Rose and Grimes, 2001)

Indicator	Advantages	Concerns
Total coliforms	Conservative indicator, easily detected	Also found in waters not impacted by faecal pollution
Faecal coliforms	Easily enumerated, more specific than total coliforms	Some faecal coliforms (i.e. *Klebsiella* spp.) have no true faecal link, which may result in false positives
E. coli	Easily enumerated, rapid	May not strengthen ability to identify risks of viruses and parasites relative to coliforms. Rapid die-off rates in certain waters, re-growth potential in tropical environments
Enterococci	Ubiquitous in wastewater effluents	There are environmental reservoirs that have no known source
Clostridium perfringens	Extremely resistant to environmental stresses (e.g. temperature, salinity, UV/sunlight) and disinfection (e.g. ozone, chlorine)	Found in low numbers, persistence makes them an unreliable indicator of recent contamination
Male-specific bacteriophage	Associated with wastes from warm-blooded animals and are a conservative model for human enteric viruses	Unstable in warm estuarine waters

E. coli, enterococci and bacteriophages are all examples of organisms that, when present, are viewed as predictive of the potential presence of enteric pathogens, since they have the same faecal source as the pathogenic organisms. Tests for coliform bacteria are standardized and relatively easy and inexpensive to use (APHA, 1995; Toranzos and McFeters, 1996). They are therefore more rapidly administered than tests determining the presence of individual pathogenic micro-organisms in water. Despite being successful in predicting possible health risks in many circumstances, there are many flaws in using microbial indicators. Research has established the inability of many of these indicators to predict the presence of disease-causing viruses (such as hepatitis A and E, coxsackie viruses, echoviruses, adenoviruses and Norwalk viruses), indigenous bacteria (such as *Legionella* and *Helicobacter*) as well as parasites (such as *Cryptosporidium* and *Giardia*) (Moe, 1996). Commonly used and alternative microbial indicators, together with the advantages and disadvantages of their use, are listed in Table 7.1.

Since it would appear that conventional detection methods do not adequately assess the risk of waterborne disease, the need for powerful new tools for the detection of pathogenic micro-organisms in water is becoming increasingly more important. Although specific rapid, sophisticated molecular methodologies are available, these technologies have only recently been applied in the field of water science and technology. Because many micro-organisms are not easily cultured or can enter a viable but nonculturable state, the current methods focus on immunological or genetic characteristics to detect the presence of specific waterborne pathogenic micro-organisms.

7.3 MOLECULAR METHODOLOGIES FOR DETECTION OF PATHOGENIC MICRO-ORGANISMS

The past decade has witnessed rapid development in areas of biological sciences, notably molecular and cell biology. As a consequence of these advances, powerful new methodologies are emerging to detect both specific pathogens and indicator organisms. Not only do these methods increase the rapidity of analysis, but also they are able to achieve a high degree of sensitivity and specificity without the need for complex cultivation and additional confirmation steps. Consequently, some of these methods permit the detection of specific culturable and/or nonculturable micro-organisms within hours, instead of the days required with the traditional methods. A summary of available technologies for molecular detection and differentiation of micro-organisms, as well as their limitations is shown in Table 7.2.

7.3.1 Immunological methods

Immunological methods are based on the specific recognition between antibodies and antigens and the high affinity that is characteristic of this recognition reaction. Consequently, many different immunoassay methods have become available for both quantitative and qualitative analysis of pathogenic bacteria in water (Bottger

Table 7.2 Emerging techniques for use in detection of pathogenic micro-organisms (adapted from Rose and Grimes, 2001)

Molecular tool	Characteristics and advantages	Limitations
Immunomagnetic capture	• Relatively specific for target organism • Can be used for recovery from complex environmental samples	Currently unable to discern viable from non-viable organisms
Nucleic acid hybridization	• Highly specific • Can be used to differentiate virulent from non-virulent strains	Only applicable to culturable microbes
Polymerase chain reaction (PCR)	• Rapid and specific • Can be used for quantitative assays • Can be applied to expressed genes (i.e., RT–PCR on mRNA) • Can be used to detect nonculturable viruses and other microbes	Unless conditions are optimized, inconsistencies in performance can produce erroneous results Currently unable to directly discern viable from nonviable organisms
Biosensors	• Can use immunoaffinity or nucleic acid hybridization, followed by detection	Currently unable to discern viable from nonviable organisms
Gene chip technology	• Rapid detection • Very sensitive • Specific technique is currently unavailable although research is ongoing	Technique is currently unavailable, although research is ongoing
Solid-state biochip	• Visionary approach currently being developed for rapid detection (min) of a number of toxins and organisms • Does not require isolation of genetic material • No lengthy incubation • No labelling or washing	Technique is currently unavailable, although progress is being made

et al., 1987; Faude and Hofle, 1997; Buswell *et al.*, 1998). These include immunocapture of cells or antigens by enzyme-linked immunosorbent assay (ELISA or EIA), or detection of targeted cells by immunofluorescence (IFA). These assays can be performed by a direct or indirect approach. In a direct immunoassay, the monoclonal or polyclonal antibodies, directed against antigens located on the surface of the target pathogen (such as capsid proteins, cell wall or flagellar antigens), are conjugated with a fluorochrome or fluorescent dye. Alternatively, secondary enzymatically or fluorescently labelled antibodies directed against the primary antibodies (now serving as antigens) can be used in an indirect immunoassay. The advantage of this procedure is that the secondary antibodies can easily be obtained from a commercial supplier with a range of conjugated fluorochromes and their use leads to signal amplification as several labelled secondary antibodies can bind to a single unlabelled primary antibody (Hoff, 1993). The antigen–antibody complex is detected and quantified by the ability of the enzyme to react with a substrate that produces either a coloured product for colorimetry or emits light for luminometry. The immunoassays are often performed on a solid phase to which the pathogen antigens have been applied, such as a membrane filter or the bottom of a microtitre plate well (De Silva *et al.*, 1992; Hasan *et al.*, 1994; Quadri *et al.*, 1994; Noah *et al.*, 1996).

Studies have shown that solid-phase enzyme immunoassays generally are too insensitive for direct detection of microbial pathogens in water, as they require a minimum of 10^3 to 10^4 target microbes (or their antigens) for detection (Faude and Hofle, 1997; Hanai *et al.*, 1997). In most situations drinking water and its sources rarely contain high enough levels of most target pathogens for direct immunoenzymatic detection. Nevertheless, enumeration of diluted specific cells can be obtained by means of immunomagnetic separation (IMS). Immunomagnetic separation, also termed immunocapture or antibody capture, is a method that uses paramagnetic synthetic beads or other magnetic particles that have been coated with monoclonal or polyclonal antibodies directed against the target microbes to recover the microbes from the sample by antigen–antibody reactions (Olsvik *et al.*, 1994). The retained microbes can be analysed directly or after they or their nucleic acids have been released or extracted from the antibody and solid phase by various physical or chemical methods. IMS methods have the advantage of selecting, separating and purifying specific target microbes from other microbes and from solutes, based on the specificity of the antigen–antibody reaction. This is a powerful approach for recovering, enriching, purifying and concentrating the target viruses, bacteria and parasites from the sample matrix (Jothikumar *et al.*, 1998; Sasaki *et al.*, 1999; Lowery *et al.*, 2000). However, it is not applicable to some pathogens because of the lack of antisera or the antigenic diversity of a large pathogen group lacking a common antigen and thus requiring many antisera.

As an alternative to the above assays, agglutination methods can be used to detect pathogens by combining dispersed micro-organisms with antibodies (on a slide, for example) and allowing for antigen–antibody reactions to produce agglutination (clumping) that can be scored as negative or various degrees of positive (Colwell *et al.*, 1992; Quadri *et al.*, 1994). One modification is latex bead agglutination in which antibodies against a specific microbial antigen are attached to latex beads. The beads are reacted with the sample and should the sample contain the specific antigen, agglutination occurs by the reaction of antigens with antibodies on the beads resulting in the beads clumping together. As with enzyme immunoassays, agglutination tests are too insensitive to directly detect and quantify most waterborne pathogens in drinking water and other aquatic samples (Ito *et al.*, 1983). The target microbes must first be cultured in order to obtain a sufficient number of them or a sufficient amount of antigen to detect and antigenically characterize them by agglutination methods.

The use of immunological methods for the detection of specific micro-organisms is a rapid and simple technique, the accuracy of which mainly depends on the specificity of the antibody (Kfir and Genthe, 1993). Nevertheless, its application to the detection of specific micro-organisms from environmental water samples is limited (Hanai *et al.*, 1997). While IFA allows specific identification and detection at a single-cell level, it does not provide information on the physiological status or viability of the detected cells. The ELISA is a rapid, simple and quite sensitive test. However, assay limitations are often associated with the specificity of the antibody used, the concentration of both antibody and antigen, and the solid matrix often leads to non-specific binding of the antigen or of the secondary antibody (Kfir and Genthe, 1993).

7.3.2 Nucleic acid-based methods

7.3.2.1 Nucleic acid hybridization techniques

The easiest way of detecting specific nucleic acid sequences is through direct hybridization of a probe to microbial nucleic acid extracts. These hybridization techniques rely on the specific binding of nucleic acid probes to complementary DNA or RNA (target nucleic acid). The probes are single strands of nucleic acid with the potential of carrying detectable marker molecules highly specific to complementary target sequences, even if these sequences account for only a small fraction of the target nucleic acid. Either DNA or RNA can serve as a nucleic acid probe, but for a number of reasons (e.g., ease of synthesis and stability), most studies have employed DNA probes (Holben and Tiedje, 1988). The probes may be used to detect genes in the bacterial genome (Southern blots) or to detect mRNA or rRNA (Northern blots) (Schneegurt and Kulpa, 1998).

According to their length, DNA probes can be grouped as either polynucleotide probes (more than 50 nucleotides) or as oligonucleotide probes (less than 20 nucleotides) (Amann *et al.*, 1997). The latter are more frequently used since they are short enough to allow for single mismatch discrimination of target nucleic acids and large quantities of oligonucleotides can be rapidly and inexpensively produced. During oligonucleotide synthesis a variety of marker or linker sequences can be introduced to the 5' end of the oligonucleotide. The principle conjugate fluorochromes are derivatives of fluorescein or rhodamine, although labels such as digoxigenin and biotin have also been used (Amann *et al.*, 1993; Zarda *et al.*, 1992; Amann *et al.*, 1995). For probes to be effective in nucleic acid hybridization, they must be specific and selective. This requires that sequence data be available. Although such data are available for some waterborne pathogens, the data are not comprehensive for all pathogens of concern. However, the use of bacterial 16S rRNA sequences to develop determinative hybridization probes is now well established (Olsen *et al.*, 1986; Pace *et al.*, 1986). The occurrence of 16S rRNA molecules in high copy numbers of usually more than 1000 in any living cell as well as the fact that 16S rRNA sequences have been determined for a large fraction of bacterial species (Olsen *et al.*, 1986; Amann *et al.*, 1995; Amann *et al.*, 1997), currently confers on these genes a very high informative value at the phylogenetic level. Comparative analyses of rRNA sequences have indicated that the rRNA molecules comprise highly conserved sequence domains interspersed with more variable regions (Gutell *et al.*, 1994; Van de Peer *et al.*, 1996). Consequently, the so-called signature sequence motifs on various taxonomic levels, from domains to subspecies, can be identified (Woese, 1987; Giovannoni *et al.*, 1988; Manz *et al.*, 1992; Amann *et al.*, 1995). Whereas species-specific probes complement the most variable regions, more general probes target more conserved regions of the molecule (Stahl and Amann, 1991; Ward *et al.*, 1992). Although several oligonucleotide probes are commercially available, new specific probes may also be designed. The design of such 16S rRNA-targeted probes is greatly facilitated by computer-aided sequence comparisons of large rRNA sequence databases (Maidak *et al.*, 1994, 1997). The principal steps involved in the design of probes are: the alignment of rRNA gene sequences; the identification of sequence idiosyncrasies; the synthesis and labelling of complementary nucleic acid probes; and the experimental evaluation and optimization of the probe specificities and assay sensitivities using cultured reference strains (Manz *et al.*, 1992; Amann, 1995).

Nucleic acid hybridization can be performed in a variety of formats. The procedure typically entails the extraction of whole-cell DNA or RNA from the sample that is subsequently fixed to a nylon or nitrocellulose membrane. Alternatively, bacterial colonies can be replicated from agar plates to membranes and their nucleic acids exposed following lysis for subsequent hybridization. Whereas the former direct approach is usually inadequate due to high detection limits and large sample volumes that are impractical for most hybridization protocols without further pathogen concentration, the latter approach is time-consuming and cannot be applied to non-culturable organisms. Consequently, the use of whole cell fluorescent *in situ* hybridization (FISH) as a method for the direct detection of bacteria in water samples has increased in popularity. The method typically combines membrane filtration, resuscitation of the bacteria on membranes and hybridization with rRNA-targeted probes. During FISH, the morphology of the cells in the sample has to be stabilized in order to maintain the morphological integrity of the cells under harsh hybridization conditions. The cell walls and membranes have to be permeabilized to allow free penetration of fluorescent oligonucleotides to the intracellular rRNA. This can be achieved with fixatives such as aldehydes and/or alcohols (De Long *et al.*, 1989; Amann *et al.*, 1995). The membranes are hybridized in hybridization solution containing a fluorescently labelled oligonucleotide. Following incubation at the hybridization temperature for one to several hours, to allow the probe to bind to complementary rRNA sequences, washing steps are used to remove unbound or the non-specifically bound fluorescent probe, and the hybridized cells are then detected by epifluorescence microscopy. Probes will only bind correctly under defined hybridization conditions and the optimization of hybridization and washing conditions is as important as the probe design (Stahl and Amann, 1991; Manz *et al.*, 1992). Several probes labelled with spectrally different fluorochromes can be simultaneously used on one sample (Lee *et al.*, 1993; Amann *et al.*, 1997), while counterstaining of the fixed cells with DAPI (4′, 6-diamidino-2-phenylindole) allows total counts to be made (Hicks *et al.*, 1992; Meier *et al.*, 1997). Alternatively, depending on the concentration of targeted cells in the sample and to increase resolution, FISH hybridization can be performed in suspension and detection can be performed by means of flow cytometry, which enables quantification of the fluorescence intensities for each target–probe hybrid (Fuchs *et al.*, 1998).

Although FISH is currently considered a highly specific cellular detection method and is relatively easy to perform, the procedure may have some limitations when applied to the detection of nutrient-starved bacterial cells disseminated in drinking waters. Generally, probes give a strong signal only if cells are metabolically active and contain large numbers of ribosomes and target rRNA. Conversely, low cellular ribosome contents may result in weak fluorescent hybridization signals (Amann *et al.*, 1995; Lebaron *et al.*, 1997). Another area of concern relates to the viability of the detected cells. Although it has been reported that the rRNA content of bacterial cells can be directly correlated with the growth rate (Amann *et al.*, 1995), more recent reports have indicated that a small number of rRNA molecules can remain for a relatively long period after the loss of culturability, thus resulting in false positives (McKillip *et al.*, 1998; Sheridan *et al.*, 1998). One possible way to overcome this issue is to couple the direct viable count techniques with FISH detection (Nishimura *et al.*, 1993; Kalmbach *et al.*, 1997; Meier *et al.*, 1997).

7.3.2.2 Polymerase chain reaction (PCR)

As a consequence of the speed, specificity and low cost of the PCR, the procedure has become one of the most widely used assays for direct detection of low levels of pathogenic microbes in environmental samples. The PCR assay can be used to selectively amplify, to detectable levels, nucleic acid sequences associated with pathogens that might be present in low numbers in water samples. PCR is a process in which target DNA, synthetic oligonucleotide primers, a thermostable DNA polymerase and the DNA subunits are combined in a microcentrifuge tube and subjected to the temperature changes needed for the DNA duplication to occur. During the PCR process, different temperatures are used to facilitate DNA denaturation, annealing of the oligonucleotide primers to the target DNA and extension of the primers across the target sequence. These cycles are repeated many times, thus resulting in increasingly greater quantities of target sequence. Under ideal conditions, the PCR can generate millions of copies of a single DNA molecule in just 20 to 30 repetitions of the temperature cycle, with each cycle requiring only a few minutes (Saiki *et al.*, 1988). The PCR-amplified products can be detected by means of techniques such as electrophoresis on agarose gels and after staining of the amplification products by a fluorochrome dye or by hybridization with a labelled probe.

There are several essential steps in the development and application of PCR for successful detection of pathogens in water samples. The key steps include identification and selection of oligonucleotide primers for target genomic sequences; testing of selected primers for sensitivity, specificity and selectivity; purification and concentration of the pathogens in environmental sample concentrates to enable efficient and reliable enzymatic amplification of low numbers of target genomic sequences; and testing of the methods for their applicability to natural pathogenic strains and actual field samples. By using PCR, a selected gene sequence specific to a group of organisms or a single species can be selectively amplified. Various primers have thus been described to amplify fragments of rRNA operons in order to detect specific organisms or groups of organisms in environmental samples (Dorsch *et al.*, 1994; Covert *et al.*, 1999). As in the case of designing oligonucleotide probes for hybridization purposes, selection of oligonucleotide primers for target pathogens requires that sequence data be available. Of particular importance are the type and function of the nucleic acid target; its length and location; and the extent to which its sequence is related to that of other, non-target but genetically related microbes. It is essential to select oligonucleotides having an appropriate length (20 to 30 nucleotides); desired sequence composition for specificity and selectivity; and appropriate melting and annealing temperatures to prevent the formation of undesirable secondary structures, primer-dimers and other artefacts that would interfere with successful PCR (Saiki *et al.*, 1988, Steffan and Atlas, 1991).

There exist many variations of the basic PCR technique. The sensitivity and specificity of the PCR may be improved by adopting a nested approach. Nested PCR involves two consecutive rounds of PCR amplification. The first round of amplification is performed with group- or organism-specific primers, whilst the second round of amplification uses the initially amplified product as the template for another round of annealing and extension with different primers (Arnheim and Erlich, 1992). The use of nested primers provides an additional level of specificity since the second round of PCR amplifcation can only be performed if the correct sequence (complementary to the nested primers) was amplified during the first round. Alternatively, nucleic acid hybridization can be performed using highly specific oligoprobes that would hybridize only with amplicons from a single pathogen type or strain. However, nested PCR permits a more rapid detection (6 to 8 h) compared to confirmation of the correct sequence amplification by probe hybridization (few days). Another variation of the PCR technique, namely multiplex PCR, allows the simultaneous detection of more than one target organism in a single PCR using multiple pairs of primers designed to be specific for different target organisms (Bej *et al.*, 1991; Rochelle *et al.*, 1997a). However, multiplex PCR may not perform well with all primer blends as the composition and length of primer oligonucleotide, as well as the size of the amplified fragments, may influence each PCR amplification (Bej *et al.*, 1991). Since the PCR method cannot be used directly for the amplification of an RNA target sequence, a complementary DNA copy (cDNA) therefore must first be synthesized. This reaction is catalyzed by the enzyme, reverse transcriptase (RT), which is able to synthesize DNA from the RNA template in the presence of specific primers and DNA subunits. The single-stranded cDNA is a suitable target for PCR amplification by making use of the same or a different set of primers. RT-PCR has emerged as a sensitive and specific approach for the detection of enteric viruses containing RNA genomes (Sobsey, 1991; Schwab *et al.*, 1996; Huang *et al.*, 2000).

Although the basic PCR method is both specific and sensitive, such standard PCR reactions are not quantitative. To obtain quantitative data from PCR-based analyses, statistical methods based on most probable number (MPN) estimations have been used (Toranzos *et al.*, 1993). In MPN-PCR, DNA extracts are diluted before PCR amplification and limits are set on the number of genes in the sample by reference to known control dilutions. Another way of quantifying PCR-amplified products for comparison is to include an internal control in the PCR reaction (Zachar *et al.*, 1993; Schwab *et al.*, 1998). Here, a known amount of target DNA is added to a PCR reaction containing DNA from the mixed microbial population. The known target DNA is complementary to the same primers and thus competes with the target sequences in the PCR reaction mixture. By preparing a dilution series of the known and unknown DNA species, it is possible to quantify the amount of product produced from the complementary gene in the extracted DNA. The known DNA target can be generated by cloning the gene of interest or purifying the PCR-amplified product after which a deletion is introduced to give a differently sized PCR product.

An alternative PCR assay for the direct enumeration of targeted cells was reported by Tani *et al.* (1998), who modified the standard PCR protocol so that nucleic acid sequences can be amplified *in situ*. This new PCR method was successfully applied to the direct enumeration of *E. coli* from a freshwater sample. With proper fixation and permeabilization conditions, the oligonucleotide primers and other reaction components are able to diffuse into the cells, and, upon thermal cycling, amplify a specific target sequence present in the cell. PCR products were labelled by digoxigenin during the amplification process and anti-digoxigenin antibodies conjugated with fluorescent dye were used for detection by epifluorescence microscopy. This approach allows a direct visualization of the fluorescent amplification products at a single-cell level and consequently a direct enumeration of cells. Even though *in situ* PCR seems promising, it has not been used for routine detection and enumeration of micro-organisms in water, as the results showed a weak fluorescence intensity signal of targeted cells. Another promising approach is real-time quantitative PCR, which consists of monitoring fluorescently labelled PCR products as they are being amplified (Heid *et al.*, 1996). This approach has also not yet been widely applied to specific detection of micro-organisms present at low concentration in environmental samples, but is likely to be developed in the future.

Although there are numerous advantages associated with PCR as a detection tool, standard PCR cannot, however, be used to detect the infectious state of an organism – only the presence or absence of pathogen-specific DNA or RNA. Yet, this viability concept is fundamental for interpreting results in terms of public health when dealing with water samples. The PCR technique must consequently be associated with a viability test. To overcome this limitation, an indirect approach has been developed for assessing the viability of PCR detected bacteria from water samples. This method is based on the analysis of each sample before and after a culture step in a non-selective medium: an increase in the PCR-amplified product after cultivation indicates the occurrence of bacterial multiplication and thus demonstrates the viability of the detected bacteria (Theron *et al.*, 2001). Another limitation inherent to PCR analysis of environmental samples is the frequent inhibition of the enzymatic reaction. Whereas humic substances are known to inhibit the DNA polymerase enzyme, colloidal matter has a high affinity for DNA (Way *et al.*, 1993; Wilson, 1997). The presence of these elements in a water sample can therefore considerably decrease the amplification yield of PCR applied to the detection of greatly diluted bacteria (Straub *et al.*, 1995). Consequently, for PCR or RT-PCR, the extracted target nucleic acid is purified by protocols utilizing, for example, Sephadex (De Leon *et al.*, 1992), Chelex (Straub *et al.*, 1995) or CTAB (Jiang *et al.*, 1995).

7.3.3 Biosensors and microarray detection

Recently, the sensitive detection of food-borne pathogens in complex environments has been achieved using biosensor technology. This technology relies on the use of microchip systems to analyse the formation of antigen–antibody complexes (Malmqvuist, 1993; Hirmo *et al.*, 1998; Mandrell and Wachtel, 1999) and may hold great promise for adaptability to monitoring drinking water quality. For most biosensor-based methods, an optical fibre is coated with a 'capture' antibody for the antigen of the micro-organism to be monitored. After the antibody has bound to the micro-organism, the optical fibre–antibody–antigen complex is incubated with a fluorophore-labelled antibody that is also specific to the antigen. The fluorophore is then excited by an evanescent wave that is transmitted through the optical fibre after which the fluorescent signal is interpreted by the detection devices of the system. These systems report rapid (minutes) detection along with a reasonably sensitive limit of detection (1×10^3 cfu/ml of bacteria) (Mandrell and Wachtel, 1999). Despite their potential as analytical tools, further research is, however, needed to address issues such as advanced sample concentration, elimination of non-specific detection, and improved detection limits.

Recent advances in the detection of sequence-specific nucleic acid hybridization have been achieved using microarrays, also known as biochips. Microarrays are glass microslides or nylon membranes containing a high density of immobilized nucleic acids (genomic DNA, cDNA or oligonucleotides) in an ordered two-dimensional matrix (Epstein and Butow, 2000). Microarrays can be prepared by synthesizing DNA *in situ* on a glass surface using combinational chemistry (Pease *et al.*, 1994) or by robotic microdeposition of cDNAs (0.5–2 kb) amplified by PCR (Zammatteo *et al.*, 2000). The sample DNA, usually bound to a fluorescent or enzyme label, is exposed to the microarray and hybridizes with the target sequences. The detection of the probe–target hybrid at each spot on the array is achieved either by direct fluorescence scanning or enzyme-mediated detection yielding a semi-quantitative result (de Boer and Beumer, 1999). Advantages of

DNA microarray technology, as compared to other techniques, include the small size of the array allowing for a higher sensitivity, the ability to simultaneously detect diverse individual sequences in complex DNA samples, and the capacity to do comparative analysis of a large number of samples (de Boer and Beumer, 1999; Van Hal *et al.*, 2000). However, as with biosensors, obstacles such as sample size, matrix-associated inhibitors, non-specific binding, and cross-hybridization must be overcome before microarrays can be used for the detection and differentiation of pathogens in environmental samples.

7.4 DETECTION OF ENTERIC MICROBIAL PATHOGENS IN WATER

Estimates indicate that about 90% of the illnesses associated with domestic water supplies are related to microbial agents; only about 10% are due to chemical agents (Craun, 1988). The commonly recognized waterborne pathogens consist of several groups of enteric bacteria, viruses and protozoa. Viruses and protozoa differ from bacterial contaminants in important ways. Because they are environmentally inert, they do not replicate in water and environmental samples. Furthermore, unlike bacterial pathogens, human enteric viruses and protozoal parasites are environmentally stable, are resistant to many of the traditional methods used to control bacterial pathogens, and have notably low infectious doses (Jaykus *et al.*, 1994; Du Pont *et al.*, 1995). Moreover, a number of problems are encountered in determining the presence of microbial pathogens in water. For each group of microbes, whether protozoa, viruses or bacteria, the method must cope with a different set of conditions or characteristics that can complicate the task of identifying particular microbes. Since the concentrations of enteric organisms are low in water, their detection in water typically starts with some type of concentration process such as filtration. This is followed by a process to recover the pathogen from the filter and then by methods to assay and characterize the pathogenic microbes.

7.4.1 Enteric bacteria

7.4.1.1 Conventional detection techniques

Bacteria can be recovered and concentrated from water by a variety of filtration methods. The most widely used filtration method for recovering bacteria is membrane filtration using microporous membranes typically composed of cellulose esters (APHA, 1995). Following filtration, the cells recovered on the membrane filter can be directly assayed or cultured.

Various approaches for culturing the target bacteria following filtration are generally followed. The target bacteria can either be cultured by pre-enrichment and enrichment methods using broth media or the filter can be placed on differential and selective media to allow the development of discrete colonies of the target pathogens. Alternatively, the bacteria can be washed off the filter surface and reconcentrated in a small volume of suspension medium, which is plated on agar media. Quantitative results are preferably obtained by colony counts on the surface of agar media, with or without the presence of the filter used for concentration (APHA, 1995). Often, organisms may be present in water samples but are unculturable. The bacteria are still viable (exhibit low levels of metabolic activity) but they fail to develop colonies on most traditional solid culture media. Such a viable but nonculturable state (VBNC) state has been described for many pathogenic bacteria including *H. pylori*, *Campylobacter*, *E. coli*, *Vibrio cholerae*, *Vibrio* spp., and *Legionella pneumophila* (Roszak and Colwell, 1987; Byrd *et al.*, 1991; McKay, 1992; Bode *et al.*, 1993; Cappelier and Federighi, 1998; McDougald *et al.*, 1998). Also, nutrient limitations and environmental stressors, such as disinfectants used during water treatment, can produce unpredictable physiological and morphological changes in many waterborne bacterial pathogens. This makes their isolation and identification problematic and specialized handling procedures are required for their resuscitation (Ridgway, 1984). By using non-selective or less selective media, and other, less stressful culture conditions, the number of culturable cells in a population of VBNC, injured, or stressed bacteria may be increased. To confirm their identity, the cultured bacteria or bacterial colonies can be characterized by using a variety of methods, such as sub-culturing on other differential and selective media, biochemical, metabolic and other phenotype analyses, immunological analyses and nucleic acid-based analyses.

Enzyme immunoassays (EIA), which rely on the use of enzymatically or fluorescently labelled monoclonal antibodies specific for individual groups and species of bacterial pathogens, have been widely used to detect and quantify pathogenic bacteria in water. Many different EIA methods are available for both quantitative and qualitative analysis (Bottger *et al.*, 1987; Faude and Hoffle, 1997; Buswell *et al.*, 1998). However, the antibodies may exhibit cross-reactions, which compromise the specificity of the test, or they may fail if the original protein target, such as a cell surface protein, undergoes changes under environmental conditions (Vickers *et al.*, 1990). Furthermore, no information regarding the viability of the organisms is obtained (Kfir and Genthe, 1993). However, by combining microscopic examination with chemical treatments for enzymatic activity, the viability of the concentrated and purified bacterial pathogens may be assayed. For example, an approach combining fluorescent antibody and tetrazolium dye reduction, which measures dehydrogenase activity, has been used to successfully enumerate viable *E. coli* O157:H7 in water (Pyle *et al.*, 1995).

7.4.1.2 Nucleic acid-based detection methods

The discovery of a large number of bacterial toxins and other virulence factors has led to powerful methods, such as gene probes and PCR, for detecting and identifying pathogenic bacteria in water, as well as bacterial pathogens in a VBNC state which may not be detected by culture-dependent techniques. Two general types of gene probes that have been developed in recent years are DNA probes complementary to a single gene or a small region of a gene (Hazen and Jimenez, 1988; Venkatesan *et al.*, 1988; Knight *et al.*, 1991, Anderson *et al.*, 1995) and DNA probes complementary to genus- or species-specific regions of 16S rRNA (Schleifer *et al.*, 1985; Romaniuk and Trust, 1987; Rehnstam *et al.*, 1989) for use in whole cell *in situ* hybridization (FISH) (Manz *et al.*, 1993; Amann, 1995; Amann *et al.*, 1995). In contrast to the use of DNA probes complementary to genes, FISH allows for greater sensitivity and it is possible to detect a single bacterial cell, without first amplifying the bacterial population by growth. There are, however, limitations to using FISH for the detection of waterborne bacterial pathogens as stressed or starved bacteria are less reactive, bacterial debris in water may interfere with fluorescence and inanimate material may bind to probes.

PCR amplification-based methods, being more rapid and very sensitive, have become the most used molecular approach for detecting infectious bacterial agents in water (Steffan and Atlas, 1991; Alvarez *et al.*, 1993; Garret *et al.*, 1993). The major limitations to using PCR for detection are the presence of substances inhibitory to PCR (Wilson, 1997) and the possibility of amplification of non-viable cells (Josephson *et al.*, 1993). Inhibition phenomena have made bacterial DNA purification an important preliminary step for the PCR reaction. Subsequently, numerous techniques for separating and purifying the DNA have been described, resulting in varying purification levels of the DNA (Tsai and Olson, 1992; Abbaszadegan *et al.*, 1993). In recent years, the use of immunomagnetic beads attached to specific antibodies has become a popular approach in facilitating the capture, concentration and purification of target bacteria prior to DNA extraction (Olsvik *et al.*, 1994).

Several assays based on the direct detection of indicator and pathogenic bacterial cells in environmental water samples by filtration and PCR (Bej *et al.*, 1991; Toranzos and Alvarez, 1992; Oyofo and Rollins, 1993; Covert *et al.*, 1999) have been developed. The use of combinations of methods such as immunomagnetic capturing of strains, PCR, and detection of immobilized amplified nucleic acids by hybridization have been proven to be useful for the detection of waterborne pathogenic bacteria which are difficult to culture from environmental sources, e.g. *Y. enterocolitica* (Kapperud *et al.*, 1993; Sandery *et al.*, 1996), and *H. pylori* (Enroth and Engstrand, 1995; Sasaki *et al.*, 1999). A disadvantage of the above approaches is that they may detect dead as well as viable bacteria. An indirect approach is thus usually adopted for assaying the viability of bacteria from water samples by first culturing bacteria prior to PCR detection (Dorsch *et al.*, 1994; Waage *et al.*, 1999; Theron *et al.*, 2000). Not only does an enrichment procedure dilute any sample-related inhibitors present, but dead bacteria are diluted as well, thus reducing the probability of detecting them by the subsequent PCR assay. It thus becomes possible to relate detection to pathogen infectivity. In addition, allowing pathogens to multiply amplifies target nucleic acids thereby facilitating their detection. Despite these advantages, this approach still relies on culturing and so it cannot be applied to current non-cultivable pathogens.

7.4.2 Enteric parasitic protozoa

7.4.2.1 Conventional detection techniques

Parasitic protozoa are usually present in low concentrations in contaminated water and therefore must be concentrated from large volumes of water before detection. In addition, since an infection in humans can be initiated by as few as 1–10 viable cysts (Akin and Jakubowski, 1986) detection methods and viability assays need to be very sensitive. *Giardia* cysts and *Cryptosporidium* oocysts are concentrated by filtration through yarn-wound filters after which retained particulates are eluted from the filters and reconcentrated by centrifugation. The pelleted cysts and oocysts are then separated from particulate debris by flotation on Percoll-sucrose gradients followed by subsequent detection (Schaefer, 1996). While recovery efficiencies as high as 100% have been reported (LeChevallier *et al.*, 1995), recovery is generally poor and greatly affected by water quality and particulate matter (LeChevallier *et al.*, 1995; Nieminski *et al.*, 1995). Alternative methods such as calcium carbonate precipitation of particulates can concentrate *Cryptosporidium* oocysts from water with concentration efficiencies as high as 68% (Vesey *et al.*, 1993).

Conventional detection methods for parasitic agents from water sample concentrates have been based on immunofluorescent staining of filtered sample concentrates (Schaefer, 1996). The sample concentrate is filtered through cellulose acetate filters and commercial kits that use fluorescein isothiocyanate-labelled monoclonal antibodies are applied for immunofluorescent staining. The stained filters are examined under an ultraviolet microscope, and cysts and oocysts are classified according to immunofluorescence, size, shape and internal morphologic characteristics. Despite having been used successfully in surveys of raw and finished waters (LeChevallier *et al.*, 1991), the method is extremely limited as it is time-consuming and expensive, requires highly skilled and experienced personnel, and does not indicate viability of the detected cysts or oocysts. In addition, cross-reactions of monoclonal antibodies with algal cells and debris can interfere with the interpretation of results (Rodgers *et al.*, 1995). Commercially available enzyme-linked immunosorbent assays (ELISA) are becoming available for the parasitic protozoa and have the promise to be highly sensitive and rapid (Aldeen *et al.*, 1998). While presently applicable only to direct use on faecal samples, these ELISA systems may become suitable for use with environmental samples, particularly if the samples are concentrated prior to analysis.

Detection methods based on cell sorting or particle counting have been developed for the separation, purification and detection of *Giardia* and *Cryptosporidium*. The Fluorescence-Activated Cell Sorting (FACS) system is a laser-based particle counter that is able to simultaneously sort particles, sense fluorescence, and determine size (Campbell *et al.*, 1993; Vesey *et al.*, 1993, 1997). Because fluorescent antibodies are used in the FACS procedure and cross-reactions with other fluorescent objects can occur, it is recommended that sorted fluorescence objects of the right size and shape be confirmed on an epifluorescence microscope. Since the FACS processing is rapid and eliminates most of the contaminating debris and background fluorescence, microscopic observation of the sample is less fatiguing. However, general application of this technology is limited by the high cost of the instruments and the requirement of a skilled, dedicated analyst. Furthermore, infectious and noninfectious organisms cannot readily be distinguished (Schaefer, 1996).

Membrane exclusion/permeability fluorogenic dyes such as propidium iodide (Sauch *et al.*, 1991; Dowd and Pillai, 1997) and 4′,6-diamidino-2-phenylindole (DAPI) (Campbell *et al.*, 1992b) have been used to assess the viability of protozoan cysts and oocysts. Detection of these differentially stained cysts and oocysts has been greatly facilitated by microscopic and analytical imaging methods such as charge-coupled device (CCD) cameras (Campbell *et al.*, 1992a; Campbell *et al.*, 1993) and confocal laser scanning microscopy (Anguish and Ghiorse, 1997). More recently, viability assays based on nucleic acid staining of *C. parvum* oocysts with SYTO-9, SYTO-59 and hexidium have been developed and used in conjunction with immunofluorescent staining methods for the detection of *Cryptosporidium* oocysts (Belosevic *et al.*, 1997; Neumann *et al.*, 2000).

The infectivity of the spores of some of the important human microsporidia and oocysts of *C. parvum* can be assessed by culturing on susceptible mammalian host cells where spores germinate or oocysts excyst and active stages of the organisms can proliferate in the cells (Upton *et al.*, 1994; Visvesvara *et al.*, 1995). The living stages can subsequently be detected (e.g., after immunofluorescent or other staining) and quantified by counting the numbers of discrete living stages of groups (foci) of them. However, for waterborne protozoa such as *G. lamblia* and *Cyclospora cayetanensis*, culture from the cysts or oocysts recovered in a water sample is still not possible.

7.4.2.2 Nucleic acid-based detection methods

Molecular approaches based on DNA hybridization have been developed for the detection of *Giardia* (Abbaszadegan *et al.*, 1991) and *Cryptosporidium* (Lindquist, 1997). These methods are generally regarded as inadequate due to high detection limits and large sample volumes that are impractical for most hybridization protocols without further pathogen concentration. The development of amplification methods such as the PCR has greatly facilitated the direct detection of low levels of pathogens in environmental samples. Consequently, several PCR methods for the detection of *Giardia* and *Cryptosporidium* in water have been developed (Mahbubani *et al.*, 1991; Johnson *et al.*, 1995; Mayer and Palmer, 1996; Kaucner and Stinear, 1998; Wu *et al.*, 2000) as well as a multiplex PCR for the simultaneous detection of these protozoa in environmental water samples (Rochelle *et al.*, 1997a). Methods that combine genus-specific PCR with restriction fragment length polymorphism (RFLP) have been applied to detect and speciate *Cryptosporidium* oocysts and have thus provided a means whereby pathogenic and non-pathogenic species can be distinguished (Patel *et al.*, 1999; Lowery *et al.*, 2000).

While PCR methods have the potential to detect a single infectious unit and may be applied to discriminate pathogenic from nonpathogenic species, they have remained limited in their application because of enzymatic inhibition, the absence of quantitative assays, and the inability to discriminate between viable and nonviable organisms. Several approaches have been adopted to address these issues. In order to overcome inhibition of the PCR, methods combining immunomagnetic separation with PCR amplification have been developed to detect low levels of *Giardia* cysts (Mahbubani *et al.*, 1998) and *C. parvum* oocysts (Deng *et al.*, 1997; Hallier-Soulier and Guillot, 1999; Rochelle *et al.*, 1999; Lowery *et al.*, 2000). In addition, PCR methods for quantifying the number of *C. parvum* oocysts in water are being developed. Two recent quantitative PCR methods allowed the quantification of 30–50 oocysts seeded into municipal water samples (Chung *et al.*, 1999) and 258 oocysts/2 litre of sludge (Udeh *et al.*, 2000), respectively.

Determining the viability of waterborne protozoa remains a technical challenge. Early approaches in determining the viability of waterborne *C. parvum* oocysts relied on an excystation protocol before DNA extraction to allow differentiation between viable and nonviable *C. parvum* oocysts by detection based on PCR (Filkorn *et al.*, 1994; Wagner-Wiening and Kimmig, 1995). More recently, various mRNAs have been evaluated in RT-PCR assays as potential markers for cyst and oocyst viability (Kaucner and Stinear, 1998). RT-PCR assays directed at mRNA encoding the enzyme amyloglucosidase (CPAG) (Jenkins *et al.*, 2000), β-tubulin (Widmer *et al.*, 1999) and the heatshock protein (hsp70) (Stinear *et al.*, 1996) have been successful in distinguishing viable from non-viable oocysts. Integrated approaches that combine immunomagnetic separation for the recovery of *C. parvum* oocysts with *in vitro* cell culturing followed by PCR (CC-PCR) (Di Giovanni *et al.*, 1999) or RT-PCR (Rochelle *et al.*, 1997b) have demonstrated that the recovered oocysts retained their infectivity.

7.4.3 Enteric viruses

7.4.3.1 Conventional detection techniques

The low level of viruses found in environmental waters necessitates the concentration of virus by adsorption to a microporous filter (Sobsey, 1982). The adsorbed viruses are subsequently eluted by passing a small volume of beef extract, as eluent, through the filter. Alternative techniques that further concentrate and purify intact virions from the sample concentrate by physicochemical methods include polyethylene glycol (PEG) precipitation, Sephadex gel chromatography, chelation of multivalent cation impurities, and ultrafiltration (Schwab *et al.*, 1993; Hurst, 1996). However, care should be taken in applying these treatments as injury or damage of the target viruses during these processes can interfere with their subsequent detection by cultivation or other methods.

Conventional methods to directly detect viruses in water after concentration have been based on the ability of enteric viruses to infect live mammalian cells in culture. Various cell types have been used, but the sensitivity of each cell line to different viruses must be taken into account since some are highly selective for certain virus types (Dahling and Wright, 1986). Primary or secondary human embryonic and primate kidney cell cultures appear to be the most sensitive host systems for enteric virus isolations (Dahling *et al.*, 1984; Payment, 1993). While the cell culture assay can detect infectious viruses in environmental samples, without additional tests, no determination can be made as to the particular strain of virus present in a sample. Additionally, the length of time needed to detect infection in the cell culture can vary greatly, from a few days to several weeks, depending on the type and number of viruses present. In addition, when a sample tests positive for viral infectivity using cell culture, the infectious agent is not necessarily known. The BGM cell line, routinely used for enterovirus assays, is susceptible to infection by many viruses, including reoviruses such as rotavirus, a pathogen often present in environmental samples in numbers greater than enteroviruses (Puig *et al.*, 1994). Assays of viruses in cell cultures can be quantified using quantal (most probable number or 50% endpoint) or enumerative (plaque) methods. The plaque assay is generally more precise and accurate than the quantal assay because relatively large numbers of individual infectious units can be counted directly as discrete, localized areas of infection (plaques). However, conventional culture assays have been limited because some of the most important waterborne enteric viruses do not replicate (e.g., Norwalk and related viruses and some caliciviruses) or replicate poorly (e.g., rotaviruses and hepatitis A virus) in cell culture (Jaykus *et al.*, 1994). These viruses will not be detected in water unless alternative methods such as ELISA assays or nucleic acid-based detection methods are applied directly to concentrated samples.

7.4.3.2 Nucleic acid-based detection techniques

Nucleic acid-based detection methods, especially *in vitro* enzymatic amplification of specific viral genomic nucleic acid sequences by PCR and RT-PCR (for the detection of RNA viruses) followed by detection of the amplified nucleic acid by hybridization have emerged as sensitive and specific approaches for the detection of many enteric viruses (Sheih *et al.*, 1991; De Leon *et al.*, 1992; Sobsey, 1994). PCR amplification methods have since advanced to the point where they have been successfully applied to investigating waterborne outbreaks caused by non-culturable human caliciviruses and in surveying drinking water and environmental sources for the presence of enteroviruses, norovirus, rotavirus, and adenoviruses (Schwab *et al.*, 1993; Gilgen *et al.*, 1995; Girones *et al.*, 1995; Castignolles *et al.*, 1998; Lodder *et al.*, 1999; Gratacap-Cavallier *et al.*, 2000; Huang *et al.*, 2000).

Despite enormous strides in the ability to detect human enteric viruses with PCR, the technique is still limited by the absence of effective concentration methods, the presence of enzymatic inhibitors, and the inability to distinguish between infectious and non-infectious virions. Several extraction techniques have been developed to isolate and purify viral nucleic acid from sample concentrates (Yang and Xu, 1993; Sheih *et al.*, 1995). However, these techniques expose labile viral nucleic acid, often single-stranded RNA, to conditions that could lead to degradation before analysis, thereby causing a loss of target and a decrease in sensitivity. In addition, the infectivity of the viruses cannot be determined. One approach to the nucleic acid-based detection of intact and hence potentially infectious enteroviruses is virus capture mediated by antibody–antigen complexes. This approach has been used to isolate, concentrate and purify virus particles from environmental water samples for subsequent nucleic acid amplification and detection (Deng *et al.*, 1994; Schwab *et al.*, 1996; Jothikumar *et al.*, 1998). The use of an integrated cell culture–RT–PCR procedure followed by nested PCR, for enhanced sensitivity, have been used to detect infectious human astrovirus, enterovirus and adenovirus (types 40 and 41) in surface water samples, in recreational marine water samples and in sewage (Reynolds *et al.*, 1996; Murrin and Slade, 1997; Chapron *et al.*, 2000).

7.5 MOLECULAR METHODOLOGIES FOR MICROBIAL SOURCE TRACKING (MST)

Detection of pathogenic organisms provides information as to the safety and public health risks associated with a given water supply; however, it often does little to define the potential sources of the contamination. Generally, because different enteric pathogens are present in the intestines of different animals, the identification of a contamination event as being either of human or animal source would provide information as to the types of pathogens that may be expected, the risk of infection, and the treatment that may be required to control transmission of disease. In response, molecular techniques are being developed as a means to identify the source of a contaminant. DNA fingerprinting is one tool for microbial source tracking (MST) and consists of a family of techniques that are under development to identify the sources of faecal contamination in various water bodies (Parveen *et al.*, 1999; Dombek *et al.*, 2000).

The basic underlying principle of DNA fingerprinting is the generation of one-dimensional patterns of macromolecules or fragments from isolates and the reference strains the isolates are considered to be related to. The patterns should neither be too complex to prevent analysis nor too simple to obscure actual genomic dissimilarities. Given sufficient sensitivity for discrimination, isolates and reference strains that exhibit a high degree of pattern similarity can be considered related. Electrophoretic analysis of whole cell proteins have been used less extensively due to problems associated with the interpretation of the data caused by effects such as the composition of growth media on the protein banding pattern and the superposition of proteins.

The different polymorphism-based procedures are generally coupled to a PCR reaction. In the amplified ribosomal DNA restriction analysis (ARDRA) technology, PCR-amplified 16S rRNA genes are digested with restriction endonucleases and the resulting fragments separated electrophoretically. Presence or absence of the restriction site within two strains will cause differences in the length of the DNA restriction fragments. The complexity of the pattern depends upon the number of target sequences and position of restriction sites. Comparison of the generated patterns to those obtained from a database will allow assignment of isolates to species or species clusters in those cases were the banding patterns are highly similar. The separated DNA fragments may also be transferred to filters for hybridization with probes specific for an organism of interest (Parveen *et al.*, 1999; Dombek *et al.*, 2000).

Two other protocols for generating DNA fingerprints use a single primer to amplify fragments with PCR before examination on agarose gels. PCR amplification of repetitive extragenic palindromic sequences (REP-PCR) takes advantage of repetitive sequences found in the microbial genome (Versalovic *et al.*, 1994; Dombek *et al.*, 2000). In the randomly amplified polymorphic DNA (RAPD) or arbitrarily primed PCR (Hadrys *et al.*, 1992; Breen *et al.*, 1995) technology, a short oligonucleotide primer (about 10 nucleotides), usually with a random sequence that is not specific for a particular gene, is used as a primer to amplify fragments. These methods yield DNA fingerprints comprised of multiple, differently sized DNA amplification products following separation by gel electrophoresis. One of the most promising technologies for microbial source tracking is, however, amplified fragment length polymorphism (AFLP) analysis. AFLP analysis appears to have the same taxonomic range as other fingerprinting techniques, but this technology combines several advantages of these different techniques, which in most cases results in the highest power of discrimination (Blears *et al.*, 1998; Koeleman *et al.*, 1998) (Fig. 7.1). This technology is based on the selective amplification of a subset of genomic restriction fragments using PCR. For AFLP, purified genomic DNA is digested with two restriction enzymes, one with an average cutting frequency and a second one with a higher cutting frequency after which oligonucleotide adapters are ligated to the genomic DNA restriction fragments. The sequence of the adapters and the adjacent restriction site serve as oligonucleotide primer binding sites for subsequent amplification of the restriction fragments by PCR. Selective

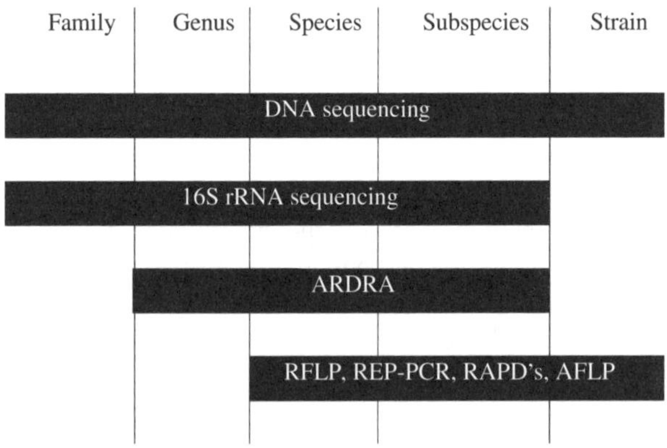

Figure 7.1 Relative applicability of various fingerprinting and DNA techniques at different levels of taxonomic resolution.

nucleotides extending into the restriction fragments are added to the 3′ ends of the adapter-specific PCR primers such that only a subset of the restriction fragments are recognized and amplified. The subset of amplified fragments is then analysed by denaturing polyacrylamide gel electrophoresis to generate the fingerprint (Vos *et al.*, 1995; Savelkoul *et al.*, 1999). Since relatively small amounts of DNA are digested and detection of AFLP fragments does not depend on hybridization, the AFLP analysis method is more reproducible and robust than other fingerprinting techniques and displays more fragments than other fingerprinting techniques (Vaneechoutte, 1996; Rademaker *et al.*, 2000).

7.6 SUMMARY AND CONCLUSIONS

Many years of experience have shown that monitoring microbiological water quality by standard indicator systems is and has been an important means for the supply of safe drinking water. Indicator methods are simple and easy to use, cheap, and cover a wide range of pathogens. However, their main limitations are lack of specificity and slow response. Furthermore, current indicators do not adequately address parasites and viruses. The recognition of new waterborne pathogens as well as re-emerging ones with increased virulence and antibiotic resistance requires examination of newer approaches to monitoring. This is critical if control and prevention of waterborne diseases are to be fully realized.

Recent developments, particularly in molecular biology research in the last few years, have resulted in a wide range of new tools, principally based on the detection of nucleic acid material and its amplification. They offer a novel, more sensitive and specific way of detecting micro-organisms. They can also identify organisms that would not be detected with current culture techniques and can be used to track new pathogenic entities, including variants of otherwise harmless micro-organisms. There are probably very few groups of micro-organisms that have not been located with these amplification techniques and several test kits have already been commercialized. Despite the success of these prototype alternative and rapid methods in detecting human enteric microbial pathogens in water, multiple barriers must be overcome before they can be used to routinely assess water quality and the microbiological safety of drinking water.

The relationship between detection by molecular approaches and the subsequent viability or infectivity of waterborne enteric pathogens remains a concern. Despite using various approaches in order to address this issue, the combination of nucleic acid methods with culture methods appears to be the most promising approach for sensitive and specific detection and quantification of infectious waterborne pathogens. However, this approach is still not ideal as it cannot be applied to the non-cultivable human enteric pathogens and it is expensive, time-consuming and not readily amenable to routine diagnostic work. In addition, methods used for purifying and concentrating the target microbes and their nucleic acids from water, so that they are free of contaminants that may interfere with the analysis, need further improvement, consolidation and simplification.

Evaluation studies and standardization of the procedures are also necessary to force the approval of new methods by health authorities, this being the basic requirement for routine application. Further research is thus needed to develop and refine the prototype protocols into collaboratively tested methods that could be routinely and expeditiously used to evaluate the microbiological safety of water. Furthermore, issues of cost, personnel efficiency and equipment needs must be evaluated to ensure that the challenge of diagnosing these emerging infections will be met. Ultimately, the true measure of the value of rapid molecular methodologies is whether they will represent improvements over conventional methodologies in terms of achieving a reduction of waterborne disease and an improvement of human health.

REFERENCES

Abbaszadegan, M., Gerba, C.P. and Rose, J.B. (1991) Detection of *Giardia* cysts with a cDNA probe and applications to water samples. *Appl. Environ. Microbiol.* **57**, 927–931.

Abbaszadegan, M., Huber, H.S., Gerba, C.P. and Pepper, I.L. (1993) Detection of enteroviruses in groundwater with the polymerase chain reaction. *Appl. Environ. Microbiol.* **59**, 1318–1324.

Akin, E.W. and Jakubowski, W. (1986) Drinking water transmission of giardiasis in the United States. *Water Sci. Tech.* **18**(10), 219–226.

Aldeen, W.E., Carroll, K., Robinson, A., Morrison, M. and Hale, D. (1998) Comparison of nine commercially available enzyme-linked immunosorbent assays for detection of *Giardia lamblia* in faecal specimens. *J. Clin. Microbiol.* **36**, 1338–1340.

Alvarez, A.J., Hernandez-Delgado, E.A. and Toranzos, G.A. (1993) Advantages and disadvantages of traditional and molecular techniques applied to the detection of pathogens in waters. *Water Sci. Tech.* **27**, 253–256.

Amann, R.I. (1995) *In situ* identification of micro-organisms by whole cell hybridization with rRNA-targeted nucleic acid probes. In *Molecular Microbial Ecology Manual* (eds. A.D.L. Akkermans, J.D. van Elsas, and F.J. de Bruijn), Kluwer Academic Publishers, Dordrecht, The Netherlands.

Amann, R., Glöckner, F.-O. and Neef, A. (1997) Modern methods in subsurface microbiology: *in situ* identification of micro-organisms with nucleic acid probes. *FEMS Microbiol. Rev.* **20**, 191–200.

Amann, R.I., Ludwig, W. and Schleifer, K.-H. (1995) Phylogenetic identification and *in situ* detection of individual microbial cells without cultivation. *Microbiol. Rev.* **59**, 143–169.

Amann, R., Ludwig, W. and Schleifer, K.-H. (1993 Identification and *in situ* detection of individual bacterial cells. *FEMS Microbiol. Lett.* **100**, 45–50.

American Public Health Association (1995) *Standard Methods for the Examination of Water and Wastewater* (ed. M.A.H. Franson), 19th edn, American Public Health Association, Washington, D.C.

Anderson, S.A., Dewis, G.D. and Pearson, M.N. (1995) Use of gene probes for the detection of quiescent enteric bacteria in marine and fresh waters. *Water Sci. Tech.* **31**(5–6), 291–298.

Anguish, L.J. and Ghiorse, W.C. (1997) Computer-assisted laser scanning and video microscopy for analysis of *Cryptosporidium parvum* oocysts in soil sediment and feces. *Appl. Environ. Microbiol.* **63**, 724–733.

Arnheim, N. and Erlich, H. (1992) Polymerase chain reaction strategy. *Ann. Rev. Biochem.* **61**, 131–156.

Bej, A.K., Mahbubani, M.H., Dicesare, J.L. and Atlas, R.M. (1991) Polymerase chain reaction-gene probe detection of micro-organisms by using filter-concentrated samples. *Appl. Environ. Microbiol.* **57**, 3529–3534.

Belosevic, M., Guy, R.A., Taghi-Kilani, R., Neumann, N.F., Gyurek, L.L., Liyanage, L.R., Millard, P.J. and Finch, G.R. (1997) Nucleic acid stains as indicators of *Cryptosporidium parvum* oocyst viability. *Int. J. Parasitol.* **27**, 787–798.

Blears, M.J., De Grandis, S.A., Lee, H. and Trevors, J.T. (1998) Amplified fragment length polymorphism (AFLP): review of the procedure and its applications. *J. Ind. Microbiol. Biotechnol.* **21**, 99–114.

Bode, G., Mauch, F. and Malfertheiner, P. (1993) The coccoid forms of *Helicobacter pylori*: Criteria for their viability. *Epidemiol. Infect.* **111**, 483–490.

Bottger, E.C., Jurs, M., Barrett, T., Wachsmuth, K., Metzger, S. and Bitter-Suermann, D. (1987) Qualitative and quantitative determination of enterobacterial common antigen (ECA) with monoclonal antibodies. *Clin. Microbiol.* **25**, 377–382.

Breen, A., Rope, A.F., Taylor, D., Loper, J.C. and Sferra, P.R. (1995) Application of DNA amplification fingerprinting (DAF) to mixed culture bioreactors. *J. Ind. Microbiol.* **14**, 10–16.

Buswell, C.M., Herlihy, Y.M., Lawrence, L.M., McGuiggan, J.T.M., Marsh, P.D., Keevil, C.W. and Leach, S.A. (1998) Extended survival and persistence of *Campylobacter* spp. in water and aquatic biofilms and their detection by immunofluorescent-antibody and rRNA staining. *Appl. Environ. Microbiol.* **64**, 733–741.

Byrd, J.J., Xu, H-S. and Colwell, R.R. (1991) Viable but nonculturable bacteria in drinking water. *Appl. Environ. Microbiol.* **57**, 875–878.

Campbell, A.T., Haggart, R., Robertson, L.J. and Smith, H.V. (1992a) Fluorescent imaging of *Cryptosporidium* using a cooled charge couple device (CCD). *J. Microbiol. Meth.* **16**, 169–174.

Campbell, A.T., Robertson, L.J. and Smith, H.V. (1992b) Viability of *Cryptosporidium parvum* oocysts: Correlation of *in vitro* excystation with inclusion/exclusion of fluorogenic vital dyes. *Appl. Environ. Microbiol.* **58**, 3488–3493.

Campbell, A.T., Robertson, L.J. and Smith, H.V. (1993) Novel methodology in the detection of *Cryptosporidium parvum*: A comparison of cooled charge coupled device (CCD) and flow cytometry. *Water Sci. Tech.* **27**(3–4), 89–92.

Cappelier, J.M. and Federighi, M. (1998) Demonstration of viable but nonculturable state *Campylobacter jejuni*. *Rev. Med. Vet.* **149**, 319–326.

Castignolles, N., Petit, F., Mendel, I., Simon, L., Cattolico, L. and Buffet-Janvresse, C. (1998) Detection of adenovirus in the waters of the Seine River estuary by nested-PCR. *Mol. Cell. Probes* **12**, 175–180.

Chapron, C.D., Balletster, N.A., Fontaine, J.H., Frades, C.N. and Margolin, A.B. (2000) Detection of astroviruses, enteroviruses, and adenovirus types 40 and 41 in surface waters collected and evaluated by the information collection rule and an integrated cell culture-nested PCR procedure. *Appl. Environ. Microbiol.* **66**, 2520–2525.

Chung, E., Aldom, J.E., Carreno, R.A., Chagla, A.H., Kostrzynska, M., Lee, H., Palmateer, G., Trevors, J.T., Unger, S., Xu, R. and De Grandis, S.A. (1999) PCR-based quantitation of *Cryptosporidium parvum* in municipal water samples. *J. Microbiol. Meth.* **38**, 119–130.

Colwell, R.R., Hasan, J.A.K., Huq, A., Loomis, L., Saeibeling, R.J., Torres, M., Galvez, S., Islam, S. and Bernstein, D. (1992) Development and evaluation of a rapid, simple, sensitive, monoclonal antibody-based co-agglutination test for direct detection of *Vibrio cholerae* O1. *FEMS Microbiol. Lett.* **97**, 215–220.

Covert, T.C., Rodgers, M.R., Reyes, A.L. and Stelma, G.N. (1999) Occurrence of nontuberculous mycobacteria in environmental samples. *Appl. Environ. Microbiol.* **65**, 2492–2496.

Craun, G. (1988) Surface water supplies and health. *J. Am. Wat. Wks. Ass.* **80**, 40–52.

Dahling, D.R., Safferman, R.S. and Wright, B.A. (1984) Results of a survey of BGM cell culture practices. *Environ. Int.* **10**, 309–313.

Dahling, D.R. and Wright, B.A. (1986) Optimization of the BGM cell line culture and viral assay for monitoring viruses in the environment. *Appl. Environ. Microbiol.* **51**, 790–812.

De Boer, E. and Beumer, R.R. (1999) Methodology for detection and typing of foodborne micro-organisms. *Int J. Food Microbiol.* **50**, 119–130.

De Leon, R., Matsui, S.M., Baric, R.S., Herrmann, J.E., Blacklow, N.R., Greenberg, H.B. and Sobsey, M.D. (1992) Detection of Norwalk virus in stool specimens by reverse transciptase–polymerase chain reaction and nonradioactive oligoprobes. *J. Clin. Microbiol.* **30**, 3151–3157.

De Long, E.F., Wickham, G.S. and Pace, N.R. (1989) Phylogenetic stains: ribosomal RNA-based probes for the identification of single microbial cells. *Science* **243**, 1360–1363.

Deng, M.Q., Cliver, D.O. and Mariam, T.W. (1997) Immunomagnetic capture PCR to detect viable *Cryptosporidium parvum* oocysts from environmental samples. *Appl. Environ. Microbiol.* **63**, 3134–3138.

Deng, M., Day, S.P. and Cliver, C.O. (1994) Detection of hepatitis A virus in environmental samples by antigen-capture PCR. *Appl. Environ. Microbiol.* **60**, 1927–1933.

De Silva, D.G.H., Candy, D.C.A., Mendis, L.N., Charts, H. and Rowe, B. (1992) Serological diagnosis of infection by *Shigella dysenteriae*-1 in patients with bacillary dysentery. *J. Infect. Dis.* **25**, 273–278.

Di Giovanni, G., Hashemi, F.H., Shaw, N.J., Abrams, F.A., LeChevallier, M.W. and Abbazadegan, M. (1999) Detection of infectious *Cryptosporidium parvum* oocysts in surface and filter backwash water samples by immunomagnetic separation and integrated cell culture–PCR. *Appl. Environ. Microbiol.* **65**, 3427–3432.

Dombek, P.E., Johnson, L.K., Zoimmerley, S.T. and Sadowsky, M.J. (2000) Use of repetitive DNA sequences and the PCR to differentiate *Escherichia coli* isolates from human and animal sources. *Appl. Environ. Microbiol.* **66**, 2572–2577.

Dorsch, M., Ashbolt, N.J., Cox, P.T. and Goodman, A.E. (1994) Rapid identification of *Aeromonas* species using 16S rRNA-targeted oligonucleotide primers: A molecular approach based on screening of environmental isolates. *J. Appl. Bacteriol.* **77**, 722–726.

Dowd, S.E. and Pillai, S.D. (1997) A rapid viability assay for *Cryptosporidium* oocysts and *Giardia* cysts for use in conjunction with indirect fluorescent antibody detection. *Can. J. Microbiol.* **43**, 658–662.

Du Pont, H.L., Chappell, C.L., Sterling, C.R., Okhuysen, P.C., Rose, J.B. and Jakubowski, W. (1995) The infectivity of *Cryptosporidium parvum* in healthy volunteers. *N. Engl. J. Med.* **332**, 855–859.

Enroth, H. and Engstrand, L. (1995) Immunomagnetic separation and PCR for detection of *Helicobacter pylori* in water and stool specimens. *J. Clin. Microbiol.* **33**, 2162–2165.

Epstein, C.B. and Butow, R.A. (2000) Microarray technology - enhanced versatility, persistent challenge. *Curr. Opin. Biotechnol.* **11**, 36–41.

Faude, U.C. and Hofle, M.G. (1997) Development and application of monoclonal antibodies for *in situ* detection of indigenous bacterial strains in aquatic ecosystems. *Appl. Environ. Microbiol.* **63**, 4534–4542.

Filkorn, R., Wiedenmann, A. and Botzenhart, K. (1994) Selective detection of viable *Cryptosporidium* oocysts by PCR. *Zentralbl. Hyg. Umweltmed.* **195**, 489–494.

Ford, T.E. (1999) Microbiological safety of drinking water: United States and global perspectives. *Environ. Health Perspect.* **107**, 191–206.

Ford, T.E. and MacKenzie, W.R. (2000) How safe is our drinking water? *Postgrad. Med.* **108**, 11–14.

Fuchs, B.M., Wallner, G., Beisker, W., Schwippl, I. Ludwig, W. and Amann, R. (1998) Flow cytometric analysis of the *in situ* accessibility of *Escherichia coli* 16S rRNA for fluorescently-labelled oligonucleotide probes. *Appl. Environ. Microbiol.* **64**, 4973–4982.

Garret, C.T., Ferreira-Centeno, A. and Nasim, S. (1993) Molecular diagnostics: Issues of utilization, regulation and organization. *Clin. Chim. Acta* **217**, 85–103.

Gilgen, M., Wegmüller, B., Burkhalter, P., Bühler, H-P., Müller, U., Lüthy, J. and Candrian, U. (1995) Reverse transcription PCR to detect enteroviruses in surface water. *Appl. Environ. Microbiol.* **61**, 1226–1231.

Giovannoni, S.J., De Long, E.F., Olsen, G.J. and Pace, N.R. (1988) Phylogenetic group-specific oligodeoxynucleotide probes for identification of single microbial cells. *J. Bacteriol.* **170**, 720–726.

Girones, R., Puig, M., Allard, A., Lucena, F., Wadell, G. and Jofre, J. (1995) Detection of adenovirus and enterovirus by PCR amplification in polluted waters. *Water Sci. Tech.* **31**(5–6), 351–357.

Gratacap-Cavallier, B., Genoulaz, O., Brengel-Pesce, K., Soule, H., Innocenti-Francillard, P., Bost, M., Gofti, L., Zmirou, D. and Seigneurin, J.M. (2000) Detection of human and animal rotavirus sequences in drinking water. *Appl. Environ. Microbiol.* **66**, 2690–2692.

Gutell, R.R., Larsen, N. and Woese, C.R. (1994) Lessons from an evolving rRNA: 16S and 23S rRNA structures from a comparative perspective. *Microbiol. Rev.* **58**, 10–26.

Guttman-Bass, N. (1987) Cell cultures and other host systems for detecting and quantifying viruses in the environment. In *Methods for Recovering Viruses from the Environment* (ed. G. Berg), pp. 195–228, CRC Press, Boca Raton, FL.

Hadrys, H., Balick, M. and Schierwater, B. (1992) Applications of random amplified polymorphic DNA (RAPD) in molecular ecology. *Mol. Ecol.* **1**, 55–63.

Hallier-Soulier, S. and Guillot, E. (1999) An immunomagnetic separation polymerase chain reaction assay for rapid and ultra-sensitive detection of *Cryptosporidium parvum* in drinking water. *FEMS Microbiol Lett.* **176**, 285–289.

Hanai, K., Satake, M., Nakanishi, H. and Venkatsewaran, K. (1997) Comparison of commercially available kits with standard methods for detection of *Salmonella* strains in foods. *Appl. Environ. Microbiol.* **63**, 775–778.

Hasan, J.A.K., Huq, A., Tamplin, R.J., Siebeling, R.J. and Colwell, R.R. (1994) A novel kit for rapid detection of *Vibrio cholerae* O1. *J. Clin. Microbiol.* **32**, 249–252.

Hazen, T.C. and Jimenez, L. (1988) Enumeration and identification of bacteria from environmental samples using nucleic acid probes. *Microbiol. Sci.* **5**, 340–343.

Heid, C.A., Stevens, J., Livak, K.J. and Williams, P.M. (1996) Real time quantitative PCR. *Genome Res.* **6**, 986–994.

Hicks, R., Amann, R.I. and Stahl, D.A. (1992) Dual staining of natural bacterioplankton with 4′,6-diamidino-2-phenylindole and fluorescent oligonucleotide probes targeting kingdom level 16S rRNA sequences. *Appl. Environ. Microbiol.* **58**, 2158–2163.

Hirmo, S., Artursson, E., Puu, G., Wadstrom, T. and Nilsson, B. (1998) Characterization of *Helicobacter pylori* interactions with sialylglycoconjugates using a resonant mirror biosensor. *Anal. Biochem.* **256**, 63–66.

Hoff, K.A. (1993) Total and specific bacterial counts by simultaneous staining with DAPI and fluorochrome-labelled antibodies. In *Handbook of Methods in Aquatic Microbial Ecology* (eds P.F. Kemp, B.F. Sherr, E.B. Sherr, and J.J. Cole), pp. 149–154, Lewis Publishers, Boca Raton, FL, USA.

Holben, W.E. and Tiedje, J.M. (1988) Applications of nucleic acid hybridization in microbial ecology. *Ecol.* **69**, 561–568.

Huang, P.W., Laborde, D., Land, V.R., Matson, D.O., Smith, A.W. and Jiang, X. (2000) Concentration and detection of caliciviruses in water samples by reverse transcription-PCR. *Appl. Environ. Microbiol.* **66**, 4383–4388.

Hurst, C.J. (1996) Detection of viruses in environmental waters, sewage, and sewage sludges. In *Manual of Environmental Microbiology* (eds. C.J. Hurst, G.R. Knudsen, M.J. McInerney, L.D. Stetzenbach, and M.V. Walter), pp. 168–175, ASM Press, Washington, D.C.

Ito, T., Kuwahara, S. and Yokota, T. (1983) Automatic and manual latex agglutination tests for measurement of cholera toxin and heat-labile enterotoxin of *Escherichia coli*. *J. Clin. Microbiol.* **17**, 7–12.

Jaykus, L.A., Hermard, M.T. and Sobsey, M.D. (1994) Human enteric pathogenic viruses. In *Environmental Indicators and Shellfish Safety* (eds C.R. Hackney and M.D. Peirson), pp. 92–153, Chapman and Hall, New York.

Jiang, X., Wang, J. and Estes, M.K. (1995) Characterization of SRSVs using RT-PCR and a new antigen ELISA. *Arch. Virol.* **140**, 363–374.

Jenkins, M.C., Trout, J., Abrahamsen, M.S., Lancto, C.A., Higgins, J. and Fayer, R. (2000) Estimating viability of *Cryptosporidium parvum* oocysts using reverse transcriptase–polymerase chain reaction (RT–PCR) directed at mRNA encoding amyloglucosidase. *J. Micriobiol. Meth.* **43**, 97–106.

Johnson, D.W., Pieniazek, N.J., Griffin, D.W., Misener, L. and Rose, J.B. (1995) Development of a PCR protocol for sensitive detection of *Cryptosporidium* oocysts in water samples. *Appl. Environ. Microbiol.* **61**, 3849–3855.

Josephson, K.L., Gerba, C.P. and Pepper, I.L. (1993) Polymerase chain reaction of nonviable bacterial pathogens. *Appl. Environ. Microbiol.* **59**, 3513–3515.

Jothikumar, N., Cliver, D.O. and Mariam, T.W. (1998) Immunomagnetic capture PCR for rapid concentration and detection of hepatitis A virus from environmental samples. *Appl. Environ. Microbiol.* **64**, 504–508.

Kalmbach, S., Manz, W. and Szewzyk, U. (1997) Isolation of new bacterial species from drinking water biofilms and proof of their *in situ* dominance with highly specific 16S rRNA probes. *Appl. Environ. Microbiol.* **63**, 4164–4170.

Kapperud, G., Vardund, T., Skjerve, E., Hornes, E. and Michaelsen, T.E. (1993) Detection of pathogenic *Yersinia enterocolitica* in foods and water by immunomagnetic separation, nested polymerase chain reactions and colorimetric detection of amplified DNA. *Appl. Environ. Microbiol.* **59**, 2938–2944.

Kaucner, C. and Stinear, T. (1998) Sensitive and rapid detection of viable *Giardia* cysts and *Cryptosporidium parvum* oocysts in large-volume water samples with wound fibreglass cartridge filters and reverse transcription-PCR. *Appl. Environ. Microbiol.* **64**, 1743–1749.

Kfir, R. and Genthe, B. (1993) Advantages and disadvantages of the use of immunodetection techniques for the enumeration of micro-organisms and toxins in water. *Water Sci. Tech.* **27**(3–4), 243–248.

Knight, I.T., Di Ruggiero, J. and Colwell, R.R. (1991) Direct detection of enteropathogenic bacteria in estuarine water using nucleic acid probes. *Water Sci. Tech.* **24**(2), 261–266.

Koeleman, J.G.M., Stoof, J., Biesmans, D.J., Savelkoul, P.H.M. and Vandenbroucke-Grauls, C.M.J.E. (1998) Comparison of ARDRA, RAPD and AFLP fingerprinting for identification of *Acinetobacter* genomic species and typing of *Acinetobacter baumannii*. *J. Clin. Microbiol.* **36**, 2522–2529.

Lebaron, P., Catala, P., Fajon, C., Joux, F., Baudart, J. and Bernard, L. (1997) A new sensitive, whole-cell hybridization technique for detection of bacteria involving a biotinylated oligunucleotide probe targeting rRNA and tyramide signal amplification. *Appl. Environ. Microbiol.* **63**, 3274–3278.

LeChevalier, M.W., Norton, W.D. and Lee, R.G. (1991) Occurrence of *Giardia* and *Cryptosporidium* spp. in surface water supplies. *Appl. Environ. Microbiol.* **57**, 2610–2616.

LeChevallier, M.W., Norton, W.D., Siegel, J.E. and Abbaszadegan, M. (1995) Evaluation of the immunofluorescence procedure for detection of *Giardia* cysts and *Cryptosporidium* oocyts in water. *Appl. Environ. Microbiol.* **61**, 690–697.

Lee, S., Malone, C. and Kemp, P.F. (1993) Use of multiple 16S RNA-targeted fluorescent probes to increase signal strength and measure cellular RNA from natural planktonic bacteria. *Marine Ecol. Prog. Ser.* **101**, 193–201.

Lindquist, H.S.A. (1997) Probes for the specific detection of *Cryptosporidium parvum*. *Water Res.* **31**, 2668–2671.

Lodder, W.J., Vinje, J., van De Heide, R., de Roda Husman, A.M., Lennen, E.J. and Koopmans, M.P. (1999) Molecular detection of Norwalk-like caliciviruses in sewage. *Appl. Environ. Microbiol.* **65**, 5624–5627.

Lowery, C.J., Moore, J.E., Millar, B.C., Burke, D.P., McCorry, K.A., Crothers, E. and Dooley, S. (2000) Detection and speciation of *Cryptosporidium* spp. in environmental water samples by immunomagnetic separation, PCR and endonuclease restriction. *J. Med. Microbiol.* **49**, 779–785.

Mahbubani, M.H., Bej, A.K., Perlin, M., Schaefer, F.W., Jakubowski, W. and Atlas, R.M. (1991) Detection of *Giardia* cysts by using polymerase chain reaction and distinguishing live from dead cysts. *Appl. Environ. Microbiol.* **57**, 3456–3461.

Mahbubani, M.H., Schaefer, F.W., Jones, D.D. and Bej, A.K. (1998) Detection of *Giardia* in environmental waters by immuno-PCR amplification methods. *Curr. Microbiol.* **36**, 107–113.

Maidak, B.L., Larsen, N., McCaughey, M.J., Overbeek, R., Olsen, G.J., Fogel, K., Blandy, J. and Woese, C.R. (1994) The Ribosomal Database Project. *Nucleic Acids Res.* **22**, 3485–3487.

Maidak, B.L., Olsen, G.J., Larsen, N., Overbeek, R., McCaughey, M.J. and Woese, C.R. (1997) The RDP (Ribosomal Database Project). *Nucleic Acids Res.* **25**, 109–111.

Malmqvuist, M. (1993) Biospecific interaction analysis using biosensor technology. *Nature* **361**, 186–187.

Mandrell, R.E. and Wachtel, M.R. (1999) Novel detection techniques for human pathogens that contaminate poultry. *Curr. Opin. Biotechnol.* **10**, 273–278.

Manz, W., Amann, R., Ludwig, W., Wagner, M. and Schleifer, K.-H. (1992) Phylogenetic oligodeoxynucleotide probes for the major subclasses of proteobacteria: problems and solutions. *System. Appl. Microbiol.* **15**, 593–600.

Manz, W., Szewzyk, U., Ericsson, P., Amann, R., Schleifer, K-H. and Stenström, T-A. (1993) *In situ* identification of bacteria in drinking water and adjoining biofilms by hybridization with 16S- and 23S rRNA-directed fluorescent oligonucleotide probes. *Appl. Environ. Microbiol.* **59**, 2293–2298.

Mayer, C.L. and Palmer, C.J. (1996) Evaluation of PCR, nested PCR, and fluorescent antibodies for detection of *Giardia* and *Cryptosporidium* species in waste water. *Appl. Environ. Microbiol.* **62**, 2081–2085.

McDougald, D., Rice, S.A., Weichart, D. and Kjelleberg, S. (1998) Nonculturability: Adaptation or devilitation? *FEMS Microbiol. Ecol.* **25**, 1–9.

McGinnis, M.J. and Foege, W.H. (1993) Actual causes of death in the United States. *J. Am. Med. Ass.* **270**, 2207–2212.

McKay, A.M. (1992) Viable but nonculturable forms of potentially pathogenic bacteria in water. *Lett. Appl. Microbiol.* **14**, 129–135.

McKillip, J.L., Jaykus, L.-A. and Drake, M. (1998) rRNA stability in heat-killed and UV-irradiated enterotoxigenic *Staphylococcus aureus* and *Escherichia coli* O157:H7. *Appl. Environ. Microbiol.* **64**, 4264–4268.

Meier, H., Koob, C., Ludwig, W., Amann, R., Frahm, E., Hoffmann, S., Obst, U. and Schleifer, K.H. (1997) Detection of enterococci with rRNA targeted DNA probes and their use for hygienic drinking water control. *Water Sci. Tech.* **35**, 437–444.

Moe, C.L. (1996) Waterborne transmission of infectious agents. In *Manual of Environmental Microbiology* (eds. C.J. Hurst, G.R. Knudsen, M.J. McInerney, L.D. Stetzenbach and M.V. Walter), pp. 136–152, ASM Press, Washington, D.C.

Murrin, K. and Slade, J. (1997) Rapid detection of viable enteroviruses in water by tissue culture and semi-nested polymerase chain reaction. *Water Sci. Tech.* **35**(11–12), 429–432.

Neumann, N.F., Gyurek, L.L., Gannie, L., Finch, G.R. and Belosevic, M. (2000) Comparison of animal infectivity and nucleic acid staining for assessment of *Cryptosporidium parvum* viability in water. *Appl. Environ. Microbiol.* **66**, 406–412.

Nieminski, E.C., Schaefer, F.W. and Ongerth, J.E. (1995) Comparison of two methods for the detection of *Giardia* cysts and *Cryptosporidium* oocysts in water. *Appl. Environ. Microbiol.* **61**, 1714–1719.

Nishimura, M., Kita-Tsukamoto, K. and Kogure, K. (1993) A new method to detect viable bacteria in natural seawater using 16S rRNA oligonucleotide probe. *J. Oceanol.* **49**, 51–56.

Noah, C.W., Poteet, S.S., Lister, M.M., Roderick, C.N., Smith D.B., Colvert, R.M., Holloand, M.A. and Cerniglia, C.E. (1996) Production and characterization of monoclonal antibodies to detect *Vibrio cholerae* serogroup O1 in a rapid enzyme-linked immunosorbent assay. *J. Food Protect.* **59**, 1153–1157.

Olsen, G.J., Lane, D.J., Giovannoni, S.J., Pace, N.R. and Stahl, D.A. (1986) Microbial ecology and evolution: a ribosomal RNA approach. *Annu. Rev. Microbiol.* **40**, 337–365.

Olsvik, O., Popovic, T., Skjerve, E., Cudjoe, K.S., Hornes, E., Ugelstad, J. and Uhlen, M. (1994) Magnetic separation techniques in diagnostic microbiology. *Clin. Microbiol. Rev.* **7**, 43–54.

Oyofo, B.A. and Rollins, D.M. (1993) Efficacy of filter types for detecting *Campylobacter jejuni* and *Campylobacter coli* in environmental water samples by polymerase chain reaction. *Appl. Environ. Microbiol.* **59**, 4090–4095.

Pace, N.R., Stahl, D.A., Lane, D.J. and Olsen, G.J. (1986) The analysis of natural microbial populations by ribosomal RNA sequences. *Adv. Microb. Ecol.* **9**, 1–55.

Parveen, S., Portier, K.M., Robinson, K., Edmiston, L. and Tamplin, M.L. (1999) Discriminant analysis of ribotyping profiles of *Escherichia coli* for differentiating human and nonhuman sources of faecal pollution. *Appl. Environ. Microbiol.* **65**, 3142–3147.

Patel, S., Pedraza-Diaz, S. and McLauchlin, J. (1999) The identification of *Cryptosporidium* species and *Cryptosporidium parvum* directly from whole feces by analysis of a multiplex PCR of the 18S rRNA gene and by PCR/RFLP of the *Cryptosporidium* outer wall protein (COWP) gene. *Int. J. Parasitol.* **29**, 1241–1247.

Payment, P. (1993) Viruses: Prevalence of disease, levels, and sources. In *Safety in Water Disinfection: Balancing Chemical & Microbial Risks* (ed. G.F. Craun), pp. 99–114, ILSI Press, Washington, D.C.

Puig, M., Jofre, J., Lucena, F., Allard, A., Wadell, G. and Girones, R. (1994) Detection of adenovirus and enteroviruses in polluted waters by nested PCR amplification. *Appl. Environ. Microbiol.* **60**, 2963–2970.

Pyle, B.H., Broadway, S.C. and McFeters, G.A. (1995) A rapid, direct method for enumerating respiring enterohemorrhagic *Escherichia coli* O157:H7 in water. *Appl. Environ. Microbiol.* **61**, 2614–2619.

Quadri, F., Chowdury, A., Hossain, J., Chodury, K., Azim, T., Shimada, T., Islam, N.K.M., Sack, R.B. and Albert, M.J. (1994) Development and evaluation of rapid monoclonal antibody based coagulation test for direct detection of *Vibrio cholerae* O139 synonym Bengal in stool samples. *J. Clin. Microbiol.* **32**, 1589–1590.

Rademaker, J.L.W., Hoste, B., Kersters, K., Louws, F.J., Swings, J., Vauterin, L., Vauterin, P. and De Bruijn, F.J. (2000) Comparison of AFLP and rep-PCR genomic fingerprinting with DNA-DNA homology studies: *Xanthomonas* as a model system. *Int. J. Syst. Evol. Microbiol.* **50**, 665–677.

Rehnstam, A.S., Norqvist, A., Wolf-Watz, H. and Hagstrom, A. (1989) Identification of *Vibrio anoillarum* in fish by using partial 16S rRNA sequences and a specific 16S rRNA oligonucleotide probe. *Appl. Environ. Microbiol.* **55**, 1907–1910.

Reynolds, K.A., Gerba, C.P. and Pepper, I.L. (1996) Detection of infectious enteroviruses by an integrated cell culture-PCR procedure. *Appl. Environ. Microbiol.* **62**, 1424–1427.

Ridgway, J.W. (1984) Bacterial recovery from water, sewage, and sewage effluents. In *The Revival of Injured Microbes* (eds. M.H.E. Andrew and A.D. Russel), p. 373, Academic Press, Orlando.

Rochelle, P.A., De Leon, R., Johnson, A., Stewart, M.H. and Wolfe, R.L. (1999) Evaluation of immunomagnetic separation for recovery of infectious *Cryptosporidium parvum* oocysts from environmental samples. *Appl. Environ. Microbiol.* **65**, 841–844.

Rochelle, P.A., De Leon, R., Stewart, M.H. and Wolfe, R.L. (1997a) Comparison of primers and optimization of PCR conditions for detection of *Cryptosporidium parvum* and *Giardia lamblia* in water. *Appl. Environ. Microbiol.* **63**, 106–114.

Rochelle, P.A., Ferguson, D.M., Handojo, T.J., De Leon R., Stewart, M.H. and Wolfe, R.L. (1997b) An assay combining cell culture with reverse transcriptase PCR to detect and determine the infectivity of waterborne *Cryptosporidium parvum. Appl. Environ. Microbiol.* **63**, 2029–2037.

Rodgers, M.R., Flanigan, D.J. and Jakubowski, W. (1995) Identification of algae which interfere with the detection of *Giardia* cysts and *Cryptosporidium* oocysts and a method for alleviating this interference. *Appl. Environ. Microbiol.* **61**, 3759–3763.

Romaniuk, P.J. and Trust, T.J. (1987) Identification of *Campylobacter* species by Southern hybridization of genomic DNA using an oligonucleotide probe for 16S rRNA genes. *FEMS Microbiol. Lett.* **43**, 331–335.

Rose, J.B. and Grimes, D.J. (2001) Re-evaluation of microbial water safety: powerful new tools for microbial risk assessment. In *Report from the American Academy of Microbiology*, American Academy of Microbiology, Washington, D.C.

Roszak, D.B. and Colwell, R.R. (1987) Survival strategies of bacteria in the natural environment. *Microbiol. Rev.* **51**, 365–379.

Saiki, R.K., Gelfand, D.H., Stoffel, S., Scharf, S.J., Higuchi, R., Horn, G.T., Mullis, K.B. and Ehrlich, H.A. (1988) Primer-directed enzymatic amplification of DNA with a thermostable DNA polymerase. *Science* **239**, 487–491.

Sandery, M., Stinear, T. and Kaucner, C. (1996) Detection of pathogenic *Yersinia enterocolitica* in environmental waters by PCR. *J. Appl. Bacteriol.* **80**, 327–332.

Sasaki, K., Tajiri, Y., Sata, M., Fujii, Y., Matsubara, F., Zhao, M., Shimizu, S., Toyonaga, A. and Tanikawa, K. (1999) *Helicobacter pylori* in the natural environment. *Scand. J. Infect. Dis.* **31**, 275–279.

Savelkoul, P.M., Aarts, H.M., de Haas, J., Dijkshoorn, L., Duim, B., Otsen, M., Rademaker, J.W., Schouls, L. and Lenstra, J.A. (1999) Amplified-fragment length polymorphism analysis: the state of an art. *J. Clin. Microbiol.* **37**, 3083–3091.

Sauch, J.F., Flanigan, D., Galvin, M.L., Berman, D. and Jakubowski, W. (1991) Propidium iodide as an indicator of *Giardia* cyst viability. *Appl. Environ. Microbiol.* **57**, 3243–3247.

Schaefer, F.W. (1996) Detection of protozoan parasites in source and finished drinking waters. In *Manual of Environmental Microbiology* (eds. C.J. Hurst, G.R. Knudsen, M.J. McInerney, L.D. Stetzenbach and M.V. Walter), pp. 153–167, ASM Press, Washington, D.C.

Schleifer, K.-H., Ludwig, W., Kraus, J. and Festl, H. (1985) Cloned ribosomal ribonucleic acid genes from *Pseudomonas aeruginosa* as probes for conserved deoxyribonucleic acid sequences. *Int. J. Syst. Bacteriol.* **35**, 231–236.

Schneegurt, M.A. and Kulpa, C.F. (1998) The application of molecular techniques in environmental biotechnology for monitoring microbial systems. *Biotechnol. Appl. Biochem.* **27**, 73–79.

Schwab, K.J., Neill, F.H., Estes, M.K. and Atmar, R.L. (1998) Improvements in the RT-PCR detection of enteric viruses in environmental samples. *Water Sci. Tech.* **38**(12), 83–86.

Schwab, K.J., de Leon, R. and Sobsey, M.D. (1993) Development of PCR methods for enteric virus detection in water. *Water Sci. Tech.* **27**, 211–218.

Schwab, K.J., De Leon, R. and Sobsey, M.D. (1996) Immunoaffinity concentration and purification of waterborne enteric viruses for detection by reverse transcriptase PCR. *Appl. Environ. Microbiol.* **62**, 2086–2094.

Sheih, Y.S.C., Baric, R.S., Sobsey, M.D., Ticehurst, J., De Leon, R. and Walter, R. (1991) Detection of hepatitis A virus and other enteroviruses in water by ssRNA probes. *J. Virol. Meth.* **31**, 1119–1126.

Sheih, Y.S.C., Wait, D., Tai, L. and Sobsey, M.D. (1995) Methods to remove inhibitors in sewage and other faecal wastes for enterovirus detection by polymerase chain reaction. *J. Virol. Meth.* **54**, 56–66.

Sheridan, G.E., Masters, C.I., Shallcross, J.A. and MacKey, B.M. (1998) Detection of mRNA by reverse transcription-PCR as an indicator of viability in *Escherichia coli* cells. *Appl. Environ. Microbiol.* **64**, 1313–1318.

Sobsey, M.D. (1982) Quality of currently available methodology for monitoring viruses in the environment. *Environ. Int.* **7**, 39–51.

Sobsey, M.D. (1994) Molecular methods to detect viruses in environmental samples. In *Rapid Methods and Automation in Microbiology and Immunology* (eds. R.C. Spencer, E.P. Wright, and S.W.B. Newsme), pp. 387–400, Intercept Ltd., Andover, England.

Stahl, D.A. and Amann, R. (1991) Development and application of nucleic acid probes in bacterial systematics. In *Sequencing and Hybridization Techniques in Bacterial Systematics* (eds. E. Stackebrandt, and M. Goodfellow), pp. 205–248, John Wiley and Sons, Chichester, U.K.

Steffan, R.J. and Atlas, R.M. (1991) Polymerase chain reaction: Application in environmental microbiology. *Annu. Rev. Microbiol.* **45**, 137–161.

Stinear, T., Matusan, A., Hines, K. and Sandery, M. (1996) Detection of a single viable *Cryptosporidium parvum* oocyst in environmental water concentrates by reverse transcription-PCR. *Appl. Environ. Microbiol.* **62**, 3385–3390.

Straub, T.M., Pepper, I.L. and Gerba, C.P. (1995) Removal of PCR inhibiting substances in sewage sludge amended soil. *Water Sci. Technol.* **31**(5–6), 311–315.

Tani, K., Kurokawa, K. and Nasu, M. (1998) Development of a direct *in situ* PCR method for detection of specific bacteria in natural environments. *Appl. Environ. Microbiol.* **64**, 1536–1540.

Theron, J., Morar, D., Du Preez, M., Brözel, V.S. and Venter, S.N. (2001) A sensitive seminested PCR method for the detection of *Shigella* in spiked environmental water samples. *Water Res.* **35**, 869–874.

Theron, J., Cilliers, J., Du Preez, M., Brözel, V.S. and Venter, S.N. (2000) Detection of toxigenic *Vibrio cholerae* from environmental water samples by an enrichment broth cultivation-pit-stop semi-nested PCR procedure. *J. Appl. Microbiol.* **89**, 539–546.

Toranzos, G.A. and Alvarez, A.J. (1992) Solid-phase polymerase chain reaction - Applications for direct detection of enteric pathogens in waters. *Can. J. Microbiol.* **38**, 365–369.

Toranzos, G.A, Alvarez, A.J. and Dvorsky, E.A. (1993) Application of the polymerase chain reaction technique to the detection of pathogens in water. *Water Sci. Tech.* **27**, 207–210.

Toranzos, G.A. and McFeters, G.A. (1996) Detection of indicator micro-organisms in environmental freshwaters and drinking waters. In *Manual of Environmental Microbiology* (eds. C.J. Hurst, G.R. Knudsen, M.J. McInerney, L.D. Stetzenbach and M.V. Walter), pp. 184–194, ASM Press, Washington, D.C.

Tougianidou D., Jacob, M., Herbold, T., Hahn, T., Flehmig, B. and Botzenhart, K. (1989) Assessment of various cell lines (including mixed-cell cultures) for the detection of enteric viruses in different water sources. *Water Sci. Tech.* **21**, 311–314.

Tsai, Y.L. and Olson, B.H. (1992) Rapid method for separation of bacterial DNA from humic acid substances for polymerase chain reaction. *Appl. Environ. Microbiol.* **58**, 2291–2295.

Udeh, P., Veenstra, J., Abraham, A.J. and John, G.H. (2000) Quantitative polymerase chain (QPCR) reaction using the MIMIC approach to estimate *Cryptosporidium parvum* oocysts, an intestinal pathogen, in municipal water treatment sludge samples. *Mol. Cell. Probes* **14**, 121–126.

Upton, S.J., Tilley, M. and Brillhart, D.B. (1994) Comparative development of *Cryptosporidium parvum* (Apicomplexa) in 11 continuous host cell lines. *FEMS Microbiol. Lett.* **118**, 233–236.

Van de Peer, Y., Chapelle, S. and De Wachter, R. (1996) A quantitative map of nucleotide substitution rates in bacterial rRNA. *Nucleic Acids Res.* **24**, 3381–3391.

Vaneechoutte, M. (1996) DNA fingerprinting techniques for micro-organisms. A proposal for classification and nomenclature. *Mol. Biotechnol.* **6**, 115–142.

Van Hal, N.L., Vorst, O., Van Houwelingen A.M., Kok, E.J., Peijnenburg, A., Aharoni, A., Van Tunen, A.J. and Keijer, J. (2000) The application of DNA microarrays in gene expression analysis. *J. Biotechnol.* **78**, 271–280.

Venkatesan, M., Buysse, J.M., Vandendries, E. and Kopecko, D.J. (1988) Development and testing of invasion-associated DNA probes for detection of *Shigella* spp. and enteroinvasive *Escherichia coli*. *J. Clin. Microbiol.* **26**, 261–266.

Versalovic, J., Schneider, F., de Bruijn, F.J. and Lupski, J.R. (1994) Genomic fingerprinting of bacteria using repetitive sequence-based polymerase chain reaction. *Meth. Mol. Cell. Biol.* **5**, 25–40.

Vesey, G., Griffiths, K.R., Gauci, M.R., Deere, D., Williams, K.L. and Veal, D.A. (1997) Simple and rapid measurement of *Cryptosporidium* excystation using flow cytometry. *Int. J. Parasitol.* **27**, 1353–1359.

Vesey, G., Slade, J.S., Byrne, M., Shepard, K. and Fricker, C.R. (1993) A new method for the concentration of *Cryptosporidium* oocysts from water. *J. Appl. Bacteriol.* **75**, 82–86.

Vickers, R.M., Stout, J.E. and Yu, V.L. (1990) Failure of a diagnostic monoclonal immunofluorescent reagent to detect *Legionella pneumophila* in environmental samples. *Appl. Environ. Microbiol.* **56**, 2912–2914.

Visvesvara, G.S., Leitch, G.L., Pieniazek, N.J., da Silva, A.J., Wallace, S., Slemenda, S.B., Weber, R., Schwartz, D.A., Gorelkin, L., Wilcox, C.M. and Bryan, R.T. (1995) Short-term *in vitro* culture and molecular analysis of the microsporidian *Enterocytozoon bieneusi*. *J. Euk. Microbiol.* **42**, 506–510.

Vos, P., Hogers, R., Bleeker, M., Reijans, M., Van de Lee, T., Hornes, M., Frijters, A., Pot, J., Peleman, J., Kuiper, M. and Zabeau, M. (1995) AFLP: a new technique for DNA fingerprinting. *Nucleic Acids Res.* **23**, 4407–4414.

Waage, A.S., Vardund, T., Lund, V. and Kapperud, G. (1999) Detection of small numbers of *Campylobacter jejuni* and *Campylobacter coli* cells in environmental water, sewage, and food samples by a seminested PCR. *Appl. Environ. Microbiol.* **65**, 1636–1643.

Wagner-Wiening, C. and Kimmig, P. (1995) Detection of viable *Cryptosporidium parvum* oocysts by PCR. *Appl. Environ. Microbiol.* **61**, 4514–4516.

Ward, D.M., Weller, R. and Bateson, M.M. (1990) 16S rRNA sequences reveal numerous uncultured micro-organisms in a natural community. *Nature* **345**, 63–65.

Way, J.S., Josephson, K.L., Pillai, S.D., Abbazadegan, M., Gerba, C.P. and Pepper, I.L. (1993) Specific detection of *Salmonella* spp. by multiplex polymerase chain reaction. *Appl. Environ. Microbiol.* **59**, 1473–1479.

Widmer, G., Orbacz, E.A. and Tzipori, S. (1999) Beta-tubulin mRNA as a marker of *Cryptosporidium parvum* oocyst viability. *Appl. Environ. Microbiol.* **65**, 1584–1588.

Wilson, I.G. (1997) Inhibition and facilitation of nucleic acid amplification. *Appl. Environ. Microbiol.* **63**, 3741–3751.

Woese, C.R. (1987) Bacterial evolution. *Microbiol. Rev.* **51**, 221–271.

Wu, Z., Nagano, I., Matsuo, A., Uga, S., Kimata, I., Iseki, M. and Takahashi, Y. (2000) Specific PCR primers for *Cryptosporidium parvum* with extra high sensitivity. *Mol. Cell. Probes* **14**, 33–39.

Yang, F. and Xu, X. (1993) A new method of RNA preparation for detection of hepatitis A virus in environmental samples by the polymerase chain reaction. *J. Virol. Meth.* **43**, 77–84.

Zachar, V., Thomas, R.A. and Goustin, A.S. (1993) Absolute quantification of target DNA: a simple competitive PCR for efficient analysis of multiple samples. *Nucleic Acids Res.* **21**, 2017–2018.

Zammatteo, N., Jeanmart, L., Hamels, S., Courtois, S., Louette, P., Hevesi, L. and Remacle, J. (2000) Comparison between different strategies of covalent attachment of DNA of glass surfaces to build DNA microarrays. *Anal. Biochem.* **280**, 143–150.

Zarda, B., Amann, R., Wallner, G. and Schleifer, K.-H. (1991) Identification of single bacterial cells using digoxigenin-labelled, rRNA-targeted oligonucleotides. *J. Gen. Microbiol.* **137**, 2823–2830.

8

Future needs and priorities

T. Ford

8.1 INTRODUCTION

It is clear from recent literature reviews, from American Academy of Microbiology Reports and from WHO expert meetings, that we are a long way from guaranteeing safe drinking water even in the most developed nations (Ford and Colwell, 1996; Ford, 1999; Rose and Grimes, 2001; Levin *et al.*, 2002). This chapter attempts to identify major knowledge gaps and summarize at least some of the needs and priorities for the future provision of safe drinking water. Obviously, the major benefits to human health are through basic hygiene and sanitation practices – still much needed areas for public health intervention in many parts of the world. However, this chapter will emphasize more research-oriented needs and priorities. These include the need for a better understanding of biofilms and their control, improvements in risk assessment methodologies, the emergence of new disease, the balance between pathogens and disinfection by-products, and the future promise of rapidly developing technologies.

The concept of safe, cheap drinking water as an inalienable human right for everyone has emerged in the more developed nations in the latter part of the twentieth century. Although the quality of our source waters, and their protection, is recognized as the single most efficient factor in determining consistently high quality drinking water, the technology now exists to take wastewater and recycle it to potable water quality.

As cities have developed, both in the United States and globally, water distribution as well as waste collection systems have become more complex and far more difficult to maintain. Thirty percent leakage from water distribution systems appears to be a common estimate, whether talking to utility personnel in Boston, USA, or Hyderabad, India. The fundamental difference between these two systems is pressure. According to the Hyderabad utility personnel, Hyderabad is able to provide a two-hour supply of water per day (Mohanty *et al.*, 2002). The system therefore remains stagnant for twenty-two hours per day and is highly susceptible to contamination through back-siphonage, cross connections, and biofilm regrowth. In contrast, Boston and other developed nation cities maintain pressure resulting in incredible wastage of treated, potable water, yet reduced risk to human health.

As we enter the twenty-first century, what are our future options? While for many regions watershed protection is still problematic, filtration and disinfection technologies are constantly being improved. A typical multibarrier approach to provide safe drinking water is presented in Table 8.1. A technology can always be developed to clean contaminated source waters, albeit at a high cost. It is harder, however, to find a technological solution to deteriorating distribution systems. Replacing pipelines is an extremely slow and disruptive process, and lining existing pipes may only provide temporary solutions. Lining materials that are currently in use, e.g., cement and epoxy, may be susceptible to deterioration, particularly in microbiologically active environments.

There are many alternatives proposed for provision of safe drinking water, including separate potable and non-potable supplies, point-of-use treatment devices

Table 8.1 A multibarrier approach to maximize microbiological quality of water.

(1) Watershed protection that minimizes anthropogenic and wildlife effects on source water, including programs to reduce the effect of waterfowl, particularly near water intake sites.

(2) A treatment system with sufficient capacity to maintain adequate pressure throughout the distribution system for 24 hours/day, and that minimizes opportunities for microbial colonization in the distribution system. This could include

- Coagulation–flocculation and sedimentation to remove colloids, associated micro-organisms, debris and macro-organisms.
- Pre-ozonation to effectively kill micro-organisms in source waters, reduce odor, taste and color, precursors for DBPs,[a] and reduce the amount of chlorine/chloramine necessary to maintain a system residual.
- Filtration to further remove particulates and micro-organisms, including granular or biological activated carbon to remove AOC.
- Cloramination to provide residual disinfection, minimize biofilm formation and reduce DBPs, with intermittent chlorination and system flushing.

(3) A rigorous program to upgrade distribution system networks and prevent interconnections through leakage, backflushing, improper hydrant use, etc.

Note: Reproduced in part from Ford (1999). Information taken primarily from Geldreich (1996).
[a]Disinfection by-products are formed by ozonation of source waters, including aldehydes and brominated by-products (discussed in Boorman *et al.*, 1999). UV disinfection, used extensively in wastewater treatment, is rapidly gaining acceptance as an alternative to ozonation.

and bottled water. None of these alternatives are without risk to the consumer and are invariably many-fold more expensive than municipal supplies. Of course, there is a strong argument that centrally supplied drinking water is dramatically undervalued, and more appropriate costing of this invaluable resource would help to address some of the problems briefly listed above.

In the foreseeable future, drinking water is likely to continue to be supplied through distribution networks. What, therefore, do we see as future needs and priorities?

8.2 RISK ASSESSMENT

To begin with, we are still unable to characterize and quantify health risks associated with drinking water meeting World Health Organization Standards (WHO, 1993). It is likely that those risks are minimal to individuals with no predisposing factors (e.g. compromised immunity). However, on a global scale it could be argued that the susceptible individuals are as common as the non-susceptible. In developed nations we think about the very young, the elderly, the pregnant and those with diseases that directly, or through treatment, compromise immunity. Even in developed nations, the burgeoning field of environmental health has shown us that exposure to pollutants in air, food and water can affect susceptibility to disease. Likewise malnutrition, stress and socioeconomic status (e.g. inner-city communities) render individuals more susceptible. In developing nations, the burden of diseases may be vast and malnutrition levels high. Rapid industrialization also exposes populations to uncontrolled pollution, leaded petrol is still accepted as the norm, and smoking, drug and alcohol abuse and prostitution are rampant.

However, although exposures appear to be vast, diseases may not be apparent due to multiple prior exposures resulting in population immunity. This then is the paradox. By all the above criteria, these populations are highly susceptible, yet immunity results in lower than expected incidence of many waterborne diseases. This immunity must come at some cost to the individual – which provides some validation to the Disability-Adjusted Life Years (DALY) approach to estimate burden of disease, which could take into account the reduced quality of life (and lifespan) from exposure to multiple infectious agents (and toxins).

A recent study by Arie Havelaar and colleagues in the Netherlands used the DALY approach to compare the risks of disinfection by-products with infectious disease (Havelaar *et al.*, 2000). They conducted a hypothetical case study involving a drinking water system typical of the Netherlands. Their goal was to compare the reduction in risk of infection with *Cryptosporidium parvum* from ozonation of the water source, with the potential risk of cancer from ingestion of bromate (formed by reaction of ozone and bromine compounds in source water). Net health benefits (in DALYs) were calculated on the basis of published clinical, epidemiologic, and toxicologic data on morbidity and mortality. Although bromate was produced in their model at concentrations exceeding US EPA guidelines, they concluded that net benefits from ozonation outweighed risks by more than an order of magnitude, with a net benefit of approximately 1 DALY/million years. The DALY approach allowed the authors to consider life and health expectancy, including evaluation of the distribution of population susceptibilities. This approach provides a far more appropriate estimate of disease burden than can be obtained solely from annual mortality rates. Even so, considerable assumptions are made in

(1) the exposure assessment, e.g., the median number of infectious *Cryptosporidium* oocysts, the median concentration of bromate, and the volume of water ingested
(2) in the hazard characterization, e.g., the shape of the dose–response curves at low exposures, the applicability of rodent models to humans, and the distributions and models used to produce median parameter values and confidence intervals
(3) and in the risk characterization, e.g., assumptions made in calculating life years lost, years with disability and weightings for different population susceptibilities.

Paul Gale from the WRc-NSF Ltd, UK, has provided a thoughtful analysis of the risk assessment process and has argued that distribution of pathogens, and in particular the protozoa, is extremely heterogeneous in drinking water (Gale, 2001). In other words, most consumers ingest zero *Cryptosporidium* oocysts and most water samples measure zero oocysts. However, a few individuals could consume a large number of oocysts. Gale argues that risk assessments based on median values obtained from spot sampling will underestimate risk as most samples measure zero. Number of organisms present in a drinking water sample should be more accurately modeled as a distribution (in Gale's example for *Cryptosporidium*, a Poisson-log-normal distribution is used). Daily risks of infection are then calculated for this distribution using Monte Carlo simulation. However, Gale also reports that risks predicted simply by using the arithmetic mean are very similar to those using Monte Carlo simulation. The arithmetic mean of pathogen density may be a better predictor of risk than the median value, as it provides a weighting to any positive samples based on the actual number of oocysts.

Gale also argues that spot sampling is inappropriate, as even during outbreak conditions most spot samples are zero, and that continuous monitoring, as is currently recommended in the UK for *Cryptosporidium*, is necessary. Similar arguments could be applied to pathogenic viruses and, in fact, to any pathogens with low infective doses and/or a tendency to adsorb to particles/biofilms contributing to a heterogeneous distribution. Gale has constructed a large number of risk assessments based on the arithmetic mean, rather than the median, including risk assessments for *Cryptosporidium parvum*, *Escherichia coli* O157, rotavirus and bovine spongiform encephalitis (BSE; see later).

Assumptions are also clearly present in estimating consumption of drinking water and it is argued that these too should be modeled on a distribution. The dose–response relationship between number of pathogens ingested and infection is highly variable on the basis of individual susceptibility (including immunity from prior exposures). Infectious doses measured in healthy volunteers may bear little relationship to the range of infectious doses in an average population.

Where else does uncertainty arise? For any pathogen, its presence in drinking water may not be enough to characterize risk. Organisms may lose their infectivity/virulence in the drinking water distribution system or after exposure to disinfection; or conversely, they may increase or change in virulence and in their ability to resist antibiotics.

Our risk assessment approach examines individual organisms. We are in fact exposed to complex mixtures of both microbes and chemicals. Questions that arise from this are:

(1) What are the synergistic effects (both in terms of infectious dose and disease outcome) of exposure to mixtures of pathogens, opportunistic pathogens and non-pathogenic microbes?
(2) What are the synergistic effects of exposure to mixtures of microbes and chemicals? For example, could simultaneous exposure to high concentrations of a contaminant chemical and an infectious agent affect the pathogen's infectious dose? Certainly, there is an argument that long-term exposure to chemical contaminants may increase susceptibility to infection.
(3) How is a pathogen's infectivity and exposure route altered by association with biofilms?
(4) How is a pathogen's infectivity and exposure route altered by intracellular survival within protozoa (King *et al.*, 1988)? (For example, it has been argued that the disease outcome from exposure to *Legionella pneumophila* could be related to mode of transmission; within biofilms, within protozoa or free-living (Kramer and Ford, 1994).)

8.3 THE PATHOGENS

Although a wide range of diseases are caused by waterborne pathogens (Table 8.2), the most common outcome, and the one that most frequently remains undiagnosed, is acute gastrointestinal infection (AGI). AGI can be caused by viruses, bacteria or protozoa, or symptoms similar to AGI may be caused by chemical contaminants. Increasingly, however, viruses are being suspected as major causative agents due in part to the difficulties inherent in both their specific diagnosis and measurement of the agents in drinking water. Of well over 100 known viruses that can potentially be transmitted in drinking water, the caliciviruses and rotavirus are most commonly identified. However, types of poliovirus, coxsachievirus, echovirus, reovirus, adenovirus, HAV, astrovirus, coronavirus, and hepatitis E have been implicated in waterborne outbreaks and there may be many further, as yet uncharacterized, groups of viruses that are responsible for AGI and other disease manifestations.

The bacteria include both true and opportunistic pathogens and are the usual list (included in Table 8.2). Campylobacteriosis remains the most common form of

Table 8.2 Pathogens in drinking water, their infectious doses, diseases and additional comments (adapted from Ford, 1999)[a]

	Infectious dose[b]	Diseases	Comments
Vibrio cholerae	10^8	Cholera	New toxigenic serogroups/AB resistance
Salmonella spp.	10^{6-7}	Salmonellosis	AB resistance
Shigella spp.	10^2	Shigellosis	AB resistance
Toxigenic *E. coli*,	10^{2-9}	Diarrhoeal diseases	Major identified cause of diarrhoeal disease
e.g., *E. coli* O157		Haemolytic–uremic syndrome	Enteropathogenic, enterotoxigenic, and enterohaemorrhagic strains identified; include multiple AB-resistant strains
Campylobacter spp.	10^6	Campylobacteriosis	AB resistance
Leptospira spp.	3	Leptospirosis	Increase with flooding events
Francisella tularensis	10	Tularemia	Significance in DW unknown
Yersinia enterocolitica	10^9	Yersiniosis	Significance in DW unknown
Aeromonas spp.	10^8	Skin and respiratory infections, gastritis	?
Helicobacter pylori	?	Gastric ulcers/cancer	Essentially, exposure route unkown
Legionella pneumophila	>10	Legionellosis, pontiac fever	Underestimated cause of pneumonia
Mycobacterium avium	?	Disseminated infections	Increasing in healthy populations

Protozoa			
Giardia lambia	1–10	Giardiasis	Underdiagnosed
Cryptosporidium parvum	1–30	Cryptosporidiosis	Underdiagnosed, extreme chlorine resistance
Naegleria fowleri	High?	Primary amoebic meningo-encephalitis	Disease very rare, yet exposures common
Acanthamoeba spp.	?	Encephalitis + others	Transmission of bacterial pathogens?
Entamoeba histolica	10–100		High rates of infection and associated mortality
Cyclospora cayetanensis	?	Cyclosporidiosis	Most outbreaks associated with contaminated produce
Isospora belli	?		Significance in DW unkown
The Microsporidia	?	Microsporidiosis	May be widespread
Ballantidium coli	25–100		Significance in DW unkown
Toxoplasma gondii	?	Toxoplasmosis	Significance in DW unkown
Viruses			
	1–10	Diarrhoeal disease, meningitis, heart disease, liver disease, etc.	Incidence probably dramatically underestimated. Known groups include caliciviruses, poliovirus, coxsachievirus, echovirus, reovirus, adenovirus, hepatitis A, hepatitis E, rotavirus, astrovirus, and coronavirus. However, many additional viruses may remain to be discovered.

[a]Data compiled from WHO (1993), Hazen and Toranzos (1990) and Geldreich (1996).
[b]Infectious dose is number of infectious agents that produce infection (asymptomatic or symptomatic) in 50% tested volunteers and is therefore not useful for risk estimates for disease.

bacterial dysentery, followed by pathogenic *E. coli*, salmonellosis and shigellosis. Under the umbrella of new pathogens, *Legionella pneumophila* and the non-tuberculous mycobacteria occupy a unique niche in their ability to proliferate in hot water systems, their environmental ubiquity and their resistance to disinfection. In addition, they can apparently survive and in some cases proliferate within protozoan cells. As with the viruses, there are many unkowns. Water may or may not be a significant route for dissemination of *Helicobacter pylori* – the research community is divided. How much do the opportunistic pathogens *Aeromonas* spp., *Pseudomonas* spp., *Klebsiella* spp., etc. contribute to morbidity and mortality through consumption of drinking water? Certainly they are a major cause of hospital-acquired infections with high associated mortality risks.

Aeromonads isolated from water, for example, contain a wide range of virulence and antibiotic resistance factors (Kincombe *et al.*, 1999; Bondi *et al.*, 2000). They are known to cause a number of non-GI infections, including wound infections from exposure to water, but their true involvement in gastrointestinal disease would appear to be unkown. They can be isolated from patients with diarrhoeal disease, but only recently have researchers working in Bangladesh shown close associations between a number of *Aeromonas* enterotoxin genes and severe diarrhoea (Albert *et al.*, 2000). This research team was also able to demonstrate that a small number of children with diarrhoea were only infected with *Aeromonas* spp., strongly suggesting that the aeromonads were the causative agent of the disease. Given the comparative ease of detection of virulence and antibiotic resistance genes in *Aeromonas* spp., and the ubiquity and ease of isolation of these opportunistic pathogens in drinking water, aeromonads should be a useful model organism for studying changing virulence and antibiotic resistance patterns.

The protozoa receive most attention in the media due to the size of recent outbreaks – partially due to low infectious doses and high resistance to water treatment (Marshall *et al.*, 1997; Egorov *et al.*, 2002). *Cryptosporidium parvum* has attracted the most attention, with *C. parvum* replacing *Giardia lamblia* as the most common cause of waterborne disease outbreaks in the UK, and the second most common cause in the US. The protozoa that are currently receiving increased scrutiny are the *Cyclospora* and *Toxoplasma*, although a waterborne route of transmission is far from proven. It is likely we will also see increased scrutiny of the routes of exposure to microsporidia. Smaller than are the other protozoans, the microsporidia group are recognized increasingly as causative agents of both human and animal diseases. They are also more likely to penetrate filtration systems than are the larger protozoa, hence it is reasonable to suspect a waterborne route of exposure. An interesting question is why are diseases caused by the pathogenic amoebae not more prevalent? A recent risk assessment for *Naegleria fowleri* showed that many surface water bodies are positive for this pathogen, yet only a few cases of primary amoebic meningoencephalitis occur each year? In fact, seroprevalence studies indicate that infection may be fairly common, although disease is absent (Cabanes *et al.*, 2001).

It would be remiss not to at least mention other potentially infectious agents; the fungi, prions and viroids. Most research is focused on the viruses, bacteria and protozoa in drinking water and very little information is available on these other infectious agents. A wide range of fungal species are, however, present in drinking water. A number of fungal species do grow submerged in water or at the air–water interface. Fungi are readily isolated from drinking water, yet are seldom implicated in disease.

Recent studies have indicated that fungal species of *Aspergillus*, *Cladosporium*, *Epicoccum*, *Penicillium* and *Trichoderma* are frequently islolated from treated drinking water (Kinsey *et al.*, 1999; Arvanitidou *et al.*, 1999). *Candida* yeasts are also occasionally isolated from drinking water, and apparently correlate with the indicator organisms, total and faecal coliforms. A number of fungi and yeasts isolated from treated drinking water are potential pathogens, or at least can produce toxic metabolites and can readily spoil foods. However, more research is clearly necessary to establish a role, if it exists, for these microbes in transmission of human diseases through consumption of drinking water.

As for viroids and prions, to my knowledge no attempts have been made to look for these infectious agents in drinking water. Viroids, single-stranded RNA, are thought to only cause plant diseases. Together with similar infectious agents known as satellite RNAs which are dependent upon a helper virus for replication, it is unlikely that these agents represent a serious threat to human health through drinking water. Of course, the absence of any information linking these agents to human disease does not mean that in future linkages will not emerge. For example, the Hepatitis Delta agent is essentially a viroid encapsulated in a hepatitis B coat.

In contrast, prions – an infectious proteinaceous material – have risen to prominence following the devastating economic and perceived human health threat from the BSE outbreak in the UK. Although prions have not been isolated from drinking waters, it is reasonable to at least consider the risks of contamination from, for example, rendering wastes, abattoirs and landfills. As previously mentioned, Paul Gale from the WRc-NSF Ltd., UK, conducted a provisional risk assessment on drinking-water-related infections with BSE (Gale *et al.*, 1998). The risk assessment was based on mice infectivity studies, where the oral ID_{50} was taken as 1 gram of BSE-infected bovine brain, equivalent to $\sim 10^{13}$ BSE prion protein molecules. Taking into account the rapid dispersion of contaminants in the aquatic environment, Gale's research group estimated that an individual would be exposed to only minute sub-fractions of an $ID_{50}$from lifetime consumption of drinking water. In addition, the inherent stickiness of prions suggests that they would be rapidly sequestered to particles and easily removed in drinking water treatment. In fact, worst-case assumptions suggested that "an individual would have to consume 2 l/day of tap water for 45 million years to have a 50% chance of infection through drinking water drawn from the aquifer". Of course, there is still considerable uncertainty here, not least of which is the absence of dose–response information for BSE.

8.4 THE PHENOMENON OF "NEW DISEASE"

It has been difficult to define what is truly the emergence of a new disease (Levins *et al.*, 1995). Disease may emerge through creation of unique ecological niches, for example the hot-water systems that support growth of *Legionella.* Factors such as population density and increasing numbers of susceptible individuals (very young, elderly, pregnant women and those immunocompromised) could provide an extensive human reservoir for opportunistic pathogens and promote changes in virulence patterns, even in developed countries. Increased adaptation to the human host might be responsible for increased infection rates in populations with no underlying susceptibility (e.g. mycobacterial diseases).

Legionella pneumophila, *Escherichia coli* O157, *Vibrio cholerae* O139, *Helicobacter pylori*, *Cryptosporidium parvum*, and Hepatitis E are all examples of micro-organisms categorized as 'new' or 'newly recognised' pathogens. It is arguable that all known pathogens developing resistance to antibiotics and/or changing virulence patterns should be added to this list. Research is clearly needed to better understand the ecology of the drinking water environment that may promote new disease emergence. This author would argue that the microbial films that form on the surfaces of pipe, the biofilms, provide an opportunity for horizontal gene transfer both within and between species. The biofilm environment may also promote expression of plasmids through exposure to chemical stressors such as metals (Ford, 1993). There is an increasing body of research linking metal resistance with multiple antibiotic-resistance determinants, presumably expressed on the same plasmids (Alonso *et al.*, 2001).

What of the classic agent of waterborne disease, *Vibrio cholerae*? Cholera represents one of the oldest known diseases to be transmitted by water, responsible for seven pandemics over the past 200 years. The causative agent, *V. cholerae*, serogroup O1, is remarkable in that it is highly environmentally resistant, multiplying in association with other organisms such as zooplankton (Tamplin *et al.*, 1990). Cholera will no doubt be with us for the foreseeable future; emerging whenever hygiene and sanitation are compromised. It is, however, changing – or at least has the potential to change. The emergence of *V. cholerae* O139 in India in an epidemic form was the first time that a non-O1 serogroup of *V. cholerae* was shown to cause epidemic cholera. There is now molecular evidence that O139 strains were derived from O1 strains through genetic modification (Faruque *et al.*, 1998).

A research team from the National Institute of Cholera and Infectious Diseases in Calcutta has examined the ubiquity of virulence genes in a number of environmental strains of *V. cholerae* cultured from surface water (Chakraborty *et al.*, 2000). Molecular techniques demonstrated the presence and expression of critical virulence genes in many of these strains. They have also examined antimicrobial susceptibilities of *Vibrio cholerae* strains isolated from cholera patients over a 6-year period (Garg *et al.*, 2000). Although the O1 and O139 strains were increasingly resistant to a wide range of drugs, they remained largely susceptible to gentamycin and tetracycline. In contrast, *V. cholerae* non-O1 and non-O139 strains exhibited high levels of resistance to virtually every class of antimicrobial

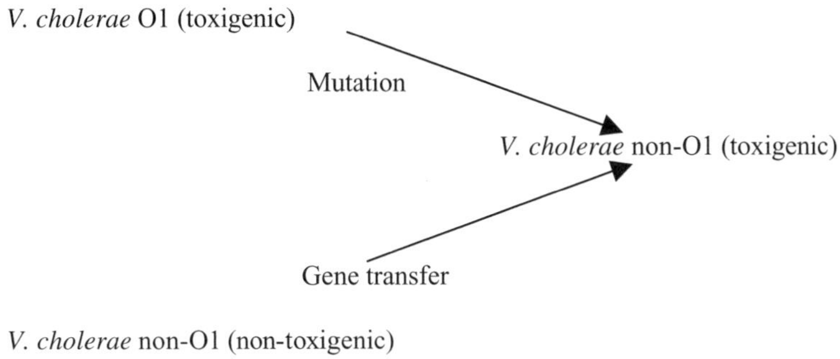

Figure 8.1 Emergence of new epidemic serogroups of *Vibrio cholerae*.

agent tested in the study. There was also an apparent transfer of resistance patterns between the O139 serogroup and the other serogroups.

From these two studies, it appears that the risk of transmission of virulence genes is high, and multiple-antibiotic resistance in non-O1 or non-O139 would make these strains of particular public health significance. It is important to learn more about the conditions that resulted in emergence of the toxigenic O139 serogroups, and could therefore result in many more, perhaps environmentally hardier, serogroups of this pathogen. The emergence of epidemic strains could occur through mutation of existing epidemic strains or through gene transfer. This is diagramatically represented in Figure 8.1; although the example is *V. cholerae*, it could equally apply to other pathogens such as the toxigenic *E. coli*. In the case of gene transfer, virulence factors could also be transferred between species. Both mutation and gene transfer would appear reasonable within the drinking water distribution system, where organisms are likely to be exposed to a variety of stressors such as chlorine and metal ions.

8.5 BIOFILMS

There is an increasing trend to see biofilms as critical risk factors in themselves. Certainly, the role of biofilms in human disease is well recognized and comparisons have been made with the drinking water distribution system. Biofilms provide a protective, nutrient-rich environment for a wide range of micro-organisms. They may enhance survival of pathogens in drinking water systems by providing protection from disinfection, and may even allow certain pathogens to proliferate (e.g. *Legionella*, Mycobacteria, and others). Biofilms are also known to contribute to pipeline degradation, 'dirty water', odor and blockage (Ford and Mitchell, 1990).

Transfer of antibiotic and virulence factors in drinking water biofilms has already been discussed and should be an area for future research. In addition, genes for polysaccharide synthesis, conveying increased resistance to chlorine and preference for biofilm formation, could also be transferred in drinking water biofilms.

Returning to the question of cholera, it is argued that the ability of *V. cholerae* to attach to surfaces and form biofilms is important in the environmental survival of this pathogen. Roberto Kolter and colleagues from Harvard University have studied important determinants in biofilm formation (Watnick and Kolter, 1999). Using simple flow cells, they examined a large number of *V. cholerae* transposon insertion mutants to determine genes involved in defective biofilm formation. For *V. cholerae* El Tor, important determinants included those responsible for (1) the biosynthesis and secretion of the type IV pili, (2) flagella motility, and (3) synthesis of extracellular polysaccharides (EPS). Their model for *V. cholerae* biofilm formation is shown schematically in Figure 8.2.

Understanding the molecular basis for biofilm formation is an important step forward in addressing biofilm control. Directly targeting specific molecules that play pivotal roles in the attachment process, e.g., pili, flagella and EPS, or targeting the expression of genes involved in their biosynthesis, presents a potentially effective way to prevent biofilm formation, whether in disease or in a drinking water pipeline (Costerton and Stewart, 2001).

In recent years, perhaps one of the most innovative advances in our understanding of biofilm dynamics has been the importance of cell-to-cell signalling. This has come to be known as quorum sensing (Bassler, 1999; Nilsson *et al.*, 2001) and was first implicated in biofilm formation using strains of *Pseudomonas aeruginosa* that lacked genes for a specific acylated hormoserine lactone (HSL). In the absence of these genes, David Davies of Binghampton University and colleagues have shown that biofilms never develop their characteristic three-dimensional structures (Davies *et al.*, 1998). These structures, that include water channels and 'pillars' of microbes, have been identified following application of confocal microscopy in the late 1980s (Lawrence *et al.*, 1991). The important role for cell-to-cell signalling in biofilm formation has provided yet another target for biofilm control. If the signals are disrupted, theoretically, a functional biofilm cannot form. Considerable research has now been conducted in this area, focused on algal products that prevent fouling of the algal fronds (Maximilien *et al.*, 1998). These chemicals, which appear to be substituted furanones, are similar to signal molecules, and are thought to effectively block their binding sites.

8.6 CHEMICALS

Although the focus of this article is on infectious agents, an assessment of health effects from consumption of treated drinking water is incomplete if the complex mixtures of microbes and chemicals are not all considered. The future trend will be for complete treatment of water that removes chemicals, toxins and microbes in as efficient a manner as possible. The number of chemicals is vast and the lists (and standards) are driven to some extent by our ability to measure chemicals with increasing precision and sensitivity. Micro-organisms and chemicals interact. Some of the chemicals of concern are of course microbial products themselves – bacterial, fungal and even algal toxins. Some are produced as a result of our water

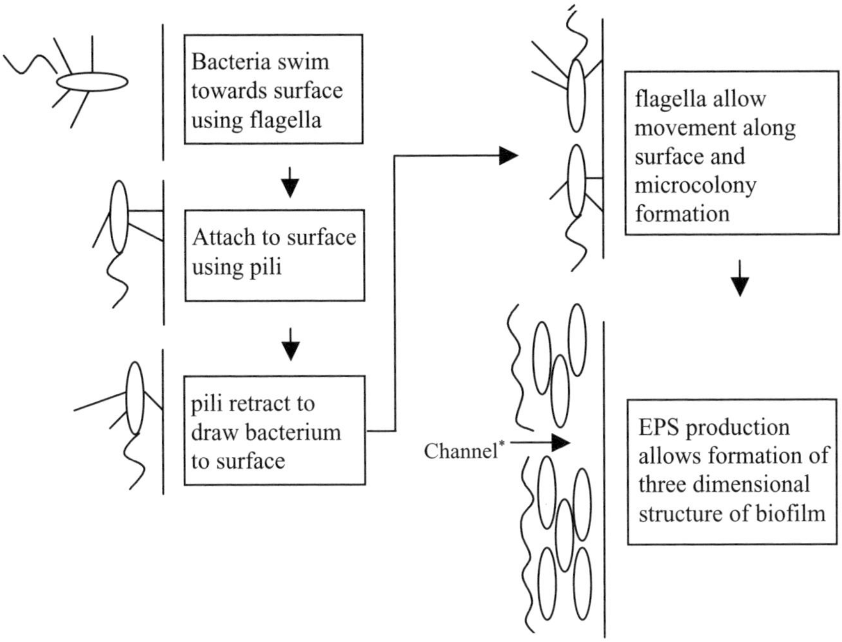

*biofilms are thought to form complex structures that include water channels, pillars, etc. (Lawrence et al. 1991)

Figure 8.2 Model for *shape V. cholerae* El Tor biofilm adapted from Figure 7, Watnick and Kolter (1999).

treatment practices to minimize microbial contamination, the so-called disinfection by-products (DBPs). Some may be produced within the distribution system pipeline owing to the activity of certain groups of organisms – the sulfate-reducing bacteria produce sulfides and other sulfur-containing chemicals; nitrifying bacteria produce nitrites and nitrates from ammonia compounds (either in source water or from chloramination).

Microbial activity is important in the fate and transport of all anthropogenic contaminants released to the environment. Although drinking water may contain the parent compounds of, for example, pesticides and herbicides, it is likely that a range of degradation products are also present. Even metals such as arsenic that have natural geologic sources may be released to surface waters through a number of microbially mediated mechanisms. There has been much discussion on the health risks from arsenic in drinking water following the dramatic health crisis in Bangladesh, and subsequent US EPA recommendations for the United States. There is little debate amongst the public health community about the health effects of exposure to arsenic. However, in the United States, enforcing stricter regulations has met political barriers because of the fact that arsenic is a naturally occurring compound and is expensive to remove from drinking water, particularly for small-scale utilities. The major anthropogenic use of arsenic is in wood preservation (as a component of chromated copper arsenate). That usage has been massive and worldwide. It is difficult to believe that the current and future legacy from use of these compounds in treated wood – leaching from existing structures and from landfills – will not become a critical concern for water utilities in the future.

DBPs are the other topic of contentious debate, particularly given the fact that health risks from exposure to these compounds remain poorly defined. Individual compounds have been shown to cause cancers in animal models and epidemiologic studies suggest slightly elevated risks from drinking water consumption (Boorman *et al.*,1999). However, we still have much to learn about the toxicity of the vast array of disinfection by-products, and, in particular, the potential for synergistic effects between these products. Relatively recent research has implicated one specific chlorinated furanone, known as mutagen X or MX, as a major contributor to drinking water mutagenicity in bacterial assays and as a carcinogen in rats (Komulainen *et al.*, 1997). Further information will no doubt emerge on other DBPs as increasingly sophisticated assays are used to assess toxicity of mixtures and the appropriate dose–response relationships.

8.7 GENE ARRAYS

Do we see a point at which gene arrays will become useful for monitoring for waterborne pathogens? A recent American Academy for Microbiology report begins with the imaginary scenario that a ‘gene chip’ placed in the flow of water will one day detect each pathogen, the fluorescence response will then trigger an alarm system that will promote an appropriate treatment response (Rose and Grimes, 2001). Gene chip technology is here to stay and the data it can potentially generate is vast. Our ability to interpret this data is, however, extremely limited and until the emerging field of bioinformatics has shown us how to effectively analyze this data, a monitor for drinking water pathogens remains elusive. In the interim, chips are being developed for specific indicator organisms:

- For a gene chip to be developed and results understood, the target organisms must be characterized and a highly specific gene segment recognized.
  Most organisms in drinking water have yet to be characterized.
- Once developed, the organism’s DNA has to come in contact with the chip surface in order to hybridize to the probe.
  Given the heterogeneous distribution of microbes in drinking water and the need to lyse the cell, this dramatically reduces the potential sensitivity of the technique.
- Further developments will undoubtedly follow.

8.8 FUTURE NEEDS AND PRIORITIES – CONCLUSION

- Valuation of water – if changes in source water protection, water treatment, distribution system upgrading and monitoring are to be effectively implemented, realistic valuation of this priceless resource needs to be taken, with appropriate subsidies for low-income communities.

- Surveillance systems – policy decisions are based on grossly inadequate data. In fact it is widely recognized that the burden of waterborne disease is underestimated by orders of magnitude. Population surveys for unreported disease incidence will be necessary to provide a clearer understanding of the disease burden from contaminated water in both developed and developing countries.
- Water treatment – approaches are needed that minimize selection for treatment resistant pathogens, biofilm formation and production of disinfection by-products.
- Monitoring – cost-effective pathogen-specific monitoring is needed to begin to evaluate risk in both developed and developing countries.
- Risk assessment – improved risk assessment methodologies are necessary to better model exposure scenarios and provide realistic estimates of pathogen infectivity. For risk assessment to be effective, accurate dose–response relationships need to be developed to concentrations relevant to drinking water. Synergy between pathogens, other microbes and chemicals needs to be better understood, and population susceptibilities more effectively modeled.

In conclusion, this discussion would not be complete without reference to wastewater. A vital step to provision of adequate, safe drinking water is to begin to understand that wastewater is a valuable resource. As we move into the twenty-first century, we will see greater and greater emphasis, not only on water reuse for non-potable applications, but also the use of recycled water to augment diminishing supplies of drinking water. The barriers to wastewater recycling are probably more issues of public perception than of cost. Just as many arid nations increasingly rely on desalination to supply drinking water, we must realize that treatment technologies are more than capable of recycling wastewater to a potable quality. The predicted increases in numbers of water-scarce countries for the twenty-first century makes education in this area critical. However, water recycling alone will not be sufficient without a concerted effort to conserve the available remaining resources.

An idealized scheme for future provision of safe drinking water is presented in Figure 8.3.

However, a word of caution. How safe should our drinking water be? Arguably for the immunocompromised it can never be too safe. However, it is conceivable that filtration technologies can provide water, at least at the level of the treatment plant, that is 100% free of not only infectious agents, but of all micro-organisms. Perhaps we can eventually provide distribution pipes that are lined with materials that effectively prevent biofilm build-up. However, is this in fact optimal for the immunocompetent? This is a question for the immunologists to address, but it may be important that we continue to think of the role of exposure to infectious agents in developing and maintaining healthy immune systems.

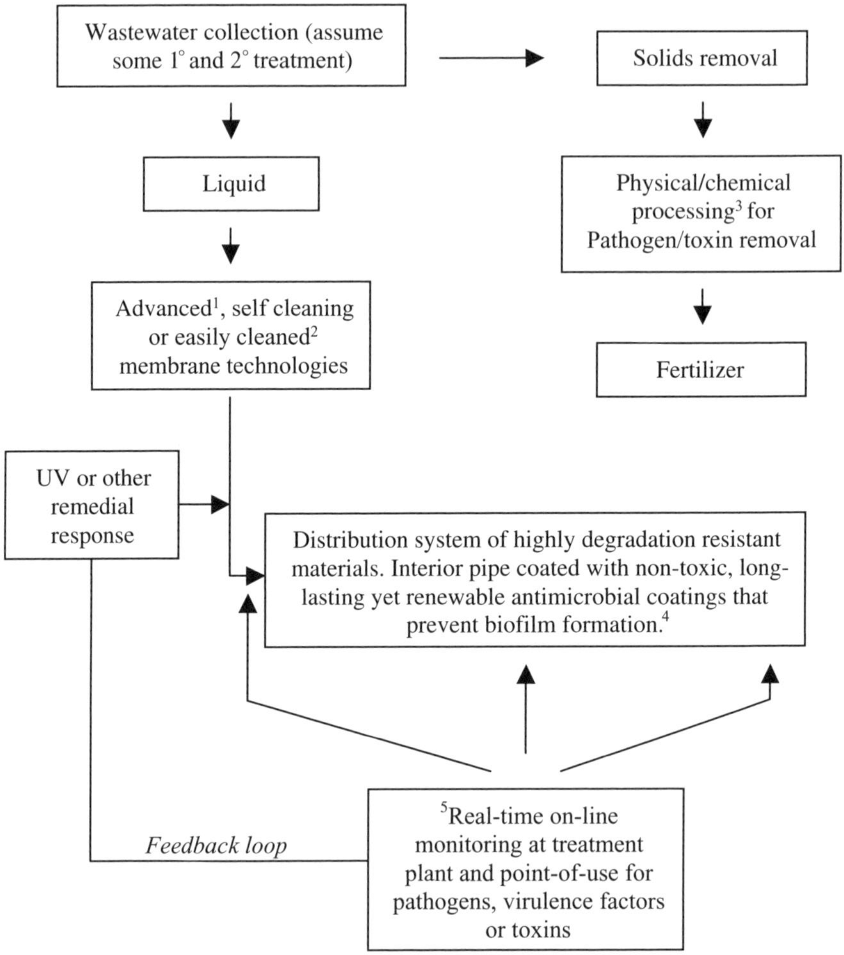

[1]Many companies now invest heavily in micro-filtration, ultra-filtration, nano-filtration and Reverse Osmosis membrane technologies. Hollow-fiber filtration technologies, for example, are allowing filtration capacities adequate for municipal water systems.
[2]For example, patented gas backwash systems for USFilters Memcor® microfiltration systems.
[3]There is currently considerable debate on appropriate criteria for land application of sewage sludges. Known as biosolids, treatment ranges from production of Class B biosolids, with a consequent risk to surface waters, to Class A biosolids, which are essentially pathogen free.
[4]For example, the substituted furanones discussed in the text.
[5]The ideal monitoring tool depicted in the recent American Academy of Microbiology Report (Rose and Grimes, 2001).

Figure 8.3 Idealized scheme for future provision of safe drinking water.

REFERENCES

Albert, M.J., Ansaruzzaman, M., Talukder, K.A., Chopra, A..K., Kuhn, I., Rahman, M., Faruque, S.G., Islam, M.S., Sack, R.B. and Mollby, R. (2000) Prevalence of enterotoxin genes in *Aeromonas* spp. isolated from children with diarrhoea, healthy controls, and the environment. *J. Clin. Microbiol.* **38**, 3785–3790.

Alonso, A., Sanchez, P. and Martinez, J.L. (2001) Environmental selection of antibiotic resistance genes. *Environ. Microbiol.* **3**, 1–9.

Arvanitidou, M., Kanellou, K., Constantinides, T.C. and Katsouyannopoulos, V. (1999) The occurrence of fungi in hospital and community potable waters. *Lett. Appl. Microbiol.* **29**, 81–84.

Bassler, B. (1999) How bacteria talk to each other: regulation of gene expression by quorum sensing. *Curr. Opin. Microbiol.* **1**, 582–587.

Bondi, M., Messi, P., Guerrieri, E. and Bitonte, F. (2000) Virulence profiles and other biological characters in water isolated *Aeromonas hydrophila*. *Microbiologica* **23**, 347–356.

Boorman, G.A., Dellarco, V., Dunnick, J.K., Chapin, R.E., Hunter, S., Hauchman, F., Gardner, H., Cox, M. and Sills, R.C. (1999) Drinking water disinfection byproducts: review and approach to toxicity evaluation. *Environ. Health Perspect.* **107**, 207–217.

Cabanes, P.A., Wallet, F., Pringuez, E. and Pernin, P. (2001) Assessing the risk of primary amoebic meningoencephalitis from swimming in the presence of environmental *Naegleria fowleri*. *Appl. Environ. Microbiol.* **67**, 2927–2931.

Chakraborty, S., Mukhopadhyay, A.K., Bhadra, R.K., Ghosh, A.N., Mitra, R., Shimada, T., Yamasaki, S., Faruque, S.M., Takeda, Y., Colwell, R.R. and Nair, G.B. (2000) Virulence genes in environmental strains of *Vibrio cholerae*. *Appl. Environ. Microbiol.* **66**, 4022–4028.

Costerton, J.W. and Stewart, P.S. (2001) Battling biofilms. *Sci. Amer.* **July**, 75–81.

Davies, G.D., Parsek, M.R., Pearson, J.P., Iglewski, B.H., Costerton, J.W. and Greenberg, E.P. (1998) The involvement of cell-to-cell signals in the development of a bacterial biofilm. *Science* **280**, 295–298.

Egorov, A.I., Paulauskis, J., Lubov, P., Tereschenko, A., Drizhd, N. and Ford, T. (2002) Contamination of water supplies with *Cryptosporidium* and *Giardia lamblia* and diarrhoeal illness in selected Russian Cities. *Int. J. Hyg. Environ. Health* **205**, 281–289.

Faruque, S.M., Albert, M.J. and Mekalanos, J.J. (1998) Epidemiology, genetics, and ecology of toxigenic *Vibrio cholerae*. *Microbiol. Mol. Biol. Rev.* **62**, 1301–1314.

Ford, T.E. (1999) Microbiological safety of drinking water: United States and global perspectives. *Environ. Health Perspect.* **107**, 191–206.

Ford, T.E. (1993) The microbial ecology of water distribution and outfall systems. In *Aquatic Microbiology – an Ecological Approach* (ed. T.E. Ford), pp. 455–482, Blackwell, Boston, MA.

Ford, T.E. and Colwell, R. (1996) *A Global Decline in Microbiological Quality of Water: a Call for Action*, 40 pp., American Academy of Microbiology, Washington, D.C.

Ford, T.E. and Mitchell, R. (1990) The ecology of microbial corrosion. *Adv. Microb. Ecol.* **11**, 231–261.

Gale, P. (2001) Developments in microbiological risk assessment for drinking water. *J. Appl. Microbiol.* **91**, 191–205.

Gale, P., Young, C., Stanfield, G. and Oakes, D. (1998) Development of a risk assessment for BSE in the aquatic environment. *J. Appl. Microbiol.* **84**, 467–477.

Geldreich, E.E. (1996) *Microbial Quality of Water Supply in Distribution Systems*, 504 pp., CRC Press, Boca Raton, FL.

Garg, P., Chakraborty, S., Basu, I., Datta, S., Rajendran, K., Bhattacharya, T., Yamasaki, S., Bhattacharya, S.K., Takeda, Y., Nair, G.B. and Ramamurthy, T. (2000) Expanding multiple antibiotic resistance among clinical strains of *Vibrio cholerae* isolated from 1992–7 in Calcutta, India. *Epidemiol. Infect.* **124**, 393–399.

Havelaar, A.H., De Hollander, A.E.M., Teunis, P.F.M., Evers, E.G., Van Kranen, H.J., Versteegh, J.F.M., Van Koten, J.E.M. and Slob, W. (2000) Balancing the risks and benefits of drinking water disinfection: disability adjusted life-years on the scale. *Environ. Health Perspect.* **108**, 315–321.

Hazen, T.C. and Toranzos, G.A. (1990) Tropical source water. In *Drinking Water Microbiology* (ed. G.A. McFeters), pp. 32–53, Springer, New York, N.Y.

King, C.H., Shotts, E.B., Wooley, R.E. and Porter, K.G. (1998) Survival of coliforms and bacterial pathogens within protozoa during chlorination. *Appl. Environ. Microbiol.* **54**, 3023–3033.

Kingombe, C.I.B., Huys, G., Tonolla, M., Albert, M.J., Swings, J., Peduzzi, R. and Jemmi, T. (1999) PCR detection, characterization, and distribution of virulence genes in *Aeromonas* spp. *Appl. Environ. Microbiol.* **65**, 5293–5302.

Kinsey, G.C., Paterson, R.R. and Kelley, J. (1999) Methods for the determination of filamentous fungi in treated and untreated waters. *J. Appl. Microbiol.* **85**, 214S–224S.

Komulainen, H., Kosma, V.M., Vaittinen, S.L., Vartiainen, T., Kaliste-Korhonen, E., Lotjonen, S., Tuominen, R.K. and Tuomisto, J. (1997) Carcinogenicity of the drinking water mutagen 3-chloro-4-(dichloromethyl)-5-hydroxy-2(5*H*)-furanone in the rat. *J. Natl. Cancer Inst.* **89**, 848–856.

Kramer, M.H.J. and Ford, T.E. (1994) Legionellosis: Ecological factors of an environmentally 'new disease'. *Z. Bacteriol. Mikrobiol. Hyg.* **195**, 470–482.

Levins, R., Awerbuch, T., Brinkmann, U., Eckardt, I., Epstein, P., Ford, T., Makhoul, N., Albuquerque de Possas, C., Puccia, C., Spielman, A. and Wilson, M.E. (1995) The Emergence and spread of new diseases. *The Ecologist* **25**, 21–26.

Levin, R., Epstein, P., Ford, T.E., Harrington, W., Olson, E. and Reichart, E. (2002) US drinking water challenges in the 21st century. *Environ. Health Perspect.* **110**, 43–52.

Maximilien, R., de Nys, R., Holmstrom, C., Gram, L., Kjelleberg, S. and Steinberg, P.D. (1998) Bacterial fouling is regulated by secondary metabolites from the red alga *Delisea pulchra. Aquat. Microb. Ecol.* **15**, 233–246.

Marshall, M., Naumovitz, M., Ortega, Y. and Sterling, C.R. (1997) Waterborne protozoan pathogens. *Clin. Microbiol. Rev.* **10**, 67–85.

Mohanty, J.C., Ford, T.E., Harrington, J.J. and Lakshmipathy, V. (2002) Microbiological health risks of drinking water in Hyderabad City; a Preliminary Study. *J. Water Suppl.: Res. Tech-AQUA* **51**, 239–251.

Nilsson, P., Olofsson, A., Fagerlind, M., Fagerstrom, T., Rice, S., Kjelleberg, S. and Steinberg, P. (2001) Kinetics of the AHL regulatory system in a model biofilm system: how many bacteria constitute a "quorum"? *J. Mol. Biol.* **309**, 631–640.

Rose, J.B. and Grimes, D.J. (2001) *Reevaluation of Microbial Water Quality: Powerful New Tools for Detection and Risk Assessment.* American Academy of Microbiology, Washington, D.C.

Tamplin, M., Gauzens, A., Huq, A., Sack, D.A. and Colwell, R.R. (1990) Attachment of *Vibrio-cholerae* serogroup-O1 to zooplankton and phytoplankton of Bangladesh waters. *Appl. Environ. Microbiol.* **56**, 1977–1980.

Lawrence, J.R., Korber, D.R., Hoyle, B.D., Costerton, J.W. and Caldwell, D.E. (1991) Optical sectioning of microbial biofilms. *J. Bacteriol.* **173**, 6558–6567.

Watnick, P.I. and Koulter, R. (1999) Steps in the development of a *Vibrio cholerae* El Tor biofilm. *Mol. Microbiol.* **34**, 586–595.

WHO. (1993) *Guidelines for Drinking-Water Quality, Recommendations*. 2nd ed, Vol. 1, World Health Organization, Geneva.

9

Converting hindsight into foresight – understanding the causes of waterborne outbreaks

Steve E. Hrudey and Elizabeth J. Hrudey

9.1 INTRODUCTION

An overview is provided in this chapter derived from a comprehensive analysis of documented drinking water disease outbreaks over the past 30 years in 15 affluent nations (Hrudey & Hrudey, 2004). Our analysis of over 70 case studies acknowledges the relatively high level of drinking water safety that residents of these rich nations normally enjoy in contrast to the depressing toll of death and disease found in developing nations. Yet these case studies almost universally demonstrate that the outbreaks involved were eminently preventable.

A consistent case is made for better focus on a comprehensive risk management approach to drinking water safety that considers the sources of danger and the effective operation of multiple barriers from the raw water source to the consumers' taps. The range of cases documented, some in substantial detail, offers a perspective on the unfolding character of a waterborne outbreak and provides an opportunity for those charged with providing drinking water, regulating its safety or protecting public health to learn from a large number of documented negative events without

 Microbial Waterborne Pathogens edited by T.E. Cloete, J. Rose, L.H. Nel and T. Ford. ISBN: 1 84339 055 8. Published by IWA Publishing, London, UK.

having to experience first-hand the trauma of these failures. The documented recurrence of common failure modes on four continents over the past 30 years suggests that achieving a preventive approach to assuring drinking water safety requires more effective means for learning from the numerous past mistakes and oversights and then implementing appropriate measures to avoid repeating these errors.

Converting hindsight into foresight became the theme for our book. This was originally inspired by involvement with Justice Dennis O'Connor's landmark public inquiry (O'Connor 2002a,b) into the Walkerton tragedy, a thoroughly preventable outbreak occurring in the richest, most highly developed province of Canada. This disaster killed 7 residents, including a 2-year-old child, and made 2,300 seriously ill in a community of 5,000 in May 2000. The public inquiry revealed failures at virtually every level of the operational, regulatory and political system, frequently as a result of widespread complacency about the importance of drinking water safety to a community's public health and well-being.

9.2 EVIDENCE FROM CASE STUDIES

The evidence for our analysis was drawn from the open, refereed English-language literature, reports of public inquiries or official outbreak investigations and other credible information sources. The more than 70 case studies (some cases involved more than one outbreak) are summarized in Table 9.1 by decade, presenting the countries of occurrence, the water source and treatment types, the pathogens involved and the main failure elements or factors.

Table 9.2 summarizes the case studies by the size of population served by the water system where the outbreak occurred. This summary shows that there were many outbreaks in medium- to large-scale systems (25 serving 10,000–99,999 and 11 serving >100,000). The total numbers in this summary do not represent the total numbers of outbreaks occurring in the various size categories over this period because only those outbreaks were reviewed that had sufficient detail about causes and outcomes provided in accessible documents.

There are clearly many small outbreaks that are either poorly documented or not reported at all, and the conventional wisdom that small systems are the ones at greatest risk is not challenged by our selection of case studies. However, our summary does show that many medium- to large-size systems in affluent nations have failed.

Each case study account was modelled on our analysis of the initial detailed record provided for the Walkerton outbreak, consisting of an overall description of the location and scope of the outbreak, a description of the drinking water system, an account of the outbreak experience as it unfolded, an analysis of the direct causes of the outbreak and a summary commentary on noteworthy aspects of each outbreak. The enormous detail provided by the Walkerton Inquiry, including access to major exhibits and transcripts of the evidence, allowed for presentation of a wide variety of human elements contributing to this disaster. Despite the readily accessible and readable inquiry reports (O'Connor, 2002a,b), the overwhelming

Table 9.1 Summary characteristics of outbreak case studies over the past three decades (Hrudey & Hrudey, 2004)

Decade	Countries	Water source type	Treatment provided	Pathogens involved	Major failure factors
1974–1979 No deaths attributed	U.S.A. (7) Note: finding all cases coming from the U.S. for this period merely reflects the availability of the best documented case studies for the 1970s.	Groundwater (1) Surface water (6)	Chlorination only (3) Chloramination only (1) Chlorination–filtration (3)	*Giardia* (4) *Campylobacter* (1) Enterotoxigenic *E. coli* O6:H16 (1) *Shigella* (1)	Sewage contamination (5) Inadequate knowledge of water source (5) Heavy rainfall–run-off (2) Disinfection failure (3) Process changes (1) Filtration failure (1) Cross connections (1) Wildlife contamination (1)
1980–1989 2 deaths	U.S.A. (9) Canada (4) Sweden (4) England (2) Norway (2) Australia (1) Scotland (1)	Groundwater (10) Surface water (13)	None or nothing effective (5) Chlorination only (8) Filtration only (2) Chlorination–filtration (7) Chloramination–filtration (1)	*Giardia* (7) *Campylobacter* (5) *Cryptosporidium* (4) Norwalk virus (3) Hepatitis A (1) Rotavirus (2) *Entamoeba* (1) Not identified (2)	Sewage contamination (12) Disinfection failure (11) Inadequate knowledge of water source (7) Cross connections (5) Heavy rainfall–run-off (5) Livestock contamination (4) Wildlife contamination (3) Process changes (3) Filtration failure (2)

(*continued*)

Table 9.1 (*continued*)

Decade	Countries	Water source type	Treatment provided	Pathogens involved	Major failure factors
1990–1999	U.S.A. (7)	Groundwater (17)[b]	None or nothing effective (12)	*Cryptosporidium* (11)	Sewage contamination (18)
	Canada (6)		Chlorination only (5)	*Campylobacter* (3)	Inadequate knowledge of water source (17)
Deaths	England (6)	Surface water (13)[b]	Chloramination only (1)	Norwalk virus (6)	
15 plus	Denmark (2)		Filtration only (1)	*E. coli* O157:H7 (4)	Heavy rainfall–run-off (7)
50–70[a]	Finland (2)	Rain water (2)	Chlorination–filtration (9)	*Giardia* (2)	
	Japan (2)		Chlorination–filtration–Ozonation (2)	Enteropathogenic *E. coli* (1)	Filtration failure (5)
	Australia (1)				Wildlife contamination (5)
	Ireland (1)			*Shigella* (1)	Cross connections (4)
	Scotland (1)			*Salmonella* (1)	Livestock contamination (2)
	Sweden (1)			*Toxoplasma* (1)	Process changes (2)
	Switzerland (1)			Not identified (3)	Disinfection failure (1)
2000–	Finland (3)	Groundwater (7)	None or nothing effective (6)	*Cryptosporidium* (5)	Sewage contamination (5)
	N. Ireland (3)	Surface water (6)	Chlorination only (2)	*Campylobacter* (5)	Heavy rainfall–run-off (2)
7 deaths	Canada (2)		Chlorination–filtration (4)	Norwalk virus (3)	
confirmed	Sweden (2)		Filtration–UV (1)	*E. coli* O157:H7 (1)	Filtration failure (4)
	England (1)				Livestock contamination (4)
	Italy (1)				Process changes (1)
	New Zealand (1)				Disinfection failure (1)
					Knowledge of source (1)
					Cross connection (1)

[a]Estimated number of AIDS victims to die in the two years following the Milwaukee outbreak.
[b]Two systems had combined surface and groundwater.

Table 9.2 Distribution of outbreak case studies reviewed by size of population (Hrudey & Hrudey, 2004)

Size of population	Transient or institutional supply	Community supply
<500	5 outbreaks Oakcreek Canyon (1989), **Saitama*** (1990), Yukon (1995), Hawke's Bay (2001), Stockholm County (2001)	2 outbreaks Skjervøy (1988), Alpine (1998),
500–9,999	4 outbreaks Crater Lake (1976), Moama (1989–90), Bermuda (1998), Gulf of Taranto (2000)	24 outbreaks Richmond Heights (1974), Camas (1976), Red Lodge (1980), Eagle-Vail (1981), Mjövik (1982), **Drumheller*** (1983), Greenville (1983), Braun Station (1984), Alsvåg (1984), **Cabool*** (1989–90), Creston – Erickson (1990), Uggelose (1991–92), Jackson County (1992), Noormarkku (1994), Temagami (1994), Fife (1995), Klarup (1995–96), Stromsund (1996), Heinävesi (1998), La Neuveville (1998), **Walkerton*** (2000), Asikkala (2000), eastern Finland (2001), Transtrand (2002).
10,000–99,999	0 outbreaks	25 outbreaks Rome, NY (1974–75), Berlin, NH (1977), Bennington (1978), Bradford, PA (1979), Bramham (1980), Georgetown (1980), Rome, GA (1980), Grums (1980), Orangeville (1985), Pittsfield (1985–86), Penticton (1986), Sälen (1986), Carrollton (1987), Sunbury (1988), Boden (1988), Saltcoats and Stevenston (1988), Naas (1991), Warrington (1992–93), **Gideon*** (1993), Ogose (1996), Cranbrook (1996), Brushy Creek (1998), Clitheroe (2000), southern Finland (2001), North Battleford (2001).
>100,000	1 outbreak **Washington County Fair*** (1999)	10 outbreaks Edmonton (1982–83), Swindon and Oxon (1989), Isle of Thanet (1990), Bradford, Yorks (1992), Kitchener–Waterloo (1993), **Milwaukee*** (1993), Victoria (1995), South Devon–Torbay (1995), NW-London & W Herts (1997), Belfast (2000, 2001)

*Outbreaks with **deaths** attributed.

media coverage and countless other summaries that have been written, many Canadians mistakenly believe that the whole tragedy can be blamed on the misdeeds of the operators or that "tougher" regulations would have prevented the outbreak.

Justice O'Connor was very clear on his findings about the first premise, stating (O'Connor, 2002a, p.24), "It is simply wrong to say, as the government argued at the Inquiry, that Stan Koebel or the Walkerton PUC were solely responsible for the outbreak or that they are the only ones who could have prevented it." Regarding the second premise, if the Walkerton water supply had satisfied the treatment requirements that had been specified for it, i.e. providing a chlorine residual of 0.5 mg/l for 15 minutes' contact time, this level of disinfection would have been more than adequate to have inactivated the pathogens responsible for the outbreak, *E. coli* O157: H7 and *Campylobacter* spp. The technical requirements did not need to be more stringent. However, Justice O'Connor found that if the Ontario policy that had been in place since 1994, requiring continuous chlorine residual monitors for unfiltered groundwater supplies under the influence of surface water, had been implemented for Walkerton, this measure costing an estimated $8,000 would have prevented the outbreak. In this case, the entire regulatory and public health oversight that should have ensured that these treatment conditions were being consistently maintained clearly failed to do so.

The experience of drinking water providers failing to perform the measures that we have learned over the past 150 years are needed to prevent drinking water from transmitting pathogens appears repeatedly throughout the case studies reported. Seeing the human interactions that contributed to these failures at Walkerton allows for some degree of "reading between the lines" for the subsequent case studies that do not have the extensive documentation available for Walkerton.

In addition to the 70 full-outbreak case studies, a summary and references describing another 25 outbreaks over the past 25 years are provided for situations where no treatment or no effective treatment was provided. Likewise, a full case study is included on the Sydney water crisis of 1998 even though no illness occurred in the community. This incident was included because of the number of relevant experiences of systemic breakdown and the challenges of public health risk-management decision making under uncertainty that were documented by the Sydney Water Inquiry (McLellan, 1998). A detailed account of a cryptosporidiosis outbreak was provided using the North Battleford water inquiry (Laing, 2002), which showed a severe failure on the part of the municipal and provincial governments responsible. In this case, Justice Laing concluded (Laing, 2002, p. 198): "There was a systematic failure on the part of the City of North Battleford to recognize its responsibility to produce safe drinking water. This failure was brought about by the City's collective lack of knowledge on what it takes to produce safe drinking water, and policies that discouraged the possibility that it might acquire such knowledge." Finally, the Edmonton giardiasis outbreak of 1982–83 (found to be the second largest giardiasis outbreak reported in the developed world – based on laboratory-confirmed cases) is documented as an example of taking a poor water quality management situation and turning it around over an extended period, given strong commitment and dedicated leadership. This account shows the interplay with local politics and presents an initial story of denial despite a mounting body of evidence that the giardiasis outbreak was waterborne.

9.3 RECURRING THEMES

The overall theme that emerges in reviewing these case studies is a need to convert hindsight into foresight. Presumably, if more of the key players who are engaged in providing drinking water were afforded an effective opportunity to learn from past mistakes, greater capacity to anticipate problems might emerge, resulting in the adoption of more effective preventive measures.

Although there were countless ways to look at the recurring themes that emerge from the outbreak case studies, we chose to categorize them according to six major themes based on the guiding principles that have been adopted to introduce the current revisions to the Australian Drinking Water Guidelines (ADWG). These were developed by a working party of the Australian National Health and Medical Research Council (NHMRC) as they undertook to restructure the ADWG to accommodate the NHMRC Framework for Management of Drinking Water Quality (Rizak *et al.*, 2003). This approach takes a comprehensive perspective focussing on managing drinking water quality from source to consumers' taps with an emphasis on achieving continuously effective performance of key process elements rather than just narrow compliance with guideline numbers. The focus is maintained on identifying and satisfying the most important elements of operations required to assure water quality and safety.

9.3.1 Pathogens pose the greatest risk to drinking water safety

Anyone engaged in producing safe drinking water needs to know that microbial pathogens represent the most pervasive and consistent proven risk to drinking water safety. While there are cases where chemical contamination of water has caused severe and life-threatening disease (e.g. naturally high levels of arsenic in groundwater in countries like Bangladesh and Taiwan), these cases arise from specific local conditions and do not represent a pervasive risk wherever drinking water is supplied. Because microbial pathogens come from humans, pets, livestock and other animals, these proven disease-causing agents represent a threat to any human water supply because they can be found wherever humans or animals are found. Waterborne outbreaks caused by pathogens can occur in any geographic setting. A drinking water supply that is not explicitly and effectively secured from potential pathogen contamination is inherently unsafe; an outbreak may not occur until an appropriate combination of conditions coincide, but for a system that fails to protect against pathogen contamination the questions to ask are: '*when*,' '*which pathogen*' and '*how severe*,' not '*if*' an outbreak will occur.

The case studies reviewed included 23 where no treatment or no effective treatment was provided. Another 25 outbreaks involving similar risky systems where literature reports were accessible were presented in summary form only. Given the overwhelming and ongoing evidence of risk to drinking water safety posed by pathogens, what seems painfully obvious to microbiologists and public health professionals does not seem to be recognized by drinking water providers and customers in many regions of affluent nations. In some cases, this misunderstanding was linked to the use of remote or apparently pristine water supplies, yet outbreaks occurred because wildlife or livestock were able to introduce faecal contamination to those sources. In others, apparently high quality, secure groundwater supplies were used, but faecal contamination was introduced by cross-connections or other distribution system contamination.

9.3.2 Robust multiple barriers are essential

Because disease-causing agents get efficiently disseminated via drinking water throughout a community, effective public health protection should involve more than one barrier to disease transmission by this route. These barriers need to be robust to withstand the inevitable vagaries of human error and the challenges by nature and circumstance. A multiple barrier approach has been advocated by public health professionals for more than a century, yet 25 of the outbreak case studies involved systems with only a single treatment barrier, most commonly, chlorination or chloramination only. Operators of systems that presumed they had secure watersheds often discovered after an outbreak that this presumption was based more on faith than evidence because there had been no routine program in place to conduct sanitary surveys to identify and manage watershed contamination hazards existing for those water supplies.

Case studies reviewed illustrated failures in one or more of the elements of the multiple barrier concept as elaborated by the Walkerton Inquiry (O'Connor, 2002b), namely, robust and effective measures involving the following elements:

- *source* protection and selection to keep the raw water as clean as possible, to reduce the risk of contamination breaching the drinking water system;
- *treatment*, often involving more than one process, to remove or inactivate contaminants, to be effectively designed, operated and maintained;
- *distribution* system security to protect against intrusion of contaminants and disinfectant residual use to assure delivery of safe water to consumers;
- *monitoring* to control treatment processes and detect contamination in a timely manner to inform risk management responses; and
- *response* capabilities to adverse conditions that are well-conceived, thorough and effective.

9.3.3 Trouble is preceded by change

A key characteristic of water treatment processes familiar to those who are experienced is that such processes are best suited to stable conditions, where water quality characteristics do not change substantially or rapidly over time. The corollary is that when conditions do change rapidly or severely, water treatment processes are vulnerable to failure. Change does not guarantee that failure will occur, but most cases of treatment process failure can be shown to involve some conditions that are abnormal. This reality means that an effective operator must understand the water treatment system and source conditions well enough to know what is the range of normal conditions, otherwise there is no capacity to recognize the unusual.

Extreme or unusual weather conditions were also a striking feature of several outbreaks, including many of the most serious: Cabool (4 deaths, severe cold leading to mains breaks), Gideon (7 deaths, severe cold leading to water turn-over in a contaminated storage tank), Milwaukee (50–70 deaths over the following 2 years, severe winter storms), Walkerton (7 deaths, heavy spring rains, 1-in-60-year rainfall) and Washington County Fair (2 deaths, heavy rains following prolonged drought). In other cases, a lightning strike caused the outage of a supply pump house, increasing reliance on a river source subject to contamination (South Devon – Torbay) and even a volcanic eruption of Mt. St. Helens, causing rapid snow melt, which led to unusually heavy run-off (Red Lodge). Observations of the impact of severe weather have been made repeatedly, being noted in outbreak reviews dating back to the 1920s (Gorman & Wolman, 1939).

This influence seems eminently generalizable: when unusual conditions in the process or the water source occur, be wary and look for resulting vulnerability. At the same time, there is a need to recognize that most changes will not be sufficient to trigger complete system failure. If this reality is not acknowledged, then repeated alerts that result in no evident failures may simply foster operator complacency. Events that cause problems but do not lead to outright failure should be documented as close-call incidents for future training purposes.

9.3.4 Operators must be capable and responsive

The human element is central to water system failures and resulting outbreaks. In one sense, this should come as no surprise because normal human behaviour involves making mistakes (Kletz, 2001). The challenge for public health officials, operators and regulators of drinking water systems is to ensure that when inevitable human errors do occur catastrophic system failure does not follow. There must be sufficient redundancy and fail-safe character to the system that individual and inevitable human errors do not accumulate to trigger a disaster.

The operator may be the most important element in the water supply system for any community, as noted in a report commissioned for the Walkerton Inquiry (Geldreich & Singley, 2002): "The ultimate protection of public health lies in the hands of the plant operator. The more knowledgeable the operator, the greater the public's protection." In many jurisdictions, training for water treatment operators is marginal at best. In smaller communities, water treatment responsibilities are assigned to individuals who have little stature, minimal training and limited support. When failures occur, it is too easy to blame the operator simply because he or she is closest to the final failure. We wonder how long the travelling public would accept hearing after every investigation of a plane crash that pilots were inadequately trained or that operational controls were inadequate to fly the plane through foreseeable weather conditions. Yet, many communities will entrust a service as vital to their health as their drinking water supply to personnel who are neither adequately trained nor supported by adequate resources.

9.3.5 Drinking water professionals must be accountable

The burden of assuring safe water does not lie solely on the operators nor is there an expectation that operators must be perfect or superhuman (Hrudey & Hrudey, 2003). Rather, there must be an informed culture of caring about the burden of responsibility that those involved in providing drinking water must assume, supported by a system that can tolerate inevitable human errors without allowing catastrophic failure. There were outbreak case studies where key individuals did not merely make mistakes, but clearly failed to discharge their duties in a responsible manner (Eagle-Vail, La Neuveville, Walkerton). In others, specific warnings went unheeded (Bennington, North Battleford, South Devon, Ugglelose, Walkerton) and in others, consumers tolerated an outbreak without securing necessary improvements, leading to a subsequent outbreak (Creston – Erickson, Georgetown, Isle of Thanet, La Neuveville, North Battleford, South Devon). The case studies also illustrate the important role that consumers can and should play in assuring safe drinking water because there were several examples where the earliest warnings of problems were evident with consumer complaints that were not acted upon effectively (Gideon, La Neuveville, Milwaukee, Naas, Pittsfield).

9.3.6 Risk management – making sensible decisions under uncertainty

Assuring drinking water safety is an exercise in risk management, which inevitably demands making important decisions without complete or compelling evidence. Some key elements of a risk management approach to drinking water safety were summarized by the Walkerton Inquiry (O'Connor, 2002b, p. 75):

- "being preventive rather than reactive
- distinguishing greater risks from lesser ones and dealing first with the former
- taking time to learn from experience
- investing resources in risk management that are proportional to the danger posed."

The resulting decision-making process must balance between making false positive and false negative errors (Hrudey & Leiss, 2003). While provision of safe drinking water calls for a precautionary approach, it should be clear that taking substantial precautionary actions on any indication of danger will lead to many false positive decisions (taking action when none is required). In a public health context this may be preferred to false negative decisions (failing to take action when action is required). However, chronic false positives will carry their own costs and may ultimately result in failures arising from subsequent complacency.

The Sydney water crisis of 1998 was presented as a full case study of what has been characterized as a false positive error (Clancy, 2000). While there was considerable controversy about whether apparently extensive contamination of Sydney treated water by *Cryptosporidium* and *Giardia* was based on accurate monitoring evidence, all parties agreed that there was no excess illness in the population (McClellan, 1998). Yet this water quality crisis was estimated to have cost the Sydney Water Corporation over $37 million in direct costs, with contingency costs of over $100 million (Quill, 1999). There was also a huge loss in consumer confidence in the safety of Sydney water.

There were a number of other examples illustrating the importance of a preventive rather than a strictly reactive approach to assuring drinking water safety. In particular, greater attention to watershed protection and to real time process control, rather than reliance solely on treated-water monitoring has been clearly shown to be a more effective approach to assuring drinking water safety. Reactive approaches are limited to measures such as issuing boil water advisories, which often are not issued quickly enough to prevent community illness, with many examples where the advisory was only issued late in the outbreak or when the outbreak was effectively over. Furthermore, the effectiveness of compliance with boil water advisories for eliminating hazardous exposure has also been shown to be limited (BGOSHU, 2000; Angulo *et al.*, 1997; O'Donnell *et al.*, 2000; Willocks *et al.*, 2000). These realities are now reflected in a number of recent risk management approaches to assuring drinking water safety: the New Zealand Ministry of Health Public Health Risk Management Plans (www.moh.govt.nz/water), the Australian NHMRC Framework for Management of Drinking Water Quality (Rizak *et al.*, 2003), the World Health Organization Drinking Water Guideline Water Safety Plans (WHO, 2003) and the Bonn Charter – A Drinking Water Quality Framework for the 21st Century (Rouse, 2004).

9.4 SUMMARY

A review of drinking water outbreaks in affluent countries over the past 30 years provides a rich body of evidence for drinking water providers who are committed to ensuring the delivery of safe drinking water (Hrudey & Hrudey, 2004). The good news is that such outbreaks are now remarkably rare in affluent countries, particularly considering the pervasive nature of microbial pathogens occurring wherever humans, their animals and wildlife reside combined with the proven capability of these pathogens to cause human disease. This reality should mean that residents of affluent countries would value the safety of their drinking water supply as one of their best public investments and most vital services. Yet, there was ample evidence in several case studies that complacency and a failure to value safe drinking water have played a role in allowing outbreaks to occur.

A greater role for consumers who value safe drinking water in demanding better from their drinking water providers, regulators and public health professionals would be beneficial. Accordingly, a list of questions to pose to water providers was offered for consumers wishing to become involved in assuring drinking water safety. (Hrudey and Hrudey, 2004)

Ultimately, this analysis seeks to promote the conversion of hindsight into foresight by allowing those who are committed to providing safe drinking water to learn from the documented failures of others without having to experience similar painful lessons first-hand.

Acknowledgements

The guidance and leadership of Justice Dennis O'Connor provided the inspiration for pursuing the book that is described in this chapter. The logistical and technical support of the Canadian Water Network and the Australian Cooperative Research Centre for Water Quality and Treatment and the financial support from Health Canada, Alberta Health and Wellness and the Natural Sciences and Engineering Research Council of Canada is gratefully noted.

REFERENCES

Angulo, F.J., Tippen, S., Sharp, D.J., Payne, B.J., Collier, C., Hill, J.E., Barrett, T.J., Clark, R.M., Geldreich, E.E., Donnell, H.D. and Swerdlow, D.L. (1997) A community waterborne outbreak of salmonellosis and the effectiveness of a boil water order. *Am. J. Public Health* **87**(4), 580–584.

BGOSHU (2000) *The Investigative Report of the Walkerton Outbreak of Waterborne Gastroenteritis May–June 2000*. Owen Sound, Ontario, Bruce-Grey-Owen Sound Health Unit, 58 pp., plus appendices.

Clancy, J. (2000) Sydney's 1998 water quality crisis. *J. Am. Water Works Assoc.* **92**(3), 55–66.

Geldreich, E.E. and Singley, J.E. (2002) *Ontario Water Suppliers: Two Expert Assessments*, Toronto, The Walkerton Inquiry, Attorney General of Ontario, 38 pp.

Gorman, A.E. and Wolman, A. (1939) Waterborne outbreaks in the United States and Canada, and their significance. *J. Am. Water Works Assoc.* **31**, 225–373.

Hrudey, S.E. and Hrudey, E.J. (2004) *Safe Drinking Water – Lessons from Recent Outbreaks in Affluent Nations*, 512 pp., IWA Publishing, London.

Hrudey, S.E. and Hrudey, E.J. (2003) Comments on the paper 'Sociotechnical systems, risk management, and public health: comparing the North Battleford and Walkerton Outbreaks' by Woo and Vicente. *Reliabil. Eng. System Safety* **82**, 343–345.

Hrudey, S.E. and Leiss, W. (2003) Risk management and precaution: insights on the cautious use of evidence. *Environ. Health Perspect.* **111**, 1577–1581.

Kletz, T.A. (2001) *An Engineer's View of Human Error*, 3rd edn, 281 pp., Institution of Chemical Engineers, Rugby, England.

Laing, R.D. (2002) *Report of the Commission of Inquiry into Matters Relating to the Safety of the Public Drinking Water in the City of North Battleford, Saskatchewan*, 372 pp., Department of Justice, Government of Saskatchewan. Web site: http://www.northbattlefordwaterinquiry.ca

McClellan, P. (1998) *Sydney Water Inquiry*, New South Wales Premier's Department, Sydney. Web site: http://www.premiers.nsw.gov.au/our_library/archives/sydwater/5threp/r5vol1.htm

O'Connor, D.R. (2002a) *Report of the Walkerton Inquiry. Part 1. The Events of May 2000 and Related Issues*, 504 pp., The Walkerton Inquiry, Toronto. Web site: http://www.attorneygeneral.jus.gov.on.ca/english/about/pubs/walkerton

O'Connor, D.R. (2002b) Report of the Walkerton Inquiry. Part 2. A Strategy for Safe Water, 582 pp., The Walkerton Inquiry, Toronto. Web site: http://www.attorneygeneral.jus.gov.on.ca/english/about/pubs/walkerton

O'Donnell, M., Platt, C. and Aston, R. (2000) Effect of a boil water notice on behaviour in the management of a water contamination incident. *Commun. Dis. Public Health* **3**(1), 56–59.

Quill, R. (1999). The 1998 Sydney water crisis: view from the inside. *1999 Annual Conference, American Water Works Association*, Chicago.

Rizak, S., Cunliffe, D., Sinclair, M., Vulcano, R., Howard, J., Hrudey, S.E. and Callan, P. (2003) Drinking water quality management: a holistic approach. *Water Sci. Technol. (GB)* **47**(9), 31–36.

Rouse, M. (2004) IWA and global water management. *Water 21*. April 2004, 68–69.

WHO (2003). Water Safety Plans, Chapter 4. *WHO Guidelines for Drinking Water Quality*, pp. 54–88, World Health Organization, Geneva.

Willocks, L., Sufi, F., Wall, R., Seng, C. and Swan, A.V. (2000) Compliance with advice to boil drinking water during an outbreak of cryptosporidiosis. *Commun. Dis. Public Health* **3**(2), 137–138.

Index